Metropolitan Science

Metropolitan Science

London Sites and Cultures of Knowledge and Practice, c. *1600–1800*

Rebekah Higgitt and Jasmine Kilburn-Toppin with Noah Moxham

BLOOMSBURY ACADEMIC
LONDON • NEW YORK • OXFORD • NEW DELHI • SYDNEY

BLOOMSBURY ACADEMIC
Bloomsbury Publishing Plc
50 Bedford Square, London, WC1B 3DP, UK
1385 Broadway, New York, NY 10018, USA
29 Earlsfort Terrace, Dublin 2, Ireland

BLOOMSBURY, BLOOMSBURY ACADEMIC and the Diana logo are trademarks of Bloomsbury Publishing Plc

First published in Great Britain 2024

A catalogue record for this book is available from the British Library.

A catalog record for this book is available from the Library of Congress.

ISBN: HB: 978-1-3504-1703-8
ePDF: 978-1-3504-1704-5
eBook: 978-1-3504-1705-2

Typeset by Integra Software Services Pvt. Ltd.
Printed and bound in Great Britain

Contents

Figures and plates

Acknowledgements

This book is the product of the research project Metropolitan Science: Places, Objects and Cultures of Practice and Knowledge in London, 1600–1800, which was funded by The Leverhulme Trust from April 2017 to July 2020 and based at the University of Kent in partnership with the Science Museum. As well as the Trust, we would like to thank the project's Co-Investigator, Alison Boyle, and Charlotte Sleigh and Tim Boon, who joined its Management Committee. Above all, we are enormously grateful to Jim Bennett, who acted as project consultant, programming our stimulating opening conference, co-editing the resulting collection of essays published in the *British Journal for the History of Science* (2019), offering advice throughout the project and reading drafts of chapters. We are deeply saddened that he died before this book reached print and he could receive these heartfelt thanks.

All of those who gave papers at the project's events, shared our conference panels and joined us in conversation have helped shape our thinking and this book. There are too many to name, but particular thanks go to Alexandra Rose and Jane Desborough, for explorations of links with the developing Science City 1550–1800 gallery at the Science Museum; Larry Stewart, for acting as commentator for our session at the 2017 History of Science Society; Andrew Mendelsohn, for his time discussing themes of mutual interest with the project team; Pamela H. Smith, for giving the keynote at our 2020 conference; and Daniel Belteki, for helping us quickly turn that conference into an online one as the pandemic spread and for his work as research assistant. We are grateful for access to archives at the various repositories we have made use of, especially Eleni Bide and Sophia Tobin at the Goldsmith's Company Library and Archive, and Victoria West at the Worshipful Company of Barbers. We also thank our editor at Bloomsbury Academic, Emily Drew, the anonymous referees for their encouraging and insightful comments and Caroline Higgitt for her time and attentive proofreading.

Abbreviations

BM	British Museum
EIC	East India Company
GHA	Goldsmiths' Hall Archive, London
HBC	Hudson's Bay Company
LMA	London Metropolitan Archives
NMM	National Maritime Museum, Greenwich
NMS	National Museums Scotland.
ODNB	*Oxford Dictionary of National Biography*
RMA	Royal Military Academy, Woolwich
SM	Science Museum, London
TNA	The National Archives

1

Introduction

London, *c.* 1600–1800

Early modern London's institutional buildings, workshops, laboratories, warehouses and wharves were hives of productive, networked, industrious activity. A snapshot of such spaces at the turn of the seventeenth century would reveal artisans, practitioners and institutional officials engaged in the testing and evaluation of coinage, molten metals, gunpowder and spices among many other products and substances. Members of London's corporations were producing weapons, measuring instruments and medicines; they were mastering practical mathematics, surveying the navigable channels of the Thames and even sea-routes throughout Britain's rapidly expanding global empire.

While British science often traces its roots to the foundation of the Royal Society of London for Improving Natural Knowledge in 1660, this book explores a different set of institutional sites to understand how knowledge and skilful, experimental practices developed in early modern London. These were the collective, industrious, practical and commercial contexts of a city that lacked a university but combined a growing population, manufacture, overseas trade, government and court. They included new incorporations – joint stock enterprises like the East India Company or guilds representing emergent crafts – and others that expanded as London's international significance and imperial dominance grew, such as the Royal Mint and the Office of Ordnance. Across the city, such institutions collected information from nearby and around the world, and developed techniques to meet individual and institutional objectives. Moving beyond histories of institutions or of natural and experimental philosophy, this book shows how these collective cultures shaped knowledge production and the city.

Understanding how, why and where Londoners gained knowledge about nature and put it to use between the end of the sixteenth century and dawn of the nineteenth can helpfully begin with a rehearsal of the city's extraordinary demographic and geographic changes. These were both actual and relative: a population more than doubling in the seventeenth century, from about 200,000 to 575,000, moved London from being the third most populous city to the largest in Europe. Geographically, it expanded in all directions, into the counties of Essex, Middlesex, Surrey and Kent. It came to incorporate the centre of royal and parliamentary power in Westminster as well as maritime parishes along the banks of the Thames.[1] In the eighteenth century, London was a centre of manufacturing, culture and consumption, as well as of finance, government, law and court.[2] Visitors inevitably commented on its size, wealth and variety, and even citizens had the sense by the middle of our period that London had become ungraspable, incomprehensible even to itself.[3] By 1800 the population had reached 900,000.[4]

London was also a port – exporting, importing and re-exporting goods from overseas and internally – and was increasingly seen as both national capital and an imperial metropolis. In terms of European centres of trade, it shifted from minor player to competitor and then undeniable winner.[5] It has been estimated that at the end of the seventeenth century, 'about 75 per cent of England's overseas trade passed through the city, and almost half of the kingdom's merchant shipping was based there'.[6] An often-repeated, though hard to confirm, statistic suggests that around 1700 the shipping industry directly or indirectly employed a quarter of the population.[7] By the later eighteenth century other ports in Britain were growing and competing, but London remained dominant. Around 1800 it still handled around 70 per cent of the nation's imports and 56 per cent of its exports, while, in the three previous decades, the value of that trade had tripled.[8]

As is well known, given London's mortality rates, its huge growth in population depended on migration. Migrants came from every corner of the country and from overseas. The term 'foreigners' referred to London inhabitants born elsewhere in England who were not citizens, while emigrants, largely French and Dutch Protestants, were known as 'aliens' or 'strangers'. Jews, Muslims and sub-Saharan Africans were also part of the heterogeneous metropolitan population. The arrival of newcomers sparked mixed reactions from Londoners (depending upon the wealth, social status and nationality of the migrant, as well as the season and economic climate), but migrants certainly diversified London neighbourhoods and the pool of available skills.[9] Strangers were engaged in a broad range of crafts and occupations, but were proportionally most important and conspicuous in metalwork, the clothing industry

1 Jeremy Boulton, 'London 1540–1700', in *The Cambridge Urban History of Britain, Volume 2: 1540–1840*, ed. Peter Clark (Cambridge: Cambridge University Press, 2000), 315–45 (pp. 315–17); Mark Jenner and Paul Griffiths, 'Introduction', in *Londinopolis: Essays in the Cultural and Social History of Early Modern London* (Manchester: Manchester University Press, 2000), 1–23.

2 Leonard Schwarz, 'London 1700–1840', in *The Cambridge Urban History of Britain*, 641–72.

3 Vanessa Harding, 'City, Capital, and Metropolis: The Changing Shape of Seventeenth-Century London', in *Imagining Early Modern London: Perceptions and Portrayals of the City from Stow to Strype, 1596–1720*, ed. J. F. Merritt (Cambridge: Cambridge University Press, 2001), 117–43 (pp. 118–21, 140).

4 Schwarz, 'London 1700–1840', 650.

5 David Sacks and Michael Lynch, 'Ports 1540–1700', in *The Cambridge Urban History of Britain*, 377–424 (p. 386). See also Margarette Lincoln, *London and the Seventeenth Century: The Making of the World's Greatest City* (New Haven and London: Yale University Press, 2021).

6 Sacks and Lynch, 'Ports 1540–1700', 390.

7 Boulton, 'London 1540–1700', 320.

8 Andrew Saint, 'The Building Art of the First Industrial Metropolis', in *London – World City 1800–1840*, ed. Celina Fox (New Haven and London: Yale University Press, 1992), 51–76 (p. 51).

9 Jacob Selwood, *Diversity and Difference in Early Modern London* (Farnham: Ashgate, 2010), 1–3.

and trade.[10] As Deborah Harkness has written, 'a healthy proportion' of immigrants to London also 'possessed skills that brought them to the attention of London citizens interested in the properties of nature'. Expertise and knowledge of English and foreign physicians, surgeons, midwives, apothecaries, alchemists and scientific instrument makers were combined, negotiated, commoditized and exchanged.[11]

While some neighbourhoods could be identified with migrant communities, early modern London could also be experienced as an overlapping patchwork of governing and corporate structures, such as parishes, wards and craft and trade associations. Corporation membership was a route to and expression of citizenship, creating, according to Phil Withington, 'a framework, or structure, for continuous and systematic public activity' by citizens within their communities and in relation to the state. Such structures provided multiple drivers of public activity creating a 'civic public sphere' of middling sorts that underpinned the development of the early modern state and laid the way for Habermas's eighteenth-century 'bourgeois public sphere' of scholars and professionals, capitalists and entrepreneurs.[12] London was also a city increasingly aware of its place in and interactions with the wider world, following and then overtaking Antwerp in its ability to concentrate wealth, especially within merchant livery companies, and to dominate overseas trade, with the creation of a new kind of corporation, the joint-stock company.[13]

As London was the nation's and, later, empire's capital it was, inevitably, shaped by changes to the nature of the state. If the early modern English state was 'an amalgamation of jurisdictions, offices, privileges, communities, and corporations encompassing, often on semi-autonomous and competitive bases, a range of public participants, languages and interests', in the eighteenth century it was increasingly centralized and bureaucratized.[14] Warfare with France underpinned the most significant changes to the English and, from 1707, British state. As the size of the army and navy trebled in the century after 1680, they, together with war debts, accounted for 75–85 per cent of an annual expenditure that drove taxation. The state was increasingly represented by professional, centrally appointed, officials, who worked within new or reformed departments, including a growing excise operation.[15] Alongside this, the joint-stock enterprise that created the English East India Company had become an imperial power, acquiring the trappings of state overseas while managing and supporting its affairs

10 Lien Bich Luu, *Immigrants and the Industries of London* (Aldershot and Burlington, VT: Ashgate, 2005), 120.

11 Deborah Harkness, '"Strange" Ideas and "English" Knowledge: Natural Science Exchange in Elizabethan London', in *Merchants and Marvels: Commerce, Science, and Art in Early Modern Europe*, ed. Pamela Smith and Paula Findlen (London and New York: Routledge, 2002), 137–60 (pp. 139, 155).

12 Phil Withington, 'Public Discourse, Corporate Citizenship, and State Formation in Early Modern England', *The American Historical Review* 112, no. 4 (2007): 1016–38 (pp. 1017–18, 1021).

13 Stephen Alford, *London's Triumph: Merchant Adventurers and the Tudor City* (London: Allen Lane, 2017).

14 Withington, 'Public Discourse', 1024–5.

15 John Brewer, 'The Eighteenth-Century British State', in *An Imperial State at War: Britain from 1689 to 1815*, ed. Lawrence Stone (London: Routledge, 1994), 52–71 (p. 57).

from London.[16] Wealth accruing from trade, manufacture and commerce encouraged private and civic investment in new industrial and infrastructural projects, physically reshaping the city and experiences of its inhabitants.

Developments such as these had significant impacts on older corporations. The extent to which guilds, or livery companies, exercised control over the crafts and trades for which they were responsible declined significantly, especially from the late seventeenth century. Regulatory searches of workshops by guild officers, as a means of market regulation, became evermore challenging as London's suburbs expanded.[17] While guild leadership was increasingly wealthy and of rising social status, their connection to and influence over their ordinary membership and household economies decreased. Socially and culturally some of their roles were replaced or influenced by the arrival of new institutions, from voluntary associations to educational establishments. However, historians have also emphasized continuity. Guilds in London as elsewhere were not replaced but adapted in order to survive or thrive in new circumstances.[18] Corporate citizenship became more rather than less significant during the seventeenth-century Commonwealth and, into the eighteenth century, royal charters were used to standardize practices and confirm the status and privileges of corporations and their members.[19] Guilds and other civic structures remained a source of urban and occupational identity, while new institutions and associations were often modelled at least in part on older corporations.[20]

Similarly, despite fires, plagues, civil war and political change, historians of London have tended to stress stability and continuity over disorder and disruption.[21] Devastating as it was, the Great Fire destroyed 'only a fifth of London's housing stock'. Other fires were frequent but had severe effects only on more limited areas. Trades continued and civic structures were retained or reformed even as political ruptures took place and new forms of finance and trade were developed.[22] Stability was, of course, in the interests of the trading elites who were gaining positions of power and influence. These interests, Valerie Pearl has suggested, were rendered more broadly

[16] P. J. Stern, *The Company-State: Corporate Sovereignty and the Early Modern Foundation of the British Empire in India* (Oxford: Oxford University Press, 2011), 3–11, 15.

[17] Patrick Wallis, 'Controlling Commodities: Search and Reconciliation in the Early Modern Livery Companies', in *Guilds, Society and Economy in London 1450–1800*, ed. Ian Anders Gadd and Patrick Wallis (London: Centre for Metropolitan History, 2002), 85–100 (p. 87); John Forbes, 'Search, Immigration and the Goldsmiths' Company: A Study in the Decline of its Powers', in Gadd and Wallis, *Guilds, Society and Economy*, 115–25 (pp. 115, 118–19); Michael Berlin, 'Guilds in Decline? London Livery Companies and the Rise of a Liberal Economy, 1600–1800', in *Guilds, Innovation and the European Economy*, ed. S. R. Epstein and Maarten Prak (Cambridge: Cambridge University Press, 2008), 316–41 (pp. 327–8).

[18] On their 'flexibility and fluidity' see Gadd and Wallis, 'Introduction', in *Guilds, Society and Economy*, 1–14 (pp. 9–10); Berlin, 'Guilds in Decline?', 330–9.

[19] Withington, 'Public Discourse', 1026–7.

[20] See e.g. Joseph Ward, *Metropolitan Communities: Trade Guilds, Identity, and Change in Early Modern London* (Stanford, CA: Stanford University Press, 1997); Noah Moxham, 'Natural Knowledge, Inc.: The Royal Society as a Metropolitan Corporation', *British Journal for the History of Science* 52, no. 2 (2019): 249–71.

[21] On sixteenth-century civic stability see Ian Archer, *The Pursuit of Stability: Social Relations in Elizabethan London* (Cambridge: Cambridge University Press, 1991).

[22] Jenner and Griffiths, 'Introduction', in *Londinopolis*, 3–4.

palatable by being equated with those of the city, underpinning its wealth and the promise of opportunities for its inhabitants. Private gain was represented as public good.[23] Increased knowledge and improved practice, pursued for commercial gain or institutional advantage, were increasingly brought into this equation.

By 1800 London was a city of 'enormous wealth, population and technical skill'. This was crucial to its being the locus of a nineteenth-century 'revolution in the sciences' that saw the arrival of new disciplines, techniques and '[m]ost of the institutions associated with modern science'.[24] Our period, then, saw enormous change not only in the size and significance of London but also in the nature and social place of science. Part of this story, and the one traditionally encountered in histories of science, was its institutionalization. The most common starting point for surveys of British science is the foundation of the Royal Society, which may be linked with the beginnings of another long-lived foundation, the Royal Observatory (f. 1675), down the river from the city at Greenwich. Thereafter we might point to educational establishments that emphasized mathematics, including the Royal Mathematical School at Christ's Hospital (f. 1673), and institutions reflecting the ideology of Improvement, such as the Society for the Encouragement of Arts, Manufactures and Commerce (f. 1754) and Royal Institution (f. 1799). These new bodies, however, drew on many existing networks and practices, connecting to older institutions and interests in ways not usually appreciated or examined.

To understand the knowledge and practices that became known as science and industry by the end of our period, we need to explore these other institutions and networks. Our case studies reflect both new and pre-existing corporations, and ones that were embedded within the contexts of civic governance, state and military organization and maritime activity, including overseas trade. Knowledge and the manipulation of the natural world were in different ways essential to each in meeting their aims of supporting the commercial advantages of their members, the reputations of their governing bodies and officials, and the fiscal, military and imperial ambitions of the state. The chapters will consider a Crown body (to 1688), the Royal Mint; a governmental department, the Office of Ordnance; two London livery companies, the Goldsmiths' Company and the Barber Surgeons' Company; a maritime guild that was founded outside the city but which acted within and well beyond it, Trinity House; and a joint-stock long-distance trading company that became an imperial government, the English East India Company (EIC).

In the following sections, we will introduce the scholarship on the history of science, technology and knowledge within which we situate this book and through which we identify our object of study. Firstly, we will consider how science – and other kinds of natural and practical knowledge – has been understood as something that co-evolved with cities. Secondly, we explore the literature that has reshaped approaches to early modern knowledge and practice, seeking to escape retrospective definitions

23 Valerie Pearl, 'Change and Stability in Seventeenth-Century London', *The London Journal* 5, no. 1 (1979): 3–34.

24 Iwan Morus, Simon Schaffer and Jim Secord, 'Scientific London', in Fox, *London – World City*, 129–42 (p. 129).

of what counts, by expanding the types of places, people and practices taken into consideration. We then turn to the role of institutions and corporations in relation to such practices, and as creating, containing or participating in knowledge communities, before outlining the spatial and material elements of our approach to these histories and their sources. Finally, we introduce the chapters and case studies, which show institutions as places and communities of natural knowledge and artificial practice in early modern London – or what we have called, with less careful attention to contemporary terminology, metropolitan science.

Metropolitan science

Science in the city

The scale and diversity of cities generally, and London in particular, have seen them cast as machines for the creation of knowledge, learning and innovation. Knowledge production is 'to a large degree an urban phenomenon', not least because cities contain a significant number of spaces of communal activity.[25] Likewise, both the city and knowledge in the city are shaped by their connectivity, including, crucially in our period, through their role as hubs for long-distance trade.[26] We agree with the call by the editors of *Science in the City* to link the history of science and urban history by studying science as local practice, and with their view 'that science and the city were coproduced within the same historical processes'.[27] Their collection of essays is perhaps typical in focusing on the nineteenth- and twentieth-century city, by which time many of the locations and structures of scientific careers and exchange that are still with us today had become apparent. Likewise, many more urban centres were recognizable as metropoles, increasingly understood as the product of human ingenuity.[28] Proliferating city institutions – many now largely dedicated to natural knowledge – are key to many of these later histories. Although often controlled spaces, supposedly separated from the disturbances of the urban environment, they are shown to have been closely engaged with, influential to and impacted by surrounding networks and spaces.[29] Growing cities of the nineteenth century not only encroached on previously

[25] Bert De Munck and Antonella Romano, 'Knowledge and the Early Modern City: An Introduction', in *Knowledge and the Early Modern City: A History of Entanglements*, ed. Bert De Munck and Antonella Romano (London: Routledge, 2020), 1–30 (p. 1).

[26] Karel Davids, 'Cities, Long-Distance Corporations and Open Air Sciences: Antwerp, Amsterdam and Leiden in the Early Modern Period', in De Munck and Romano, *Knowledge and the Early Modern City*, 126–48.

[27] Sven Dierge, Jens Lachmund and Andrew Mendelsohn (eds), 'Introduction: Toward an Urban History of Science', *Science in the City, Osiris* 18 (2003): 1–19 (p. 3). On co-production in science and technology studies, see Sheila Jasanoff, 'The Idiom of Co-production', in *States of Knowledge: The Co-production of Science and Social Order* (London and New York: Routledge, 2004), 1–12.

[28] Other recent examples are Tanya O'Sullivan, *Geographies of City Science: Urban Lives and Origin Debates in Late Victorian Dublin* (Pittsburgh: Pittsburgh University Press, 2019) and Mitchell G. Ash (ed.), *Science in the Metropolis: Vienna in Transnational Context 1848–1914* (London: Routledge, 2020).

[29] For example, David Aubin, 'The Fading Star of the Paris Observatory in the Nineteenth Century: Astronomers' Urban Culture of Circulation and Observation', *Osiris* 18 (2003): 79–100.

distinct suburbs and open space, they were also increasingly crowded, encouraging new scientific institutions to find ways to present themselves as distinctive, separated and specialized spaces.[30] City and science shaped each other, both physically and representationally.

Nevertheless, the editors of *Science in the City* recognize that urban cultures supported many activities that were 'not scientific' in the modern sense, 'yet involve[d] the creation or transmission of knowledge'.[31] Such activities included the mapping, recording and archiving that were essential to civic governance as well as trade, commerce and manufacture. More recently, Andrew Mendelsohn has considered such practices in the early modern period and, in particular, the extent to which physicians working within and for cities were 'central to the rise of shared, organized written information about the human and natural world'.[32] Urban institutions connected to governance, law and religion were also responsible for creating professional communities of practice, as revealed, for example, by scholars of literary culture who have brought the activities of clerks, scriveners and stationers out of the archival shadows. The role of clerks in institutional record-keeping has been shown to be creative and varied, supportive of corporate ends as well as individual ambitions.[33]

Much of the work on urban knowledge cultures has emphasized 'thick description' as a means of representing the 'flow of intellectual activities and the proliferation of expert knowledges' within and between cities and their local, national and transnational contexts.[34] As well as institutions, cities can be considered as exemplifying or containing 'trading zones'. This is a term that Pamela Long, drawing on Peter Galison's exploration of twentieth-century sites of collaboration across paradigms in physics, has used to characterize early modern sites that allowed 'the complex interactions among those trained in a practice or skill and those from learned traditions, as well as those who combined both practice and learning'.[35] Exchanges, such as those required and enabled by technologically invested sites like mines and arsenals, were not exclusively urban

30 Sophie Forgan, '"But Indifferently Lodged … ": Perceptions and Place in Building for Science in Victorian London', in *Making Space for Science: Territorial Themes in the Shaping of Knowledge*, ed. Jon Agar and Crosbie Smith (Basingstoke: Macmillan, 1998), 195–215; Graeme Gooday, 'The Premisses of Premises: Spatial Issues in the Historical Construction of Laboratory Credibility', in ibid., 216–45.

31 Dierge, Lachmund and Mendelsohn, 'Introduction', 16.

32 J. Andrew Mendelsohn, 'Introduction: Civic Medicine', in *Civic Medicine: Physician, Polity, and Pen in Early Modern Europe*, ed. Annemarie Kinzelbach and Ruth Schilling (Abingdon and New York: Routledge, 2020), 1–6 (p. 1).

33 Matthew Davies, '"Monuments of Honour": Clerks, Histories and Heroes in the London Livery Companies', in *Parliament, Personalities and Power: Papers presented to Linda S. Clark*, ed. H. Kleineke (Woodbridge: Boydell, 2011), 143–66; Jennifer Bishop, 'The Clerk's Tale: Civic Writing in Sixteenth-Century London', *Past & Present* 230, no. 11 (2016): 112–30.

34 Dorothee Bantz, 'Metropolitan Natural Histories: Inventing Science, Building Cities, and Displaying the World', in *Science in the Metropolis*, 25–42 (pp. 25–8); Oliver Hochadel, 'Periphery and Metropolis: Some Historiographical Reflections on the Urban History of Science' (ibid., 43–63).

35 Pamela O. Long, 'Trading Zones in Early Modern Europe', *Isis* 106, no. 4 (2015): 840–7 (p. 847).

but their initiation usually was.[36] Cities brought together people who offered or sought the requisite range of skills, knowledge and social, political and financial capital.

For Deborah Harkness, London in the late sixteenth century, with its range of people and interests, generated an atmosphere that was 'creative and competitive – the ideal environment for cultural and intellectual exchange'.[37] She describes 'an urban sensibility toward the natural world', classified as vernacular science, that was 'based in urban ways of knowing and evaluating nature that came from the densely overlapping social worlds of the City' and the combination of collaboration and competition that they inspired.[38] In this sense, she pictures the whole of London as 'a house of science and a prototype of a modern laboratory', inspiring Francis Bacon's vision of Salomon's House and giving impetus to empirical virtues of observation, experiment, pragmatism and (relative) openness.[39] Hal Cook, too, has presented urban commerce, this time of the Dutch Golden Age, as a driver of close engagement with the natural world and materials, placing emphasis on observation and experimentation. He argues that the financial and political capital produced gave 'the values of the urban merchants, including their intellectual values' increasing status.[40]

Looking forward to the eighteenth century, the shops of scientific instrument makers form another kind of site that Long includes among trading zones, facilitating interactions and discussions between scholars, gentlemen, artisans and practitioners. Jim Bennett's comparison of these sites in Paris and London indicates the livelier and more open, commercial and international nature of the trade and the exchanges it enabled in the latter.[41] This sense of London as representing, at least by the turn of the eighteenth century, a particularly open and increasingly public culture of knowledge and practice is supported too by Larry Stewart's account of 'public science' in contexts that were less elite but strikingly entrepreneurial, from coffee houses to practitioner- and artisan-led societies and schools.[42] Liliane Hilaire-Pérez has likewise put the mobility, public culture and networks of London artisans at the heart of her reading of

36 For a brilliant recent study of the 'vernacular culture of geometry' in the mining cities of the Holy Roman Empire – involving complex interactions between practitioners, humanists, state administrators and investors – see Thomas Morel, *Underground Mathematics: Craft Production and Knowledge Production in Early Modern Europe* (Cambridge: Cambridge University Press, 2023).

37 Deborah Harkness, *The Jewel House: Elizabethan London and the Scientific Revolution* (New Haven and London: Yale University Press, 2007), 2.

38 Harkness, *The Jewel House*, xvii, 6. On vernacular science, and its role in a broader understanding of the Scientific Revolution, see Pamela H. Smith's conclusion to *The Body of the Artisan: Art and Experience in the Scientific Revolution* (Chicago: University of Chicago Press, 2004), 237–41.

39 Harkness, *The Jewel House*, 8–9.

40 Harold J. Cook, *Matters of Exchange: Commerce, Medicine, and Science in the Dutch Golden Age* (New Haven and London: Yale University Press, 2007), 40.

41 Jim Bennett, 'Shopping for Instruments in Paris and London', in Smith and Findlen, *Merchants and Marvels*, 370–95.

42 Larry Stewart, *The Rise of Public Science: Rhetoric, Technology, and Natural Philosophy in Newtonian Britain, 1660–1750* (Cambridge: Cambridge University Press, 1992), and *idem*, 'Other Centres of Calculation, or, Where the Royal Society Didn't Count: Commerce, Coffee-Houses and Natural Philosophy in Early Modern London', *British Journal for the History of Science* 32, no. 2 (1999): 133–53.

the 'industrial Enlightenment'.[43] Institutions were important foci and drivers, but the formal and informal networks that city life supported were layered and overlapping. Many historians have thus chosen to emphasize entanglements and hybrid practices, blurring older distinctions that previously characterized the history of science and technology. In the process, the city has been revealed as a productive assemblage through which knowledge and its own built environment were shaped and mutually justified.[44]

Early modern knowledge and practice

Many of the recent works discussed above have drawn on and contributed to a body of scholarship that has sought to reframe what it is that historians of early modern science study. It is perhaps unnecessary to rehearse again the historiographical discussions that have contrasted the approach of those like Robert K. Merton, Edgar Zilsel and Paolo Rossi, who drew on sociology and social history to emphasize the commercial and military drivers of knowledge and skill, with that of Alexandre Koyré and A. Rupert Hall, who rejected the idea that craft skill played a significant role in shaping natural and experimental philosophy.[45] These divisions were the legacy of earlier attempts to separate, and create hierarchies between, the work of the hand and the mind. Such histories have since been interrogated, and the argument that historians of science and knowledge should pay attention to both, and that contemporaries collapsed or did not recognize such distinctions, is now widely accepted.[46] Practical and material concerns have been placed centrally rather than brushed to one side: 'Like commerce,

[43] Hilaire-Pérez, 'Technology as Public Culture', which responds to Joel Mokyr's *The Gifts of Athena: Historical Origins of the Knowledge Economy* (Princeton and Oxford: Princeton University Press, 2002). See also Liliane Hilaire-Pérez, *La Pièce et le Geste. Artisans, Marchands et Savoir Technique à Londres au xviiie Siècle* (Paris: Éditions Albin Michel, 2013).

[44] De Munck and Romano, *Knowledge and the Early Modern City*. On 'hybrid activity' see Lissa Roberts, 'Workshops of the Hand and Mind: Introduction', in *The Mindful Hand: Inquiry and Invention from the Late Renaissance to Early Industrialisation*, ed. Lissa Roberts, Simon Schaffer and Peter Dear (Amsterdam: Royal Netherlands Academy of Arts and Sciences, 2007), 1–8 (p. 3).

[45] Robert K. Merton, 'Science, Technology and Society in Seventeenth-Century England', *Osiris* 4 (1938): 360–632; Edgard Zilsel, 'The Sociological Roots of Science', *American Journal of Sociology* 47, no. 4 (1942): 544–62; Paolo Rossi, *Philosophy, Technology, and the Arts in the Early Modern Era*, trans. Salvator Attansio (New York: Harper and Row, 1970); Alexander Koyre, 'Galileo and Plato', *Journal of the History of Ideas* 4, no. 4 (1943): 400–28; A. R. Hall, *Ballistics in the Seventeenth Century* (Cambridge: Cambridge University Press, 1952). See discussion in e.g. Harold J. Cook, Pamela H. Smith and Amy R.W. Meyers, 'Introduction: Making and Knowing', in *Ways of Making and Knowing: The Material Culture of Empirical Knowledge*, ed. Pamela H. Smith, Amy R. W. Meyers and Harold J. Cook (Ann Arbor: University of Michigan Press, 2013), 1–16 (pp. 4–5); De Munck and Romano, *Knowledge and the Early Modern City*, 4–6, and Lesley B. Cormack, 'Handwork and Brainwork: Beyond the Zilsel Thesis', in *Mathematical Practitioners and the Transformation of Natural Knowledge in Early Modern Europe*, ed. Lesley B. Cormack, Steven A. Walton and John A. Schuster (Cham: Springer, 2017), 11–35.

[46] See Roberts, Schaffer, and Dear, *The Mindful Hand*; Pamela O. Long, *Artisan/Practitioners and the Rise of the New Sciences, 1400–1600* (Corvallis: Oregon State University Press, 2011); Pamela H. Smith, *From Lived Experience to the Written World: Reconstructing Practical Knowledge in the Early Modern World* (Chicago: The University of Chicago Press, 2022).

science arose not from liberating the mind from the world but from keenly interested engagement with it.'[47]

Our approach, focused on corporations and corporate sites and, as discussed below, attending to spaces and material goods, as well as the people associated with them, means we are necessarily exploring knowledge and practice that was generated by 'activity in the world' and 'doing things with purpose'.[48] We show the frequent and productive connections that existed between artisans and philosophers, the real-world questions that prompted experimental and ingenious solutions, and some of the ways in which the worlds of commerce and craft practice crossed over with and incorporated those of mathematics, experimental philosophy and learned knowledge. Despite our title, we are cautious about using terms, including 'science' and 'experiment', that not only mean something different today than they did in the past but that also changed significantly during our period.[49] Indeed, they were concepts that were emerging and being shaped, in part by just such contexts as those we explore here.[50] Thus, while Harkness and others, writing about the early modern period, have embraced such terms with due glossing and justification, we pay close attention to those we encounter in the archive and contemporary print, while also drawing more broadly on terms such as natural knowledge and artificial practice – recognizing their intertwined and multidirectional relationships.[51]

As well as the productive skill of workshop-based artisans, historians have explored the practical contexts within which mathematical techniques were developed and put to work. These included the accounting needs of merchants, bookkeepers and bankers, as well as the technical practices of draughtsmen, surveyors and navigators.[52] Nearly three decades ago Frances Willmoth noted the close connection between practical mathematicians – those who made a living from applying, writing about and teaching mathematics – and the Office of Ordnance. There she revealed 'a continuum of social and intellectual relationships' between mathematicians, natural philosophers and 'the practical skills of craftsmen'.[53] This linking of and relatively easy movement between the categories of scholar and craftsman revealed through attention to problems and

[47] Cook, *Matters of Exchange*, 2; see also Smith and Findlen, *Merchants & Marvels*, e.g. p. 3.

[48] Smith, Meyers and Cook, 'Introduction: Making and Knowing', 2.

[49] On experiment as equivalent to trial, test and adaptation, see Simon Werrett, *Fireworks: Pyrotechnic Arts and Sciences in European History* (Chicago: University of Chicago Press, 2010), 56, 81–3 and *Thrifty Science: Making the Most of Materials in the History of Experiment* (Chicago and London: University of Chicago Press, 2019), 3–5, 85–6.

[50] On the late emergence of 'science', placed around 1760–1848, see Andrew Cunningham and Perry Williams, 'De-Centring the "Big Picture": "The Origins of Modern Science" and the Modern Origins of Science', *The British Journal for the History of Science* 26, no. 4 (1993): 407–32.

[51] Harkness, *The Jewel House*, 2, xv–xvii; on natural knowledge and artificial practice see Jim Bennett and Rebekah Higgitt, 'London 1600–1800: Communities of Natural Knowledge and Artificial Practice', *The British Journal for the History of Science* 52, no. 2 (2019): 183–96 (p. 186). Roberts, Schaffer and Dear, *The Mindful Hand*, xviii–xix.

[52] Pamela H. Smith and Paula Findlen, 'Commerce and the Representation of Nature in Art and Science', in *idem Merchants and Marvels*, 1–25 (pp. 16–18); Harkness, *The Jewel House*, Ch. 3.

[53] Frances Willmoth, 'Mathematical Sciences and Military Technology: The Ordnance Office in the Reign of Charles II', in *Renaissance and Revolution: Humanists, Scholars and Craftsmen*, ed. J. V. Field and Frank James (Cambridge: Cambridge University Press, 1993), 117–32 (p. 117). This 'continuum' has been more recently explored in Cormack, Walton and Schuster, *Mathematical Practitioners*.

instruments held in common has been classed by Jim Bennett as 'the mechanics' philosophy'. This he has shown to have developed within practical contexts for individual and collective benefit by a self-conscious London community towards the end of the sixteenth century.[54]

The expansion of attention to a pluralized range of knowledges and practices has, then, served to blur dichotomies between science and technology, scholars and craftsmen, the work of the head and of the hand, and across the history of science and technology. One approach has been to refer to 'technoscience', although this carries the danger of anachronism and risks hiding contemporary efforts to articulate distinctions that went on to shape the categories defined in the following centuries.[55] Conversely, a concept coined as a way avoiding separation between science and its context, or debates about the relative importance of intrinsic and extrinsic factors, in the twentieth century – Jon Agar's problem-generating 'working worlds' – has proved helpful.[56] Our institutional foci were part of wider, and overlapping, working worlds of navigation, military power, metalworking, medicine and more that generated problems and solutions. These resulted from craft practice that could include processes of abstraction – models, mathematics or experiments – that were addressed within that world or crossed into locations of more elite learning. These processes helped generate an increasingly public world of discourse and exchange that was of use to the overlapping groups of artisans, institutional elites and scholars.

However, it would be wrong to suggest that this circulation of problems and concepts was simple or unhindered. While Harkness pointed to the public nature of knowledge in London as part of the reason for the growth of an 'empirical culture' at the end of the sixteenth century, our case studies explore the restrictions on access to spaces and knowledge that were inevitably a feature of production that involved commercial or military competition, or required confidence in value or authority.[57] We also show that the changing cultural position of experimental, codified or scientific knowledge – with its values of gentlemanly disinterest and openness – and the changing social mix of institutional leaders interacted in revealing but not necessarily straightforward ways to reshape how different kinds of knowledge and their relative secrecy or openness were understood or presented externally.[58]

In a city grown rich on private trade and manufacture, support for these enterprises could be and was interpreted as publicly beneficial. It was the involvement of mathematicians and natural and experimental philosophers in the problems of the

[54] J. A. Bennett, 'The Mechanics' Philosophy and the Mechanical Philosophy', *History of Science* 24, no. 1 (1986): 1–28 (pp. 6, 11).

[55] Ursula Klein, 'Technoscience avant la lettre', *Perspectives on Science* 13, no. 2 (2005): 226–66; Peter Dear 'Towards a Genealogy of Modern Science', in Roberts, Schaffer and Dear, *The Mindful Hand*, 431–2 (p. 435).

[56] Jon Agar, *Science in the Twentieth Century and Beyond* (Cambridge and Malden, MA: Polity Press, 2012), 3–6.

[57] Harkness, *The Jewel House*, 8.

[58] Steven Shapin, *A Social History of Truth: Civility and Science in Seventeenth-Century England* (Chicago and London: Chicago University Press, 1994); William Eamon, 'From the Secrets of Nature to Public Knowledge: The Origins of the Concept of Openness in Science', *Minerva* 23, no. 3 (1986): 321–47.

city's working worlds – through which they could assert the validity and usefulness of their approaches – that ensured a public culture of knowledge that might in other contexts have been associated with scholarly seclusion. Artisans, who sought to state their own claims to knowledge of the secrets of nature or valuable artificial practices by revealing them (at least partially) in print, were also seeking to project an image of increased social status.[59] Each group benefitted from being able to place a foot in the others' world, and could be brought together by the mixed interactions – across social status, levels of experience and types of knowledge – that institutional or project-focused contexts generated.[60] Following knowledge networks, historians have shown that 'hosts of people, of many different social and geographical locations, were involved in the establishment of matters of fact'.[61] It has been shown, too, that direct experience could be taken as credible testimony, widening the cast of individuals, as well as locations and practices that have been seen to count within the history of science.[62]

By the end of our period, 'industry' had shifted meaning from 'socially-distributed diligence' to 'systems of organised and mechanised manufacture' that involved increased surveillance and public accountability.[63] While some histories of the Industrial Revolution have taken the injection of scholarly ideas into artisanal worlds as causal, others have continued to emphasize the fluidity and hybridity of people and practices over unidirectional influence.[64] Ideologies of Improvement and Enlightenment developed from acknowledgement of the effective application of knowledge and practice and supported their expansion and codification. Paul Slack has argued that British successes and rising incomes by the years around 1700 provided a positive feedback loop to encourage further investment and innovation as well as certain kinds of behaviour and outlook.[65] Whatever their roots, including in the knowledge and practices of artisans, whose mechanical arts shaped experiences of the physical world, and the indigenous people of Britain's colonized territories, who knew

[59] Jasmine Kilburn-Toppin, 'Writing Knowledge, Forging Histories: Metallurgical Recipes, Artisan-Authors and Institutional Cultures in Early Modern London', *Cultural and Social History* 18, no. 3 (2021): 297–314.

[60] See discussion of the public and collaborative approach generated by scholars' attention to artisanal knowledge in relation to Paolo Rossi's modification of Zilsel's thesis in Eamon, 'From the Secrets of Nature', 336–7.

[61] David S. Lux and Harold J. Cook, 'Closed Circles or Open Networks? Communicating at a Distance during the Scientific Revolution', *History of Science* 36, no. 2 (1998): 179–211 (p. 180).

[62] Philippa Hellawell, '"The Best and Most Practical Philosophers": Seamen and the Authority of Experience in Early Modern Science', *History of Science* 58, no. 1 (2019): 28–50.

[63] Roberts, Schaffer and Dear, *The Mindful Hand*, xxvi.

[64] Paola Bertucci, *Artisanal Enlightenment: Science and the Mechanical Arts in Old Regime France* (New Haven and London: Yale University Press, 2017), 21–3; Hilaire-Pérez, 'Technology as Public Culture'.

[65] Paul Slack, *The Invention of Improvement: Information and Material Progress in Seventeenth-Century England* (Oxford: Oxford University Press, 2014).

how to identify and work with the natural products of their land, ideas of improvement were to be deeply marked by elitist and imperialist assumptions.[66]

Such shifts, which emphasized public as well as private benefit and created an increasingly public space for knowledge, had an impact even on traditionally closed corporations. Embracing new rhetoric and the moral status given to certain kinds of activity could be a useful manoeuvre for corporations and institutions, helping to justify or gain approval for activities otherwise open to charges of personal greed or profiteering.[67] Providing for training and education could also help meet corporate aims of greater efficiency or efficacy through both standardization and specialization. Individuals, too, could gain social status or opportunities by laying claim not just to experience and practical skills but to the book-learning and education that aimed to make these generalizable and more widely acceptable. These changing opportunities and accompanying self-definitions played their role in the separation of scholarly and practical, or scientific and technical, endeavours and in decreasing the fluidity of 'trading zones' and hybridity of practices evident in the early modern period.[68]

Places, spaces and things

Institutions

The hybridity and entanglements of early modern knowledge emphasized in much of the recent literature have often been explored through study of particular crafts or skills, projects or problems. In this book, the thread we follow to gain perspective on such entanglements is that of institutions and institutional sites. These are not the kinds of institutions more commonly studied by historians of science and most readily associated with the development of natural and experimental philosophy or the improvement of mechanical arts – the Royal Society, Society of Arts, British Museum or Royal Institution. Nevertheless, they are the ones in which different kinds of knowledge and practice, including these more 'elite' forms, played significant, if varied, roles. The particular, and changing, needs of these institutions produced distinct knowledge cultures but also created connections to other institutions and groups with related interests, resources and skills. They could unite a wide range of individuals, of

[66] See Richard Drayton, *Nature's Government: Science, Imperial Britain and the 'Improvement' of the World* (New Haven and London: Yale University Press, 2000), on the 'ideology of development' (p. 59) that shaped the British Empire. For a recent exploration of the deep connections between artisanal praxis, state authority and the eighteenth-century Enlightenment, see Lauren R. Cannady and Jennifer Ferng (eds), *Crafting Enlightenment: Artisanal Histories and Transnational Networks*, Oxford University Studies in the History of the Enlightenment (Liverpool: Liverpool University Press, 2021).

[67] See Ted Binnema, *Enlightened Zeal: the Hudson's Bay Company and Scientific Networks, 1670–1870* (Toronto, Buffalo and London: University of Toronto Press, 2014) on this company's use of investment in science to 'burnish its corporate reputation' as well as to directly aid its activities, and Chapter 7 on the EIC.

[68] Long, 'Trading Zones', 841–3; Liliane Hilaire-Pérez, 'Technology as a Public Culture in the Eighteenth Century: The Artisans' Legacy', *History of Science* 45, no. 2 (2007): 135–53, 137–9.

sometimes radically different social status and skillsets, who were all necessary to the function and meaning of the corporation.

Harkness's *The Jewel House* reveals London's worlds of natural knowledge and mathematical practice in the period in which our story begins, largely through individuals located within the city's overlapping social, geographic and civic contexts and networks. To Francis Bacon, she suggests, this activity was energetic but inchoate and purposeless, while her book emphasizes its 'great variety' as well as 'communal character'.[69] Harkness, however, did not explore the purposeful work that was undertaken in institutional and corporate contexts, instead anticipating the foundation of a new corporation dedicated to knowledge – the Royal Society.[70] While we readily agree that individuals frequently acted outside or across multiple corporate structures, we argue that institutional cultures are worth attending to in order to gain a new understanding of the nature of what we call metropolitan science.[71] They were important generators of and clients for knowledge and practice and, more widely, if not centres of calculation, they were centres of gravity for skills, information, materials and patronage.[72] Rather than following known individuals to institutional settings, we start with the corporation to see what knowledge meant and who was associated with it.[73]

Historians have noted that seventeenth-century London saw a proliferation of corporations. In the first four decades of the seventeenth century, for example, twenty-seven new incorporations of specialist artisans were chartered, including the Apothecaries (1617), Tobacco Pipe Makers (1619) and Clockmakers (1631). These were encouraged by the Crown, which gained financially, but also represented the city's expanding and diversifying skills, technologies and trade.[74] While guilds have traditionally been seen as inhibiting the innovation that we might associate with a thriving culture of knowledge and practice, others have noted the extent to which it

69 Harkness, *The Jewel House*, 7–8.

70 In her second chapter, Harkness discusses members of the Company of Barber-Surgeons as part of a community of expert practitioners, but in relation to disputes that took place beyond the corporation, on the streets and in the press.

71 On London's merchants working across multiple mercantile corporations, see Edmond Smith, *Merchants: The Community that Shaped England's Trade and Empire* (New Haven and London: Yale University Press, 2021), 10.

72 On centres of calculation, see Bruno Latour, *Science in Action* (Cambridge, MA: Harvard University Press, 1987), 215–20. The term is most often associated with well-networked scientific centres (individuals or institutions) that supported imperial ambitions from the second half of the eighteenth century.

73 Examples of the former include Frances Willmoth, *Sir Jonas Moore: Practical Mathematics and Restoration Science* (Woodbridge: Boydell & Brewer, 1993), 138–57; Boris Jardine, 'Instruments of Statecraft: Humphrey Cole, Elizabethan Economic Policy and the Rise of Practical Mathematics', *Annals of Science* 75, no. 4 (2018): 304–29, which includes consideration of Coles roles within the Goldsmiths and at the Mint; John Craig, *Newton at the Mint* (Cambridge: Cambridge University Press, 1946).

74 Michael Berlin, '"Broken all in pieces": Artisans and the Regulation of Workmanship in Early Modern London', in *The Artisan and the European Town, 1500–1900*, ed. Geoffrey Crossick (Aldershot: Scolar, 1997), 75–91 (p. 77); Berlin, 'Guilds in Decline?', 328–9.

was supported or adapted to, as guilds themselves changed their roles and priorities.[75] Some of these changes were prompted by the arrival of new forms of corporation, particularly joint-stock companies; the reform and increased bureaucratization of government; and the promotion of new interests, including natural and experimental philosophy at the Royal Society.

Anne Goldgar and Robert Frost's collection of studies of early modern institutional culture shows the rich histories that can be revealed. Like us, the editors and contributors are interested in the ways in which institutional identity can be established and maintained, for example, through creating their own sense of history as well as by establishing rules and creating boundaries in relation to other groups. Institutions 'operated in a complex corporate web' of overlapping privileges or interests that shaped choices about cooperation and rivalry, or publicity and secrecy, as well as providing a sense of identity and belonging, or opportunities for advantage and power. Public opinion and the market had an increasing impact, as, by the end of our period, did the changing relationships between natural knowledge, politics and business.[76] Similarly, drawing on specific and contingent stories of institutions in early modern London, Miles Ogborn has identified their roles in the city's modernization, characterized by individualization, formation of a 'public sphere', commodification, bureaucratic rationalization, state formation and a communications revolution.[77]

Naturally, within corporations, very different kinds of knowledge and practice played a variety of roles. They might be dispersed among members and/or maintained and represented centrally. Studying the workshops of artisans or productive sites like the arsenal or dockyard helps recover 'the hybrid activities involved in the intimately related processes of material and knowledge production'.[78] Focusing on institutional imperatives reveals the tensions evident between gaining new techniques, materials or skills versus maintaining control of craft secrets or commercially and militarily sensitive knowledge. This might play out in attitudes to textual versus tacit knowledge, expertise versus experience, or local versus standardized or universalized principles. There might also be tensions between individual and collective interest. Corporate attitudes to knowledge and practice, whether focused internally or externally, might be both collaborative and competitive.

Focusing on institutional histories can offer the benefit of a long series of regularly kept archives. Although there may be significant gaps, particularly due to loss by fire, institutions that continue to exist or have clear successors tend to provide rich legacies in accessible repositories. We have made use, particularly, of the papers of the Royal Mint and Office of Ordnance records at The National Archives; Trinity House at

[75] Epstein and Prak (eds), *Guilds, Innovation and the European Economy*; Sheilagh Ogilvie, 'Rehabilitating the A Reply', *Economic History Review* 61, no. 1 (2008): 175–82; Sheilagh Ogilvie, *The European Guilds: An Economic Analysis* (London and Princeton: Princeton University Press, 2019), Ch. 8, 'Innovation'. See also Hilaire-Pérez, 'Technology as a Public Culture', 139–43 on the French context.

[76] Anne Goldgar and Robert I. Frost, *Institutional Culture in Early Modern Society* (Leiden and Boston: Brill, 2004), xiii–xix, quote at p. xiv.

[77] Miles Ogborn, *Spaces of Modernity: London's Geographies 1680–1780* (New York and London: Guilford Press, 1998), 1.

[78] Roberts, Schaffer and Dear, *The Mindful Hand*, xvi.

London Metropolitan Archives; the Spectacle Makers' and Clockmakers' Companies at Guildhall Library; the East India Company archive among India Office Records at the British Library; and the Goldsmiths' and Barber Surgeons' Companies, which have remained with those bodies. Such records can, however, be unforthcoming, more obviously offering themselves to institutional chronologies than to studies of meanings and cultures. However, recent scholarship has done much to help us understand the subjectivities and instabilities of archives, and the ways in which they reveal relationships of power, knowledge and corporate identity.[79] Among other things, this encourages attention to the material nature of the archive, including the roles of the cabinets, boxes and chests – in which they were kept and which were at the heart of corporate existence – and of the clerks who assembled and often lived with the institutional record.[80] These systems and practices reveal the 'layered purposes' of documents, including 'record-keeping' and 'memory-keeping' that gave meaning to corporate activities.[81]

Thinking spatially and materially

Miles Ogborn's characterization of the modern world is 'a hybrid and cosmopolitan one forged from a multiplicity of flows and networks of people, material objects and ideas, and their power-laden and inventive conjunctions and transformations in the places and spaces that they connect and create'.[82] Our case studies, too, consider these flows and networks by focusing on the institutional spaces and places through which they passed and within which they interacted. While we draw deeply on written accounts – unpublished manuscripts and contemporary works in print – our approach is also informed by the spatial, visual and material. This means, on the one hand, using buildings, plans, maps, images and objects as sources and, on the other, looking for evidence about spaces and objects within textual sources. Both working between different types of sources and 'thinking' spatially, visually and materially can provide additional ways of understanding institutional histories. We therefore draw on scholarship that explores geographies of the city and of knowledge; shows how spaces and practices shape each other, co-defining their meanings; and uses objects to investigate practices and to understand representations of place and identity.

London has fruitfully been thought of 'not as a single world, but as a set of proximate, sometimes overlapping places and spaces'.[83] 'Thick space' has been offered

[79] Kate Peters, Alexandra Walsham and Liesbeth Corens (eds), *Archives and Information in the Early Modern World* (Oxford: Published for The British Academy by Oxford University Press, 2018).

[80] Heather Wolfe and Peter Stallybrass, 'The Material Culture of Record-Keeping in Early Modern England', in Peters, Walsham and Corens, *Archives and Information*, 179–209 and Arnold Hunt, 'The Early Modern Secretary and the Early Modern Archive', in ibid., 105–31.

[81] Wolfe and Stallybrass, 'The Material Culture of Record-Keeping', 179. On 'layered purposes' and using studies of archives to understand practices of knowledge making, see also the introduction and essays in Elizabeth Yale (ed.), 'Focus: The History of Archives and the History of Science', *Isis* 107, no. 1 (2016): 74–120.

[82] Ogborn, *Spaces of Modernity*, 20.

[83] Deborah Harkness and Jean E Howard, 'Introduction: The Great World of Early Modern London', *Huntington Library Quarterly* 71, no. 1 (2008): 1–9 (p. 1).

as a characteristic of the metropole, evoking the density not just of the physical world (built environment, people, animals, things) but also of contextual and interpretive meanings.[84] Attention to urban spaces, understood as having 'generative capacity' and 'produced in a process that was both material and discursive', has added new dimensions to our understanding of the early modern city.[85] In this book we use spatial thinking at several different levels. At the highest, we pay attention to the city's connections, particularly with its river providing access to neighbouring settlements, docks, more open spaces and, of course, the sea. We also consider the geographies of neighbourhoods, exploring proximities of particular institutions and their role as centres of more informal networks and related businesses. Focusing on those institutions, we explore the nature of their sites and buildings and seek to understand the use and meaning of space within them. Such spatial approaches can be helpful in making sense of longer chronologies such as those discussed here.

It is now some decades since historians of science took their 'spatial turn', part of a process of 'challenging traditional views of the automatic universality of scientific knowledge'.[86] They have explored the connection between space and authority and the mutual impacts of spatial arrangements and practices, representations and perceptions.[87] Equally inspired and provoked by Foucault's cultural history of space, as well as the spatial analyses of Lefebvre and others, they have shown science and knowledge to be embodied and spatially located – the products of practical activities taking place within local social, cultural and material environments. Important contributions to this field have looked chiefly at the processes and effects of the creation of 'privileged sites' for science that developed in the modern period.[88] However, these too could be heterogeneous spaces. Dorinda Outram, for example, revealed the Muséum National d'Histoire Naturelle as a mix of public and domestic spaces, which connected urban professionals with amateur or non-elite field naturalists.[89]

Spaces have, then, been understood to have 'motivated, legitimated, directed, and constrained action'.[90] It has also been argued that they entangle the physical, political and social, being 'established performatively in speech acts, migration, and other human activities'.[91] Traditions, ceremonies and displays can play a role in giving meaning to

[84] See Dorothee Brantz, Sasha Disko and Georg Wagner-Kyora, 'Thick Space: Approaches to Metropolitanism' in their edited collection of the same name (Transcript Verlag: Bielefeld, 2012), 9–27, and Brantz, 'Metropolitan Natural Histories', 25–42.

[85] Peter Arnade, Martha Howell and Walter Simons, 'Fertile Spaces: the Productivity of Urban Space in Northern Europe', *The Journal of Interdisciplinary History* 32, no. 4 (2002): 515–48 (pp. 516, 541).

[86] Jon Agar and Crosbie Smith, 'Introduction', in Agar and Smith, *Making Space for Science*, 1–23 (p. 2).

[87] Adi Ophir and Steven Shapin, 'The Place of Knowledge: A Methodological Survey', *Science in Context* 4, no. 1 (1991): 3–22.

[88] Agar and Smith (eds), *Making Space for Science*, 9–14; see Henri Lefebvre, *The Production of Space*, trans. Donald Nicholson-Smith (Oxford: Basil Blackwell, 1991). That Lefebvre's ideas responded to work that contrasted a localized, 'placed' past with a 'placeless' modernity makes their appeal to historians of science and technology clear. See Arnade, Howell and Simons, 'Fertile Spaces', 520.

[89] Dorinda Outram, 'New Spaces in Natural History', in *Cultures of Natural History*, ed. Nicholas Jardine, James A. Secord and Emma C. Spary (Cambridge: Cambridge University Press), 249–65.

[90] Arnade, Howell and Simons, 'Fertile Spaces', 517.

[91] Mitchell G. Ash, 'Metropolitan Scientific Infrastructures and Spaces of Knowledge in Vienna, 1848–1918', in Ash, *Science in the Metropolis*, 1–21 (p. 8).

the knowledge produced and used within institutional sites. The materiality of spaces is part of what defines their meanings and the behaviours appropriate to them. This includes the objects – functional or decorative – within them and the way in which the built environment works with social hierarchies to offer or hinder access, control or secrecy.

As well as exploring the unique nature of specific sites, historians of science have noted the importance of networks and different forms of communication both in creating the authority of individual sites and in overcoming the specifics of locality and locally produced authority to create knowledge claims with wider meaning.[92] The processes by which knowledge is circulated have also been shown to be creative, producing new knowledge, adaptations and appropriations.[93] Considering the early modern period, Lissa Roberts has emphasized 'geographies of skill' over 'geographies of scientific knowledge', placing focus on the material objects and people involved and offering a more multidirectional process involving hybrid knowledge and practices. However, she notes too that 'the specific places in which goods and meanings were made mattered greatly to the complex patterns of skill's ownership and control'.[94] Historians have highlighted the physical movement and labour required for knowledge to travel, including the tacit knowledge of mobile craftsmen and the practices of mediators in adapting new techniques. Considerable efforts of regulation and standardization were required for knowledge to become public, generating sufficient trust for governance, tax collection and effective mobility.[95]

Geographical approaches have helped to map the connections between the city and its knowledge communities. Harkness has painted a vivid picture of specific neighbourhoods and individual streets, where personal connections and knowledge networks could overlap and reinforce each other.[96] Patrick Wallis's work on London's medical marketplace has revealed geographical concentrations, such as those of grocers and apothecaries around Cheapside and Bucklersbury in the fourteenth century, creating a persistent pattern: by 1666, druggists and apothecaries were 22 per cent of all householders in the parishes of St Benet and St Stephen Walbrook, 91 per cent of whom were based on or near Bucklersbury.[97] Studies of scientific instrument makers have shown proximity to other trades, particularly publishing and bookselling, clustered at St Paul's Churchyard, but also goldsmiths, linked by the practices of engraving as well

92 Agar and Smith, *Making Space for Science*, 21–3.

93 David Livingstone, *Putting Science in Its Place: Geographies of Scientific Knowledge* (Chicago and London: University of Chicago Press, 2003); James A. Secord, 'Knowledge in Transit', *Isis* 95, no. 4 (2004): 654–72.

94 Lissa Roberts, 'Geographies of Skill: Introduction' and 'Mapping Steam Engines and Skill in Eighteenth-Century Holland' in Roberts, Schaffer and Dear, *The Mindful Hand*, 189–96, 197–218 (pp. 216, 192).

95 Simon Schaffer, 'Introduction' to 'Art and Industry' in Roberts, Schaffer and Dear, *Mindful Hand*, 309–23.

96 Harkness, *The Jewel House*, Ch. 1.

97 Patrick Wallis, 'Consumption, Retailing, and Medicine in Early-Modern London', *The Economic History Review* 61, no. 1 (2008): 26–53 (pp. 31–2).

as natural and mathematical book knowledge.[98] The commercial success of the trade in the eighteenth century, with increasing diversity of products and customers, meant instruments were made and sold near user communities, such as maritime Wapping, along the long shopping street formed by Fleet Street and the Strand and increasingly towards more fashionable Westminster.[99]

Several of our case studies lead us to focus on the southeastern corner of the City, the location of the Tower of London, containing the Office of Ordnance and Royal Mint, and nearby a range of sites of maritime interest: the Navy Board, Custom House, Trinity House, East India House and Muscovy House. This was an area that was, as a result, what Simon Schaffer described as 'a world containing rational mechanics texts, instrument shops, stationers and naval trainees', as well as Ordnance gunners, draughtsmen and engineers. Already a place of opportunity for practical mathematicians, in the eighteenth century it was a key location for those 'ambitious to connect Galileo and Newton's principles with the demands of navigation, shipbuilding and commercial engineering'.[100]

The scholarship that has sought to reveal the hybrid practices and knowledges of the early modern world, blurring distinctions between science and technology, bringing into view a wider range of participants and activities, has been closely associated with the 'material turn' in the history of science and knowledge. This has encouraged focus on, among other things, the objects of scientific study, the materials transformed by artificial practice, the scientific instruments that enable the exploration of nature and its application, and the material nature of knowledge circulation.[101] *Materials and Expertise*, edited by Ursula Klein and Emma Spary, uses materials 'as a route into mixed artisanal and learned practices, which contributed to artisanal innovation, the development of the consumer market, and the formation of the observational and experimental sciences of the early modern period.' Material substances offered opportunities and posed problems for knowledgeable artisans, educated merchants, polite connoisseurs and university-trained practitioners, circulating between workshops, laboratories, marketplaces, coffeehouses, salons, hospitals, dispensaries, studies and lecture halls.[102] For Pamela Smith, objects 'are records of practice as well as

98 Mary Margaret Robischon, 'Scientific Instrument Makers in London during the Seventeenth and Eighteenth Centuries' (unpublished PhD, University of Michigan, 1983), 181–3; Boris Jardine, '"Neere unto the North dore of Paules": new light on the life and times of Humphrey Cole', *Bulletin of the Scientific Instrument Society* 142 (2019): 20–4.

99 Alexi Shannon Baker, '"This Ingenious Business": the Socio-Economics of the Scientific Instrument Trade in London, 1700–1750' (unpublished D. Phil thesis, Oxford, 2010), 62–74.

100 Simon Schaffer, 'The Charter'd Thames': Naval Architecture and Experimental Spaces in Georgian Britain', in Roberts, Schaffer and Dear, *The Mindful Hand*, 278–305 (p. 287). See also Stewart, *The Rise of Public Science*.

101 Useful outlines of some of these historiographical shifts can be found in Simon Werrett, 'Matter and Facts: Material Culture and the History of Science', in *Material Evidence: Learning from Archaeological Practice*, ed. Alison Wylie and Robert Chapman (New York: Routledge, 2014), 339–52; Adam Mosley, 'Objects, Texts and Images in the History of Science', introduction to a special issue of *Studies in the History and Philosophy of Science A* 38, no. 2 (2007): 289–302.

102 Ursula Klein and E. C. Spary, *Materials and Expertise in Early Modern Europe: Between Market and Laboratory* (Chicago and London: University of Chicago Press, 2010), 1–2.

repositories of knowledge', creating a 'vernacular science of matter'.[103] Surviving objects evidence access to materials, the proximity of particular kinds of expertise and know-how, the location of production, commerce, training or trialling.

Objects and material culture can have different meanings according to the spaces in which they are used or seen, and therefore we explore them together. Lissa Roberts has, for example, shown the different meanings attached to steam engines when they were displayed or used in gardens, lecture rooms or public spaces of the city.[104] In drawing out our institutional histories we have considered, where possible, surviving objects and have looked in textual and visual sources for evidence relating to materials, objects and the physicality of our spaces. In our case studies we have focused particularly on the skills and practices surrounding work with metals, whether precious metals and coinage or the iron and brass used to make instruments, tools and guns. Gunnery also leads to knowledge of the recipes for making and practices of testing and using gunpowder. These practices were increasingly associated with scientific instruments and how-to books, as were those of surveying, metrology and navigation encountered in other case studies. Beyond this, the knowledge associated with our institutional case studies was also given physical and visual form in maps, charts, paintings, statues, decoration and displays. Together, these elements help us to shed light on the collective cultures of knowledge explored in our case studies.

Case studies

We present our institutional case studies in Chapters 3–7, where we further explore and illustrate these themes. We do this at different levels, depending on the institution, the surviving sources and the questions we seek to explore. There are overlaps of geography or type of institution that allow us to have a tighter chronological focus in some chapters, while using others to consider longer histories. The latter require us to consider changes to the types of knowledge and practice pursued and to their (re)interpretation, internally and externally. Where they could, both individuals and institutions sought to adapt to changing circumstances, including the rising status of book learning and new formulations of the public virtue of useful knowledge. The people we meet in these sites acted in their own interests that often but not always coincided with those of the institution that shaped their opportunities and interactions. Throughout, we are interested in activities undertaken to meet institutional needs and forward corporate interests, but also in those undertaken because of, or shaped by, the nature of our institutional sites. These include activities that will be recognized as artisanal practices and those associated with more elite and scholarly forms of knowledge: we suggest both hybridity and continuity between these, linked as they were by personnel, location, materials and objectives.

[103] Pamela H. Smith, 'Making as Knowing: Craft as Natural Philosophy', in Smith, Meyers and Cook, *Ways of Making and Knowing*, 17–47 (p. 20); Smith, *Body of the Artisan*, 8.

[104] Roberts, 'Mapping Steam Engines and Skill in Eighteenth-Century Holland'.

The first two institutions we explore were based within a multifaceted site of contested jurisdiction, the Tower of London at the very eastern boundary of the customary City. As well as being a royal stronghold and prison, it was the site of the Royal Mint and Menagerie and of bureaucracy, information storage, trial and manufacture. Though it was a royal institution, its successful functioning and legitimacy depended upon the labour and expertise of city dwellers. Periodically, the know-how of foreign or 'alien' craftsmen was also sought out for technical improvements. Moreover, even in the sixteenth century, the Tower was a tourist destination. Chapter 3 focuses on the early modern Royal Mint – the primary institution in England for the production and testing of silver and gold coin – and explores the Tower as a major transnational site of knowledge production and interchange. We find that, in the production of coin and the testing of bullion and specie at this site, an uneasy dynamic operated between secrecy and openness. There was in general a principle of secrecy concerning all minting activities, and yet curious gentlemen natural philosophers visited the Mint, hoping to glimpse the assay trials in which precious matter underwent separation and transformation. Moreover, the skilled practitioners who worked within these sites authored treatises which circulated throughout the city, and which claimed to reveal workshop secrets and the reports of official trials to interested parties. With an eye to the social and political dynamics of technical knowledge creation and exchange, this chapter explores the multifunctionality of sites of knowledge production, the social and professional range of those involved in trials and experimentation, and the challenge of managing embodied knowledge and 'secrets' within urban space.

Also located in the Tower was the Ordnance Office, the government department responsible for the supplies and buildings of the army and the ordnance of army and navy. Under its jurisdiction were storehouses at major dockyards, the complex that became the Royal Arsenal at Woolwich, the Royal Observatory at Greenwich, the Artillery and Engineer Corps, the Royal Military Academy and the Ordnance Survey. Chapter 4 explores how, both centrally and across its associated sites and activities, the Ordnance was an important locus for the development and use of skills in practical mathematics – including surveying, architecture and engineering – and in managing, manipulating and experimenting with materials. New spaces arriving at this period, among their offices, stores, workshops and homes, were a Drawing Room in the Tower, a foundry and laboratory at Woolwich, and fortifications that formed a toehold in overseas territories. This chapter will consider the productive exchanges between skilled individuals living and working within or close to the Tower and the mathematical and experimental approaches that were fostered there and at the Ordnance's riverside sites. It will explore the relationship of the Ordnance to skilled contractors – from Continental iron founders to London's scientific instrument makers – and the collective contexts within which knowledge was shared and put to use. The self-consciously scientific and experimental culture that existed at Woolwich by the early nineteenth century had clear precedents in earlier practices and drew on a long tradition of exchange between Ordnance officials, mathematical practitioners and experimental philosophers.

Chapter 5 focuses on the livery company hall, a key type of institutional space in London. By the turn of the seventeenth century there were approximately sixty of these within the city walls. Typically grand courtyard houses in design, these buildings were the institutional homes of London's influential artisanal and mercantile guilds. From the late medieval period these corporations had been responsible for the regulation of crafts and trades, and the adequate training and disciplining of their members. Company halls were very important urban sites of civic politics, bureaucracy, sociability and trade. Additionally, they could be spaces for memorialization and for the performance of religious devotions and philanthropy. Taking as our focus two such company buildings – Goldsmiths' Hall on Foster Lane and Barber Surgeons' Hall at the northwest boundary of the City – this chapter explores how guild buildings could operate as spaces of material knowledge production and exchange across the seventeenth and early eighteenth centuries. While considering the multipurpose nature of such buildings, particular attention will be given to two spaces that existed within them that were closely associated with the use of knowledge, technical practice, production and the maintenance of institutional authority. These were the Assay Office in Goldsmiths' Hall and the Anatomy Theatre at Barber Surgeons' Hall. The former was crucial to public confidence in the value of precious metals and closely interacted with the assaying practices at the Mint, raising issues about the location of expertise and commercial and national sensitivities. The latter, designed and constructed in the mid-to-late 1630s under the supervision of Inigo Jones, was the capital's first purpose-built facility for the public dissection of cadavers. The Anatomy Theatre established Barber Surgeons' Hall as a metropolitan centre for anatomical observation and learning (it also ignited the public imagination and indignation about institutional interference with human remains).

Another institution based in a courtyard house was the Corporation of Trinity House, a maritime guild founded to regulate pilotage in the Thames and for the charitable relief of mariners. Its rising significance reflected the growth of London's port and its maritime trade, which were evermore central to its self-presentation, as well as its increasing need for effective control over the environment of the Thames and the international shipping that navigated its channels. Our focus in Chapter 6 is on Trinity House's home, from 1661 to 1796, in Water Lane, Tower Street Ward. Here it was located close to many sites central to its role, of ever-greater relevance to the city and the merchant and royal navies. From within this building it exercised its responsibilities and monopolies, including establishing and maintaining lights, buoys and markers in the Thames and beyond; collecting fees and duties; examining and certifying pilots, Naval ships' masters and boys of the Royal Mathematical School; surveying the river's navigable channels; and dredging and selling ballast. It was also called upon, by parliament, courts of law, the City, Admiralty and Royal Society, to offer its appraisal of matters navigational, including disputes at sea, surveys, improvement schemes and new machines and instruments. Through its governing members it reflected the knowledge and interests not just of experienced mariners but also of overseas trade and naval administration, while relying on the skills and knowledge of a network of pilots, lighthouse and buoy keepers, mathematicians, surveyors, clerks and ballast men. This chapter will explore Trinity House's regulation, maintenance and representation of knowledge and practice. It highlights an emphasis on experience and collective opinion in the judgements and claims to expertise of its members, while acknowledging a tension between the increasingly wealthy, gentlemanly and aristocratic leadership and their oversight of technical activities such as mathematical examination and specialist hydrographic survey.

The EIC, the focus of Chapter 7, was one of the sources of such wealth and of navigational experience and expertise. Its headquarters, successively extended, remodelled and rebuilt from 1648 until 1800, was East India House on Leadenhall Street. It was the principal seat of authority for English commerce in the Indian Ocean and beyond, and, from 1757, it also became the administrative centre of a very large and expanding territorial empire. As such, the EIC needed to gather, store, assess and mobilize large bodies of knowledge and expertise about the sea routes to Asia, about the commodities to be traded there and about the languages, politics and cultures of Asian peoples. The building stood at the financial heart of London, while also overseeing an array of increasingly far-flung spaces of work and knowledge-making, in London, its surroundings and overseas. The Company was at once a producer and user of natural and technical knowledge, a testing ground for new equipment and ideas, and an immense human and logistical infrastructure that men of science – including, notably, the Company's own servants and officers – sought to exploit. We explore how the early experience of voyages to Asia was transformed into useable navigational knowledge, and how the EIC selected and examined its mariners and captains; how the criteria and technologies for this changed over time, highlighting evolving attitudes to secrecy and openness; and how the Company's increasingly specific requirements shaped, and were shaped by, the wares and business models of their neighbouring suppliers. It examines how knowledge accrued by the Company and its servants was organized, assessed, stored and valued in London, and how its mercantile pursuits shaped the rhetorical presentation of the Company and its headquarters as a scientific centre.

First, however, Chapter 2 provides an overview of London's sites of exchange for knowledge and practice, from the public spaces of the river, streets, taverns, coffee houses, shops, markets and the Royal Exchange to the relatively private and enclosed spaces of homes, offices and workshops. It puts our case studies in the context of the historiography of urban, social and knowledge spaces and of a broader understanding of London's geographies of exchange, consumption and display. It also attempts to grasp the range of goods and manufactures that Londoners might encounter, including exotic items imported from overseas, and which can be helpfully understood as the products or tools of natural knowledge and artificial practice. Such objects include scientific instruments, the product of a growing London trade that highlights the significance of craft skills and reveals connections between several of our institutional sites. The trade, and the mathematical, optical and philosophical instruments it produced, was a visible signifier of knowledge and practice in the city, and particularly of its application to trade and maritime or military venture. Its civic institutions, the Spectacle Makers' and Clockmakers' Companies, which lacked their own Halls, contrast with the larger and more established sites discussed in the following chapters, but their self-representations allow an exploration of the changing relationship between craft skill, natural knowledge and scientific expertise that characterizes our period.

2

City sights: Objects of knowledge and sites of exchange

The river of Thames

Despite the disruptions of the mid-seventeenth century, the sense of London's developing commercial and cultural significance that was evident in 1600 continued to increase. The city challenged and then overcame the dominance of Antwerp, reinterpreting its own history to claim this as the fulfilment of its destiny.[1] Its increasing overseas trade, and the range of people and goods that arrived from abroad, was becoming key to the city's identity, linking it evermore closely with the river through which that trade passed after navigating the high seas. The Thames was, as Joseph Ward has put it, 'a mirror for the metropolis', which increasingly reflected its commercial role.[2] The river, presented as a source of wealth and comfort, was feted in poetry and masque, depicted in paintings and sculpture. In his poem, 'On Cooper's Hill' (1668), John Denham described how the Thames, 'the most lov'd of all the Ocean Sons', brought bounty to his own shores and then

> Visits the world, and in his flying towers
> Brings home to us, and makes both *Indies* ours;
> Finds wealth where 'tis, bestows it where it wants
> Cities in deserts, woods in Cities plants.
> So that to us no thing, no place is strange,
> While his fair bosom is the World's exchange.[3]

The bounty of the Thames, and its crucial role in carrying the wealth and goods that the city could offer to the monarch and kingdom, was celebrated through the pageantry that it staged. In particular, there was the annual parade for the inauguration of the Lord Mayor on 29 October, an event that had become evermore elaborate

1 See e.g. Alford, *London's Triumph*; Lincoln, *London and the Seventeenth Century*.
2 Joseph P. Ward, 'The Taming of the Thames', *Huntington Library Quarterly* 71, no. 1 (2008): 55–75 (p. 56).
3 John Denham, *Coopers-Hill. A Poem* (London: Printed and Sold by H. Hills, in Black-Fryers, near the Water-side, 1709), 10, 11.

across the sixteenth century. It carried representatives of the City to Westminster in barges and then staged celebrations along the river for the return journey. These were complex collaborations between writers, artificers and guild worthies, and occasions of music, tableaux and masque, representing mythical figures, exotic creatures and astrological endorsement of the monarch.[4] By the early seventeenth century, the larger livery companies were routinely spending around £1,000 on each spectacle.[5]

In 1585, the Thames was represented as a nymph, who offered her fish and waterways to the City and declared her loyalty to the Queen. Beside her were a soldier and a sailor and 'this thrise reuerend honorable Dame,/ Science the sap of euery common wealth' who 'Surnam'd Mechanicall or Liberall/ Is vowed to honour London with her skill'.[6] In 1634, when Sir Robert Parkhurst of the Clothworkers' Company was chosen as Lord Mayor, his barge was filled with Company goods that had come 'from foreigne parts by sea'. Parkhurst was then addressed by an actor representing Thetis, a goddess of the sea, who spoke of how, by the Thames, 'this Child of mine', she had sent

> … silks and velvets, oyle, and wine,
> Gold, silver, Jewels, fish, salt, sundry spices,
> Fine and coarse linen, drugs of divers prices:
> What every Realme or climate can produce,
> I see it safe transported for your use.[7]

The wealth, goods and skills of the city were also made part of the *Aqua Triumphalis*, held on 23 August 1662 to celebrate the marriage of Charles II and Catherine of Braganza and to assert the loyalty of the city to the restored monarchy. John Evelyn called it 'the most magnificent Triumph ever floted on the Thames', with Samuel Pepys estimating the presence of 10,000 vessels.[8] These were 'dressd and adorned with all imaginable Pomp' and accompanied by 'Thrones, Archives, Pageants & other representations, stately barges of the Lord Mayor & Companies, with various Inventions, musique, & Peales of Ordnance both from the vessels and shore'.[9] These last would have been produced by gunners and guns supplied by the Office of Ordnance, based at the Tower of London, an industrious site we will turn to in the following two chapters. An engraving of the crowded scene shows 'Thetis *drawn in a Sea Chariot, accompanied with* Tritons, *and* Sea Nymphs', and Thames, accompanied by representations of a lion

4 Tracey Hill, *Pageantry and Power: A Cultural History of the Lord Mayor's Show, 1585–1639* (Manchester: Manchester University Press, 2010), Ch. 2.

5 Ian W. Archer, 'The City of London and River Pageantry: 1400–1856', in *Royal River: Power, Pageantry and the Thames*, ed. Susan Doran (London: Scala, 2012), 80–5 (p. 80).

6 George Peele, *The Device of the Pageant Borne before Wolstan Dixie Lord Maior of the Citie of London* (London: Printed by Edward Allde, 1585), A2r–v.

7 Jonathan Schneer, *The Thames: England's River* (New Haven and London: Yale University Press, 2005), 67.

8 *The Diary of John Evelyn*, vol. 3, *Kalendarium, 1650–1672*, ed. E. S. de Beer (Oxford: Clarendon Press, 1955), 333; *The Diary of Samuel Pepys*, 11 vols, ed. Robert Latham and William Matthews (London: Bell, 1970–1985), vol. 3, 1662 (1970), 175.

9 *The Diary of John Evelyn*, vol. 3, 333.

and a unicorn.[10] The key to unlocking the riches supplied by the river from overseas was successful navigation, and thus the printed account noted that '*Thetis* was also married to *Peleus* an excellent Navigator, and therefore is presented with gray hairs, which is partly to intimate the Antiquity of Navigation, and partly to shew the cares and fears of Seamen, who are observed to become sooner Gray hair'd then others'.[11]

Actual workers of boats and ships were not only required to manoeuvre the craft on the teeming river but were also drawn into the festivities. Watermen and seamen performed two 'Drolls', involving continuous 'dancing and singing' at different points on the river. The seamen, 'both on the Stage and the Rope', assumed 'several Antick postures and dances' to entertain the nobility and gentry as they waited for the more formal entertainment offered to the king and queen to begin.[12] Other trades of the city were also visible, particularly through the 'pageants', or floating stages, of the Twelve Great Livery Companies. Many of these – the Mercers, Drapers, Merchant Taylors, Haberdashers and Skinners – represented heraldic animals, patron saints, personified virtues and representations of luxury that included the presence of liveried Black servants. The float of the Goldsmiths' Company, one of the subjects of Chapter 5, carried a figure of Justice holding symbols of the craft's practice and trustworthiness, a balance and a touchstone, with two virgins before her, holding a golden hammer and golden cup. Unmentioned in the published account but visible in the print is a distinctly artisanal addition to this scene, with men using tools and closely scrutinizing some product of this skill (Fig. 2.1).

While the figure raising a hammer at the block exemplified the workshop endeavours of goldsmiths, the four men positioned around the table are likely to have represented the regulatory authority of the Goldsmiths' Company. Earlier in the century, the inaugural procession of goldsmith Sir James Pemberton as lord mayor of London – Anthony Munday's *Chruso-thriambos: The Triumphes of Golde* (1611) – featured a magnificent pageant sculpted like a golden mountain. Arranged in hierarchical formation on the 'Rocke or Mount of Gold' were figures representing the various labours and crafts essential to the goldsmiths' trade, including miners who found the ores of gold, refiners and 'the Mint-Maister, Coyners, Gold Smithes, Iewller … Plate-Seller'. Towards the very summit of the mountain was a figure integral to the regulation of the trade who tested the material purity of ores, plate and coin: 'to distinguish those precious Mettals of Gold and Siluer, from base adulterating or corruption, wee show there also an ingenious Say-Maister'. For spectacular effect, fireworks emerged from the assayer's furnace on the pageant.[13] We will encounter this assay master and his culture of knowledge again at the Tower in Chapter 3.

[10] Dirk Stoop's etching of the spectacle was published in London in 1662; John Tatham, *Aqua triumphalis, Being a True Relation of the Honourable the City of Londons Entertaining Their Sacred Majesties upon the River of Thames and Wellcoming Them from Hampton-Court to White-Hall Expressed and Set Forth in Severall Shews and Pageants the 23. day of August, 1662* (London: Printed for the author by T. Childe and L. Parry, 1662), 11, 6–7.

[11] Tatham, *Aqua Triumphalis*, 11. See also Schneer, *The Thames*, 62; and, on the iconography of Restoration London's street pageants, L. J. Morrissey, 'English Pageant-Waggons', *Eighteenth-Century Studies* 9, no. 3 (1976): 353–74.

[12] Tatham, *Aqua Triumphalis* [n.p.].

[13] Jasmine Kilburn-Toppin, *Crafting Identities: Artisan Culture in London, c.1550–1640* (Manchester: Manchester University Press, 2021), 188.

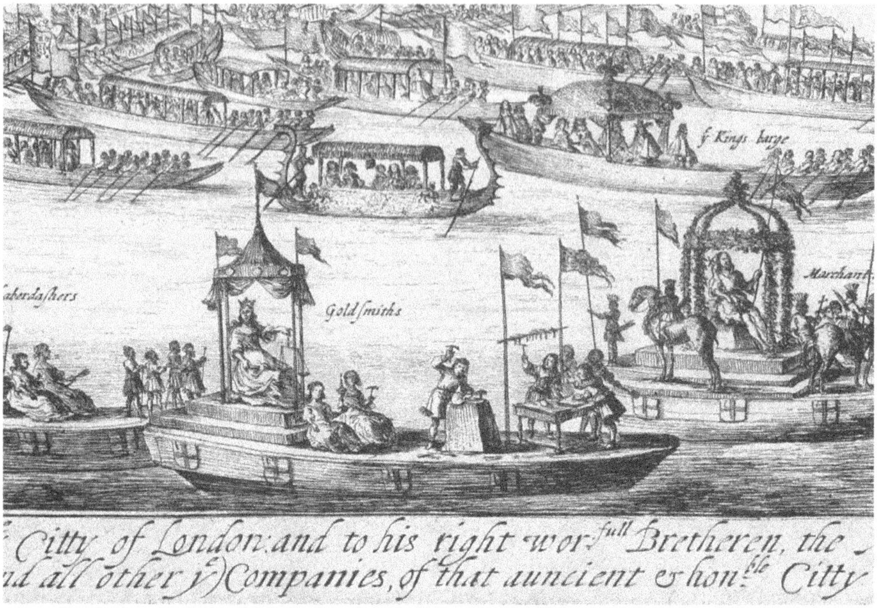

Figure 2.1 Detail from Dirk Stoop's etching of the 1662 *Aqua Triumphalis*, showing the pageant of the Goldsmiths' Company (BM: Sheepshanks.3095). © The Trustees of the British Museum.

Londoners and visitors commented on the river as a site of remarkable activity and industry even on ordinary days. Ships were built, goods unloaded, people transported, wharfs built and banks maintained. Its navigability below London Bridge was the responsibility of Trinity House which, by the eighteenth century, had responsibility for maintaining the river's depths. This was done by removing silt from the riverbed, dug up by crews working ballast lighters, and by surveying the shifting sandbanks that defined the river's navigable channels. As will be discussed in Chapter 6, the surveys carried out by Trinity House, sometimes accompanied by members of the Navy Board, had their own kind of performativity. It was apt that they, too, used the river each year when their Master was elected, borrowing barges from the Admiralty and Navy Board to transport its Elder Brethren – made up of captains and admirals of the merchant and royal navies, as well as representatives of the Admiralty and Navy Board – from the City, their place of business since 1661, to their original home in the maritime parish of Deptford. That this might be an opportunity for modest display is indicated by an order in 1779 for 'a new Silk Flag, with the Arms of the Corporation emblazon'd thereon' to be made 'preparatory to Trinity Monday next' when it would be flown on one of the barges as they travelled down and back up the river.[14]

Because the river provided transport, space, water, power and waste disposal, it was a key site of industry. While much of this was centred on the production of ships and chandlery, the river's banks supported a wide range of productive activities. London and

[14] 3 April 1779, Court Minutes, Corporation of Trinity House, MS30004/013, p. 48 (London Metropolitan Archives).

its neighbouring parishes were both the largest market and the centre of production, first in England and then in Europe, whether for luxury goods such as clothing, jewellery and furniture or for food, tools, paper, engines and building materials.[15] The open site of the river would have revealed both traditional and innovative practices, and lent itself to experimental demonstrations, especially for engines designed to pump water. In the 1580s, Peter Morris, a German immigrant, received a patent and put on a public demonstration of his water pumps by driving water through lead pipes over the steeple of St Magnus Church at the north end of London Bridge. The City gave him £50 and allowed him the lease of two arches under the bridge. Here he placed wheels that pushed water to nearby houses and up to a conduit at the east end of Cornhill. In 1600, the mathematical practitioner Edward Wright received a loan of £240 from the City to complete a pump that would draw water from the Thames to Leadenhall. The City also granted him access to St Botolph's Wharf, close to the Tower, and arranged for the lease of additional land from the Muscovy Company for the works in 1602.[16] Wright's 1598 patent was for a device that would draw water 'for the daily use of the city of London' but also 'of any other city or town, for draining and drawing water out of mines, fen grounds, marshes and other waterlogged grounds and for the grinding of corn'.[17] The City and its needs supported innovation, whether in machinery or commercial arrangements, by Londoners, Strangers or Aliens. It also became increasingly dependent on private companies and developing technological systems.[18]

Building on Morris's business, by the early eighteenth century the London Bridge Water Works ran four waterwheels, took over other water conduits and had become incorporated into a network of supply.[19] It remained a site of innovation and an impressive sight to the curious. The new wheels, designed by George Sorocold, were illustrated for the *Universal Magazine* in 1749, the scale exaggerated in comparison to a small wherry carrying sightseeing gentlemen who point to its churning motion (Fig. 2.2).[20] Just over a decade later, a steam engine was installed at the northern end of the bridge, designed to overcome the regular loss of power that came with the turn of the river's tide. Further upriver, the York Buildings Company had already installed a steam-powered pump designed by Thomas Savery, replacing it in the 1720s with a Newcomen engine, a project on which the experimental philosopher John Theophilus Desaguliers advised. A large wooden tower on the river bank below the Strand, visible in many paintings and engravings of the north bank, contained the engine, emitting such noise and smoke that it was dubbed a dragon.[21] Earlier, Savery had used the Thames to host a more temporary demonstration. He had 'made sufficient demonstration of the use of my ENGINE' – a paddleboat powered by an eight-man capstan – 'on the THAMES, of which thousands were Eye-witnesses'. He claimed that they 'seem'd to like

15 Martin Daunton, 'London and the World', in Fox, *London – World City*, 21–38 (p. 35).

16 Leslie Tomory, *The History of the London Water Industry, 1580–1820* (Baltimore: Johns Hopkins University Press, 2017), 32–4, 40.

17 Patent quoted in Jardine, 'Instruments of Statecraft', 318.

18 See Mark S. R. Jenner, 'From Conduit Community to Commercial Network? Water in London, 1500–1725', in Jenner and Griffiths, *Londinopolis*, 250–72 (quotation p. 264).

19 Tomory, *The History of the London Water Industry*, 88.

20 H. Beighton, 'A Description of the Water-Works at London Bridge', *The Universal Magazine of Knowledge and Pleasure* 5 (1749): 241–4.

21 Stewart, *The Rise of Public Science*, 344–54.

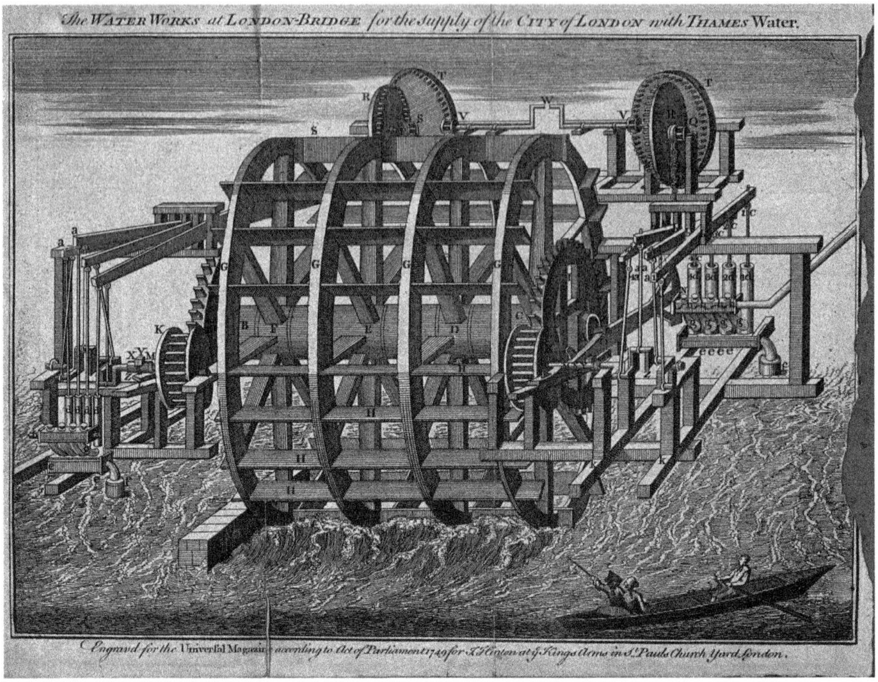

Figure 2.2 The water works at London Bridge for the supply of the City of London with Thames water, engraved for the *Universal Magazine*, 1749. Wellcome Library no. 37626i.

it', that 'the Publick News Papers' spoke well of it, although the hoped-for support from the Admiralty was not forthcoming.[22]

City streets

Neither pageantry nor experimental demonstrations were limited to the river. Within the streets of London, celebrations of what were seen as the sources of prosperity were played out, themselves drawing on, and drawing in, skills and innovation. While Charles II's marriage was celebrated on the river, his coronation the year before saw the traditional procession from the Tower of London, across the City and to Westminster Abbey. Pepys described how he and two Navy Office colleagues watched from the home of John Young, flag-maker to the Navy, on Cornhill, while other members of his household had seen the festivities from Tower Hill and Fleet Street.[23] The Cornhill location, near the Royal Exchange, placed them close to the second of four temporary triumphal arches, this one being dedicated to the Navy. This and the other arches were

[22] Thomas Savery, *Navigation improv'd: or, The Art of Rowing Ships of All Rates, in Calms, with a More Easy, Swift, and Steady Motion, than Oars Can* (London: Printed and sold by James Moxon, at the Atlas in Warwick-Lane, 1698), 18.

[23] *The Diary of Samuel Pepys*, vol. 2, 1661 (1970), 82–3.

designed by an Englishman and a Dutchman: the architect and surveyor Peter Mills, for many years the Master of the Tylers' and Bricklayers' Company, and the courtier and architect Sir Balthazar Gerbier. The designs and the pageantry were described in print by John Ogilby, a Scottish-born and London-bred cartographer and publisher with experience of mounting theatricals and revels.[24]

The Naval arch celebrated imperial ambitions that united the interests of Crown and City and drew on skills of mariners and navigators. It was surmounted with a figure of Atlas bearing a globe, topped with a ship, below which were two armillary spheres, representing astronomical navigation, and the motto Unus Non Sufficit (one is not enough – a reference to Juvenal's comment on Alexander the Great, one world is not enough). The centrepiece was a painting of the young Charles with his father, pointing to the ship *Sovereign of the Sea*. Beneath this, 'living Figures', balanced precariously on the pedestals of the upper story, represented Europe, Asia, Africa and America, 'with Escutcheons, and Pendents, bearing the Arms of the Companies, Trading into those Parts'.[25] Military and economic power were symbolized by depictions of the Tower and Royal Exchange. On the other side of the arch, a painting represented the king's brother, James Duke of York, then Lord High Admiral, as Neptune, with four female 'living Figures' within niches below, 'representing ARITHMETICK, GEOMETRY, ASTRONOMY, AND NAVIGATION'. There were also two stages erected on the street, one with a man as river Thames, crowned with London Bridge and holding a model ship. The other stage, made to look like the upper deck of a ship, supported three seamen, accompanied by drums and trumpets, who sang of ships, battles and distant lands as the procession passed.[26]

Charles's coronation procession was the last that took the traditional route from the Tower to Westminster. However, work associated with the Tower remained a feature, for the gunners and fireworkers of the Ordnance Office there were called on for salutes and displays. The specialist work of designing and constructing fireworks drew on imported skills, as well as local artisans and military manpower, and produced ingenious effects that provided popular spectacles and attracted the interest of natural philosophers.[27] While the river remained a focus for displays in 1613 (the wedding of James I's daughter, Elizabeth), 1685 (the coronation of James II) and 1713 (the end of the War of Spanish Succession), the westward spread of fashionable London meant that Green Park was chosen to celebrate the end of the war of Austrian Succession in 1749. Gun salutes were a more regular feature of Ordnance life. In the autumn of 1761, correspondence records Ordnance arrangements for salutes marking George III's wedding, coronation and entry to the city on Lord Mayor's Day. Ensuring correct timing could involve 'a Signal Rockett fired from the Great Court in St James's', 'a Flag Hoisted on the North West Tower of Westminster Abbey' or 'an Ordnance Jack Hoisted

[24] Construction was by Mills and the designs have usually been attributed to Gerbier, although drawings of the arches have been attributed to the architect and sculptor Edward Pearce: Christine Stevenson, 'Occasional Architecture in Seventeenth-Century London', *Architectural History* 49 (2006): 35–74 (p. 56). On Ogilby see ibid., 56–8 and Charles W. J. Withers, *Geography, Science and National Identity: Scotland Since 1520* (Cambridge: Cambridge University Press, 2001), 65–7.

[25] John Ogilby, *The Relation of His Majestie's Entertainment Passing through the City of London, to His Coronation* (London and Edinburgh: 1661), 12.

[26] Ogilby, *The Relation of His Majestie's Entertainment*, 11, 13–15 (quote pp. 12–13).

[27] See Werrett, *Fireworks*.

on London Bridge'. On Lord Mayor's Day, 'Guns on the Towers Wharf' were to be fired 'as in King George the First's time'.[28]

The sounds of gunpowder explosions were by no means limited to displays, with the proofing of guns, powder and shells conducted regularly at the Tower and frequent exercise of gunners, as well as trials of guns, taking place at the Artillery Garden in Spitalfields, as discussed in Chapter 4. The Ordnance was a key customer but private citizens also supported London's gun-making industry, which, from 1675 to its height in 1715, saw 450 makers active in the city, largely in areas neighbouring the Tower.[29] This could be dangerous work. A 1715 explosion at Mr Walker's gunpowder shop in Thames Street, between Custom House and Billingsgate, killed fifty people.[30] It also destroyed property, including the building of Trinity House. The following year, Matthew Bagley, the Ordnance's contracted gunfounder, and several gentlemanly witnesses lost their lives when the large-scale recasting of captured French guns at his Moorfields foundry went badly wrong.[31] This accident precipitated the Ordnance's decision to build their own brass foundry as part of their expansion of the site that became Woolwich Arsenal, combining metallurgical and chemical manufacture, trials, training and education, downriver from the City. In the City and at Woolwich, trials of new techniques or designs for guns were demonstrations, made before plausible witnesses and potential patrons.

From the 1660s, the performance of machines, engines and instruments was part of the business of the new Royal Society, but demonstrations were also witnessed by less select audiences in courtyards, buildings and streets across the city, just as Savery's paddleboat was widely seen on the Thames. As Harkness described, by the end of the sixteenth century Londoners had many examples of mechanical skill in their vicinity, whether the production of power by waterwheels and windmills, the processes carried out in glasshouses, kilns and workshops, or the exciting novelty of 'artificial motion' as demonstrated by Henrick Johnson at the Guildhall.[32] The eighteenth century saw some 136 steam engines put to work across the capital, pumping water to supply houses and businesses including foundries and breweries.[33] More amusingly, at the Sign of the George near the Royal Exchange in 1717, one could view from 9 am to 9 pm an 'Astronomical and Musical clock' which played every three hours a 'variety of … Marches, Country Dauces, and Opera Tunes'. The clock showed 'the Golden Number' and the 'Motion of the Sun with the exact Time of his Rising and Setting, the Length of

[28] Surveyor of Ordnance, Sir Charles Frederick, Letter Book – 1761–1767 (pp. 1, 2, 7): Records of the British Surveyor of Ordnance, MsColl 1075 (New York Public Library).

[29] Lois G. Schwoerer, *Gun Culture in Early Modern England* (Charlottesville: University of Virginia Press, 2016), 90.

[30] Werrett, *Fireworks*, 97.

[31] Charles Ffoulkes, *The Gun-Founders of England: With a List of English and Continental Gun-Founders from the XIV to the XIX Centuries* (Cambridge: Cambridge University Press, 2011), 55–6.

[32] Harkness, '"Strange" Ideas and "English" Knowledge', 147.

[33] John Kanefsky and John Robey, 'Steam Engines in 18th-Century Britain: A Quantative Assessment', *Technology and Culture* 21 (1980): 161–86 (pp. 175–6). The authors count seven steam engines in London for the period to 1733 and twenty to 1780.

the Day and Night at all Times also his Meridian Altitude', as well as the phases of the Moon and the times at which eighteen stars reached the meridian.[34]

Instruments associated with natural and experimental philosophy also became increasingly visible. From 1675 those who made the excursion to Greenwich would have seen the new observatory building, with its long refracting telescopes visible to those walking in the Royal Park. However, around 1664 Gresham College in Bishopsgate, a site for learning and lectures since the end of the sixteenth century, might have afforded those in the City a view of a similar telescope, supported on a high pole, made by the optical instrument maker Richard Reeve for the College's Professor of Geometry, Robert Hooke.[35] Hooke also set up a rooftop observing platform that would have been visible from neighbouring streets and buildings. Similar structures were to be found at schools specializing in mathematics for navigation, notably at Christ's Hospital in Newgate Street, or on the houses of instrument makers, including that of Jeremiah Sisson in Beaufort Buildings – recorded in a 1760s view towards the Strand by the Ordnance draftsmen Thomas and Paul Sandby (Fig. 2.3/Plate 1). In the late seventeenth century, John Marshall, adept at self-promotion, demonstrated the power of his telescopes by having 'his name placed in white on the roof of a house some twenty houses from his own'.[36] Prospective customers must have stood in, or been visible from, the street as they trained Marshall's instruments on the lettering.

The city's increasingly global trade also made its impact on London's streets. The overseas trading companies celebrated in the tableau vivant of the Naval coronation arch made some individuals very wealthy, at the expense of peoples and lands that they exploited. Their imported goods could threaten livelihoods in London, such as those of the thousands of weavers who marched through the city in 1697, but they also supplied the businesses of apothecaries, grocers and haberdashers. The EIC, the headquarters of which we consider in Chapter 7, was vital to the Ordnance and gun-makers for their supply of saltpetre, a key ingredient of gunpowder. Imported goods and materials also offered a view of the wider world to the curious, whether through objects incorporated into collections, new additions to botanical gardens or simply what could be seen by ordinary people in the streets. Arriving from Bermuda, bananas – the first seen in the city – were displayed by the apothecary Thomas Johnson in his shop window for several weeks in 1633.[37] Exotic animals were brought to public view, whether they were diplomatic gifts, like the elephant that arrived in 1675, or curios for sale and spectacle, like the rhinoceros of 1684.[38] Elephants were still attracting attention a century later, when one was landed at Tower Wharf and walked through the streets to the menagerie at Exeter Exchange on the Strand in 1785. Its journey underlines the connection

34 *Old Bailey Proceedings Online* (www.oldbaileyonline.org, version 8.0, 27 May 2022), October 1717 (a17171016-1).

35 The telescope in the centre of the courtyard is illustrated by Hooke: Classified Papers, CLP/20/61 (Royal Society).

36 Bennett, 'Shopping for Instruments', 374.

37 Wallis, 'Consumption, Retailing, and Medicine', 40.

38 Christopher Plumb, *The Georgian Menagerie: Exotic Animals in Eighteenth-Century London* (London: I.B. Tauris, 2015), 106–7; Anna Winterbottom, 'An Experimental Community: The East India Company in London, 1600–1800', *British Journal for the History of Science* 52, no. 2 (2019): 323–43 (pp. 327, 330–6).

Figure 2.3 Watercolour by Thomas and Paul Sandby showing a view of Beaufort Buildings, looking towards the Strand, 1760s. The figure and apparatus on the roof among the chimneys left of centre show the location of Jonathan Sisson's observatory (BM: 1880,1113.2854). © The Trustees of the British Museum.

See also colour version, reproduced in plate section 1.

between distant lands and seas, via the Thames, to London's military stronghold and, ultimately, sites of fashionable display and commodity.[39] Elephants' undoubtedly unpleasant perambulations around the city point us to other sites in which knowledge and novelties were exchanged and displayed. The elephant of 1675 was taken to Garraway's coffee house, while in 1720 one was shown at West Smithfield, where it was plied with tavern-supplied ale by spectators. After its death, it was dissected for the edification of philosophers and physicians on Sir Hans Sloane's lawn.[40]

If the river Thames was a mirror of the metropolis, the Royal Exchange was lauded as its microcosm: the whole world captured within the heart of the City. Modelled on Antwerp's Bourse, in Flemish style, it was founded by Sir Thomas Gresham in the 1560s and signalled his desire that London should be a rival commercial force. It came to represent the success of that ambition, and ultimately England's, and particularly London's, economic dominance, based on access to goods, people, knowledge and

[39] Christopher Plumb, *Exotic Animals in Eighteenth-Century Britain* (PhD thesis, University of Manchester, 2010), p. 39; *idem*, *Georgian Menagerie*, 157–9.

[40] Christopher Plumb, '"Strange and Wonderful": Encountering the Elephant in Britain, 1675–1830', *Journal for Eighteenth-Century Studies* 33, no. 4 (2010): 525–43 (p. 527).

ideas from around the world. The Exchange was designed as an open courtyard with a covered walkway on all sides which facilitated business discussion in all weathers; it became *the* place to hear commercial news and gossip concerning overseas trade.[41] It also included two floors of shops, 120 in all, for which, by the late sixteenth century, there was a waiting list of potential lessees.[42] One visitor estimated that the building could hold up to 4,000 merchants.[43] Often compared to the Tower of Babel, it was said to have been built to accommodate the 'bargaines, contracts and comerce' of tradesmen and the 'great store of merchants strangers from many forraine nations', becoming known as 'The Eye of London'.[44] The second, classical Exchange, built after the Great Fire, had distinct areas, or walks: 'Each nation has its own place … so that those who have trade to do with them can find them with ease' to hear 'reports from many countries and regions'.[45] It made, said Joseph Addison, 'this metropolis a kind of *Emporium* for the whole Earth'.[46] The anonymous early eighteenth-century author of *Hickelty Pickelty* could not do better than appropriate a description of St Paul's Cathedral from Bishop John Earle's *Microscosmography* eighty years earlier: it was 'the whole World's Map, which you may here discern in its perfectest Motion, justling and turning'.[47]

A trip to the Exchange might be business, pleasure or simple diversion. Visitors to London came to admire the building itself. Ambrose Walford from Essex – who ended up as an unfortunate victim of organized violent theft – had 'come to Town upon Business' in 1719, and 'being willing to satisfy his Curiosity in viewing the Publick Buildings &c. went to the Royal Exchange … gratified his Eye with that fine Structure and the Effegies of the Kings and Queens there'.[48] There was much to buy from the Exchange, if it could be afforded, from permanent shops (such as haberdashers, mercers, stationers, watchmakers and jewellers), or traders who set up stalls within or outside the courtyard. There was also much to be observed. Pepys browsed 'a great plenty of fine prints' at the Exchange in April 1666 and, a year later, bought there 'a few more prints of cittys'.[49] It attracted other businesses into its neighbourhood, offering goods, services and refreshments. Bookshops surrounded the Exchange (like that operated by Thomas Hacket at the sign of the Green Dragon, which specialized in practical works such as books on navigation), and from the second half of the seventeenth century coffeeshops also proliferated.[50] This commercial site was also part

41 Jean E. Howard, *Theatre of a City: The Places of London Comedy, 1598–1642* (Philadelphia: University of Pennsylvania Press, 2007), 33.

42 Linda Levy Peck, *Consuming Splendor: Society and Culture in Seventeenth-Century England* (Cambridge and New York: Cambridge University Press, 2005), 46.

43 Alford, *London's Triumph*, 109.

44 Edmond Howes, *The Annales, or General Chronicle of England, Begun First by Maister John Stow …* (London, 1615), 667, 868.

45 Quoted in Christine Stevenson, 'Making Empire Visible at the Second Royal Exchange, London', in *Court, Country, City: British Art and Architecture, 1660–1735*, ed. Mark Hallett, Nigel Llewellyn and Martin Myrone (New Haven and London: Yale University Press, 2016), 51–72 (p. 53).

46 *The Spectator* 1, no. 69 (19 May 1711), 392.

47 Anon, *Hickelty Pickelty: Or, a Medly of Characters* (London: 1708), 10.

48 *Old Bailey Proceedings Online*, February 1719, trial of William Wilson (t17190225-27).

49 *The Diary of Samuel Pepys*, vol. 7, 1666 (1972), 102; vol. 8, 1667 (1972), 383.

50 Alford, *London's Triumph*, 112.

of the thriving metropolitan medical marketplace. Print advertisements for medical services reveal next to the Exchange the premises of Mr Elmy 'Operator, of known Integrity, and above 25 Years Practice … who most safely and expeditiously Cures Deafness and Noise in the Ears in any of what Age soever'.[51] At Mrs Garraway's shop 'at the King's Arms at the Royal Exchange-Gate' one could purchase 'the best Water in the World … for the Itch or any Itching Humours'.[52] Under the southwest piazza of the Royal Exchange could be found bookseller Marshall Sheepey, who sold a 'most excellent volatile anti-apoplectic Elixir, so much fam'd for certainly curing convulsive Fits'.[53]

The density of crowds, affluence of visitors and diversity of luxury products also attracted 'disorderly' behaviour and criminal activity. The presence of working women such as fruit sellers could be problematic; there were suspicions about exactly what service was for sale, and women often faced harassment from men in this male-dominated mercantile space. The Exchange also became 'a dangerous and troublesome civic space, ill lit and beset on Sundays, holidays and at night by football players, "lewd boys", "rogues" and "whores"'.[54] Men and women were tried at the Old Bailey for theft from Exchange shops of perukes, a lute string, pearl and bone lace, watches, men's leather shoes and a ring set with diamonds.[55] Pickpocketing was also a frequent occurrence. Thomas Brook, 'as he stood talking against the Royal-Exchange' on Sunday 2 July 1721, 'lost his [silk] Handkerchief out of his Pocket'.[56] Daniel Evans 'was looking in at a picture-shop by the Royal Exchange' when he noticed his handkerchief was missing.[57] Timothy Malhoney 'came to the Royal Exchange to see for a captain of a ship' when he was victim to a confidence trick and persuaded by a gang to visit a nearby alehouse where he was violently robbed.[58] There was a pillory at the Exchange where defendants convicted of especially controversial crimes, such as seditious words or attempted sodomy, were punished publicly by the crowd.[59]

Shops and signs

Neither crime nor commerce was limited to the Royal Exchange. London's main streets were given colour and a sense of variety and activity by the increasing prominence of

51 *Old Bailey Proceedings Online*, May 1689 (a16890516-1).
52 *Old Bailey Proceedings Online*, December 1718 (a17181205-1).
53 *Old Bailey Proceedings Online*, September 1750, trial of Edward Newby Thomas Harrison Joseph Bradley (t17500912-38).
54 Laura Gowing, '"The Freedom of the Streets": Women and Social Space, 1560–1640', in Griffiths and Jenner, *Londinopolis*, 130–51 (p. 143).
55 *Old Bailey Proceedings Online*, t16840227-20; t16870512-42; t16890703-22; t16930426-72; t16940830-4; t17550409-27; t17640502-70.
56 *Old Bailey Proceedings Online*, July 1721, trial of Philip Jersey (t17210712-1).
57 *Old Bailey Proceedings Online*, July 1763, trial of Lawrence Miller (t17630706-53).
58 *Old Bailey Proceedings Online*, January 1772, trial of James Morgan Henry Edwards John Falmy, otherwise Felmy, otherwise Blind Jack the Kidnapper (t17720109-62).
59 Clive Emsley, Tim Hitchcock and Robert Shoemaker, 'Crime and Justice – Punishment Sentences at the Old Bailey', Old Bailey Proceedings Online.

shop window displays and the hanging street signs that identified business locations. As London's suburbs and industrial areas grew, no individual could have reasonably mastered the entire metropolis and shop signage aided street navigation.[60] In his 1716 poem *Trivia: or, The Art of Walking the Streets of London*, which captured many of the hazards, inconveniencies and sensory experiences of the 'winding Alleys' and 'long perplexing Lanes untrod before', John Gay advised: 'Be sure observe the Signs, for Signs remain, Like faithful Land-marks to the walking Train.'[61] Signs were criticized as hazards – particularly after the Great Fire, or after accidents caused by their weight that brought down house fronts and killed passers-by – but they remained common until the second half of the eighteenth century.[62] Even after their removal was ordered in 1762, many lived on in merchants' and artisans' trade cards.[63] Signs could be traditional or generic, with ready-made examples available from the market in Harp Alley off Shoe Lane, Fleet Street, described as 'a sort of Noah's Ark, in which animals of every hue, kind, and colour, might be seen'.[64] These signs were the work of painters, carvers and blacksmiths and might show considerable skill and ingenuity. Some reflected particular trades, perhaps with reference to the coats of arms of a particular guild, or became associated with them through apprenticeship lines that continued to use a particular sign or premises.[65] Company halls, and the many commercial and domestic properties owned by wealthy guilds, 'bore heraldic and other carved ornament', such as the Mercers' Company's symbol of the 'Mercers' Maiden' (a stone sculpture of the Maiden from 1669 is still extant at Corbet Court).[66] Among trades that tended to use distinctive street signs were those of optical, mathematical and philosophical instrument makers. They were in the process of creating new markets for such goods, whether presented as luxuries, associated with gentlemanly natural philosophy, or as improved tools and methods for practical work.[67]

Optical instrument makers who were free of the Spectacle Makers' Company very commonly used pairs of bow spectacles on their signs. They were easily recognizable, represented a key commodity even for those who also sold more specialized instruments, and reflected the unofficial coat of arms of the Company, showing three pairs of golden spectacles. Rendered in gilded wood and hanging before shops, these

60 Tim Hitchcock and Heather Shore, 'Introduction', in *The Streets of London: from the Great Fire to the Great Stink* (London: Rivers Oram Press, 2003), 1–9 (p. 4).

61 Quoted in Clare Brant and Susan E. Whyman (eds), *Walking the Streets of Eighteenth-Century London: John Gay's Trivia (1716)* (Oxford: Oxford University Press, 2009), 170, 179.

62 James Ayres, *Art, Artisans and Apprentices: Apprentice Painters and Sculptors in the Early Modern British Tradition* (Oxford: Oxbow Books, 2014), 203, 206; David Garrioch, 'House Names, Shop Signs and Social Organization in Western European Cities, 1500–1900', *Urban History* 21, no. 1 (1994): 20–48 (p. 20).

63 Garrioch, 'House names, Shop Signs', 37.

64 Anon, 'Paul Sandby and His Times', *Library of the Fine Arts* 2 (1831): 338–45 (pp. 338–9).

65 Ayers, *Art, Artisans and Apprentices,* 198–216; Garrioch, 'House names, Shop Signs', 25–31.

66 Christine Stevenson, *The City and the King: Architecture and Politics in Restoration London* (New Haven and London: Yale University Press, 2013), 13–14.

67 Deborah Jean Warner, 'What Is a Scientific Instrument, When Did It Become One, and Why?', *The British Journal for the History of Science* 23, no. 1 (1990): 83–93; Liba Taub, 'Introduction: Reengaging with Instruments', *Isis* 102, no. 4 (2011): 689–96.

must have attracted attention as they caught the light. John Radford worked at the Golden Spectacles without Temple Bar from at least 1668, and Thomas Ribright at the Golden Spectacles in the Poultry in the mid-eighteenth century.[68] One, two or three pairs of spectacles featured in many shop signs and names, often combined with other, more up-to-date, instruments or with figures that signalled theoretical as well as practical knowledge. Some trade cards depict the street signs themselves, revealing that elaborate ironwork was used to link the different elements (Figs 2.4–2.6).[69] John Yarwell worked from the Archimedes and Three Pair of Golden Spectacles in Ludgate Street in the 1690s, and his sign included a painted representation of Archimedes, anachronistically wielding a telescope. Samuel Scatliff was at Friar Bacon's Head, while Matthew Richardson and James Mann combined Isaac Newton and spectacles. George Willdey and Timothy Brandreth sold spectacles and a wide range of instruments and 'toys' at the Archimedes and Globe, adding to the hanging sign depicting this with two 'Posts before the Door' topped with 'a pair of large Globes' or, if their trade card depiction was accurate, a large armillary sphere on one and Archimedes, with telescope, atop a terrestrial globe on the other.[70] Choosing to advertise their wares or new designs were James Ayscough at the Great Golden Spectacles and Quadrant in Ludgate Street, John Cuff at the Reflecting Microscope and Spectacles on Fleet Street and Edward Nairne at the Golden Spectacles, Reflecting Telescope and Hadley's Quadrant in Cornill, opposite the Royal Exchange.[71] From the seventeenth century, then, those who walked the city's main thoroughfares would have had representations of scientific instruments swinging and shining above their heads.

There were ways of drawing attention to premises without hanging signs. Prescott Street, south of Goodman's Fields to the northeast of the Tower of London, seems to have been one of the first places that used numbering instead of signs to identify buildings – a novelty mentioned in Hatton's *A New View of London* in 1708.[72] One of the buildings there housed a mathematical school run by John Colson, who was also a mathematical examiner to Trinity House. The fictional *London Spy* by Ned Ward found himself at this location after a visit to the Tower and describes being struck by a door painted round with zodiac signs. It is possible that Colson, who made astronomical observations as well as teaching mathematics for navigation, may have decorated his school thus, though the story instead suggests it is the

[68] Webster Signature Database http://historydb.adlerplanetarium.org/signatures/; Gloria Clifton, *Dictionary of British Scientific Instrument Makers, 1550–1851* (London: Zwemmer in association with the National Maritime Museum, 1995), 230.

[69] For examples relating to other trades see Ayres, *Art, Artisans and Apprentices*, 200–2.

[70] Science Museum: 1951-685/88 (Yarwell), 1934-105 (Scatliff), 1934-106 (Richardson), 1951-685/51 (Mann), 1951-687/41 (Brandreth and Willdey). Several other makers, into the nineteenth century, used Archimedes as a sign, with spectacles, a globe or on his own: D. J. Bryden and D. L. Simms, 'Archimedes as an Advertising Symbol', *Technology and Culture* 34, no. 2 (1993): 387–91.

[71] Science Museum: 1934-101 (Ayscough), 1934-119 (Cuff), 1934-118 (Nairne).

[72] Edward Hatton, *A New View of London; or, An Ample Account of that City, in Two Volumes* (London, 1708), vol. 1, 65.

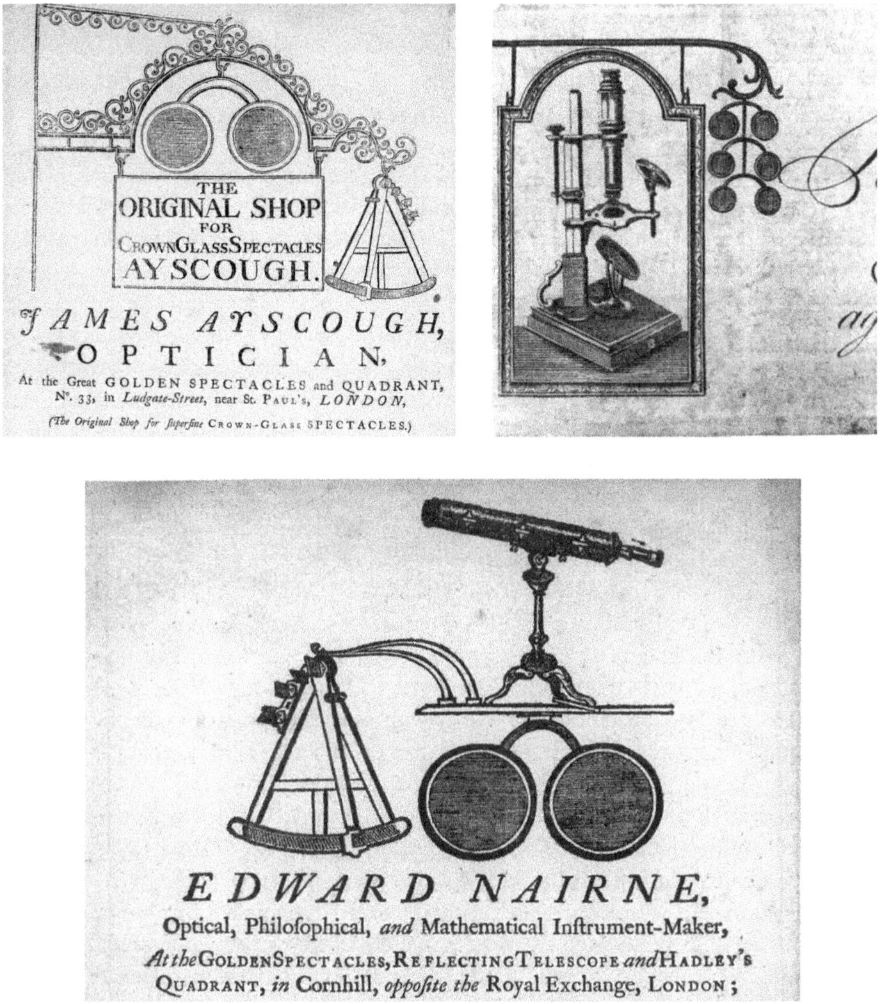

Figures 2.4–2.6 Details from the trade cards of James Ayscough, John Cuff and Edward Nairne, depicting hanging signs (SM: 1934–101; 1934–122/12; 1934–118). © Science Museum/Science & Society Picture Library.

abode of 'some strange foretelling *Star-Peeper*'.[73] By the 1820s, when numbering was the norm across London and Westminster, buildings in the main streets were said to be 'covered with names and signs, yards in length, generally gilt, and in relief'.

[73] Edward (Ned) Ward, *The London-Spy Compleat, in Eighteen-Parts. By the Author of The Trip to Jamaica* (London: Printed and sold by J. How, 1703) (p. 318). Ward may also have conflated John Colson with Lancelot Colson, a probable kinsman, physician, astrologer and chymist, who published an almanac: Rebekah Higgitt, 'The Mathematical Examiners of Trinity House: Teaching and Examining Mathematics for Navigation in London during the Long Eighteenth Century', in *Beyond the Academy: The Practice of Mathematics, 1600–1850*, ed. Philip Beeley and Christopher Hollings (Oxford: Oxford University Press, 2024), 80–111.

On top of this, 'the eye of the stranger is incessantly caught by the new and brilliant articles exposed for sale in the windows', with goods shining out, carefully arranged to maximum aesthetic advantage.[74] One sign, long in use by vendors of navigational instruments, was the figure of a naval officer or midshipman, who we will meet near East India House in Chapter 7.

Celebrated as 'no less than the great Mart and *Metropolis* of the Kingdom ... the largest as well as noblest City of the World' at the end of the seventeenth century, the range and luxury of goods available in London long remained noteworthy.[75] While this could attract censure, visitors couldn't help but be impressed:

> All that Peru and the mines of Golconda can afford, all that refinement and luxury can invent, all treasurers that the four quarters of the globe possess, – are here heaped up. It is impossible not to be astonished in seeing these riches displayed. Here the costly shawls from the East Indies, there brocades and silk-tissues from China, now a world of gold and silver plate, then pearls and gems shedding their dazzling lustre, home manufactures of the most exquisite taste, an ocean of rings, watches, ribbons, lace, bonnets, and fruits from all the zones of the habitable world – attract, tempt, astonish, and distract your eyes.[76]

Retailers used display and shop layouts to entice customers, to inform them and to generate trust in the quality of their goods or in their own skill and knowledge.[77] Goldsmiths often used large glass display cases to show off their valuable wares. The 'goldsmith's press' – substantial cupboards with open shelves and glass doors at the front – frequently covered the entire back wall of the shop, thus prominently displaying the goldsmiths' plate and providing a focus for customers. The sheer expanse of glass used for these display fittings also gestured to the financial probity and credibility of the goldsmith-retailer (and sometimes goldsmith-banker).[78] For apothecaries, as Patrick Wallis has shown, trust and legitimacy were particularly charged issues but might be addressed with high-quality displays, by advertising approval of the Society of Apothecaries, by cultivating social and intellectual networks and by separating the retail, manufacturing and consulting functions of their role. These elements played out through the use of objects within the space of their shops, such as recognizable and

[74] Heinrich Heine, *The Prose Writings of Heinrich Heine*, ed. Havelock Ellis (London: Walter Scott, 1887), available at Project Gutenberg (https://www.gutenberg.org/files/37478/37478-h/37478-h.htm). See Jon Stobart, Andrew Hann and Victoria Morgan, *Spaces of Consumption: Leisure and Shopping in the English Town, c.1680–1830* (London: Routledge, 2007), 115–16 on the importance of shop windows in giving potential customers 'a taste of what lay inside the shop' (p. 116).

[75] 1694 proposal to revise Stow's survey, quoted in Jenner and Griffiths, *Londinopolis*, 2. The claim was inaccurate: at this date, Beijing, Constantinople and perhaps Edo in Japan were larger than London.

[76] *Letters from Albion* (1814), quoted in Celina Fox, 'Introduction: A Visitor's Guide to London World City, 1800–40', in *London – World City*, 11–20 (p. 16).

[77] Wallis, 'Consumption, Retailing, and Medicine'; Claire Walsh, 'Shop Design and the Display of Goods in Eighteenth-century London', *Journal of Design History* 8, no. 3 (1995): 157–76.

[78] Claire Walsh, 'The Design of London Goldsmiths' Shops in the Early Eighteenth Century', in *Goldsmiths, Silversmiths and Bankers: Innovation and the Transfer of Skill, 1550 to 1750*, ed. David Mitchell (Stroud: Alan Sutton, 1995), 96–111 (pp. 97–100).

high-quality jars placed on purpose-built shop fittings, the use of a pill tile bearing the arms of the Society of Apothecaries on the shop counter, assigning certain tasks to different spaces of public shop and private workshop, and collecting material medica and exotic natural history specimens or instruments for display, discussion, experiment or instruction.[79]

The manufacture and retail of optical and mathematical instruments were a new trade in London, and only from the end of the sixteenth century did instrument makers begin to shift towards a public world of trade and commerce, having previously operated largely within private networks of courtly patronage or to supply specific commissions.[80] However, by the later seventeenth century and in the eighteenth, instrument makers' shops could operate in similar ways to those of apothecaries in terms of the range of possible activities and use of private, public and semi-public spaces. The instrument maker Anthony Thompson, based in Hosier Lane in Smithfield in the 1650s–60s, was evidently well integrated with practitioner and scholarly networks.[81] This brought his products to attention of readers of several mathematical texts and attracted others to his shop for opportunities to meet and speak to people with shared interests. The London mathematician John Collins described in a letter how he had discussed an edition of Euclid with 'a Cambridge scholar' at Thompson's shop.[82] In the 1630s the Strand workshop of mathematical instrument maker Elias Allen was a place of discussion and served as an address for the exchange of mathematical correspondence.[83]

As the City-based instrument trade developed, many of the more successful makers, and those offering a diverse range of goods, moved westward, along the Strand and into Westminster.[84] Such shops could lay out more on attractive displays in windows and interiors, perhaps including instruments alongside a range of other desirable goods.[85] When the French mathematician Jean Bernoulli visited London in 1768, he knew his correspondent would be aware 'of the richness and brilliance of the shops of London', but doubted he could 'imagine how much astronomy contributes to the beauty of the spectacle', with telescopes and octants 'ranged and set out with

[79] Wallis, 'Consumption, Retailing, and Medicine', 32–6, 40–1.

[80] Jardine, 'Instruments of Statecraft', 308.

[81] H. R. Calvert, 'Thompson of Hosier Lane, Instrument Makers of the Seventeenth Century', *Isis* 26, no. 1 (1936): 50–2.

[82] Philip Beeley, 'Practical Mathematicians and Mathematical Practice in Later Seventeenth-Century London', *British Journal for the History of Science* 52, no. 2 (2019): 225–48 (p. 243).

[83] Hester Higton, 'Portrait of an Instrument-Maker: Wenceslaus Hollar's Engraving of Elias Allen', *British Journal for the History of Science* 37, no. 2 (2004): 147–66 (p. 155).

[84] On the locations of London instrument makers, see Robischon, 'Scientific Instrument Makers in London' and Baker, 'This Ingenious Business', also Alexi Baker, 'Symbiosis and Style: The Production, Sale and Purchase of Instruments in the Luxury Markets of Eighteenth-Century London', in *How Scientific Instruments Have Changed Hands*, ed. Alison D. Morrison-Low, Sara J. Sechner and Paulo Brenni (Leiden and Boston: Brill, 2016), 1–20.

[85] On the instrumentalizing of curiosity for commercial opportunity see Jenni Dixon, 'The Toyshop, the Cabinet, and Eighteenth-Century Curiosity', *History of Retailing and Consumption* 5 (2019): 205–27.

care'.[86] This is quoted by Jim Bennett as evidence of the 'consumer expansion' in the instrument trade that had reached London but was not evident in Paris. His examples also reveal the role of demonstration and performance within shops, with Balthazar de Monconys introduced to the magic lantern at Richard Reeve's shop on Long Acre in 1663, and Von Uffenbach enjoying optical and pneumatic demonstrations at Francis Hauksbee's workshop in Wine-Office Court off Fleet Street in 1710.[87] These were a kind of infotainment-cum-infomercial that merged education, social opportunities, advertising and evidence of the maker's knowledge and skill.

Even the shops of less well-known instrument makers might have displays of various sorts that were visible from the street and attracted the attention of potential customers and, less desirably, from thieves. Court records, evidence very different from philosophical and mathematical books and correspondence, can be revealing of how scientific instruments might be available and desirable to a broader range of Londoners, and in less specialized contexts. Thomas Parnell, a mathematical and nautical instrument maker at St Botolph, Aldgate, nearly suffered the loss of a telescope when some men who worked on the quays attempted distraction by asking to see sail needles while one of their numbers tried, unsuccessfully, to hide the long tube under his clothes. Placing instruments where they could attract passers-by could be risky: in 1786, Thomas Sutherland, based in Piccadilly, 'lost a horizontal watch, capped and jewelled' that was 'hung up by a small brass hook, within an inch of the window'. John Marshall, who had 'a kind of bow window' to his shop in 1780, spoke of stolen products that were 'in the inside of the shop, beside my work-board, within six or seven inches of the show-glasses'. However, the optical instrument maker Philip Brock, in Church-row, Aldgate, lost a significant amount of stock from inside his shop to a putative customer who 'came in to buy a glass for a show'. This was because he had placed himself in the window, where he was 'at work' and unable to serve or intercept the incomer. This was, perhaps, for the light and suggests a small space that combined manufacture and retail, though the visible demonstration of skill might in itself have attracted attention and enhanced credibility.[88]

Institutional spaces

The institutional sites that will be explored in the following chapters, and the people who lived, worked and passed through them, coexisted with this changing, visible landscape of activity within the city. They also, as already shown, contributed to it in many different ways by partaking in civic occasions, by overseeing trade activities or as actual or potential patrons for hopeful projectors (individuals seeking support for their inventions or projects). Corporations were made up of individuals who

[86] Quoted in Bennett, 'Shopping for Instruments', 370.

[87] Bennett, 'Shopping for Instruments', 377, 371–2, 376.

[88] All of these examples are from Jasmine Kilburn-Toppin, 'Instrument Makers, Shops, and Expertise in Eighteenth-Century London', in *Spaces of Enlightenment Science*, ed. Gordon McOuat and Larry Stewart (Leiden: Brill, 2022), 74–90.

might have multiple connections to different institutions or more informal networks within the City, creating productive linkages and acting across more or less public spaces. Institutional sites themselves had clear presence within the City, often acting as centres of gravity that attracted other institutions, businesses and people into their orbit. Livery company halls were said by early seventeenth-century customs official and antiquarian Thomas Milles to be at the centre of the 'unweldy, and disorderlie Babell of buildings, which the worlde calls *London*'.[89] Buildings and their façades might make a public, corporate statement, and parts of interiors might be accessible to the curious, or the hopeful, even if other spaces were likely to be private or reached only by a few outsiders.

The following chapters focus on sites of several sorts. The Tower of London, at the eastern fringes of the City, was a multi-purpose site of many buildings and several institutions. Taking up most space were the Royal Mint and the Office of Ordnance but others were present, ranging from the Royal Menagerie to the Hall of the Company of Moneyers. Large courtyard houses in the City might suit wealthy merchants or be put to institutional use, as was the case with the site of Trinity House, in Water Lane, Tower Street Ward and East India House, which occupied the mansions of its governors or other merchants before acquiring and rebuilding premises in Leadenhall Street in the early eighteenth century. The wealthier guilds had acquired, adapted and built large halls on City sites corporately owned from the early fifteenth century, and these were among the first buildings to be reconstructed after their destruction in the Great Fire.[90] This demonstrates their importance to corporate status and the fact that although their role in craft regulation had declined, guilds remained central to city life.[91] However, not all corporations, or associations, had halls, or even rooms of their own. Many made use of the semi-public spaces of taverns and coffee houses, which could set aside rooms for meetings, private dining, lectures or discussion. These are well known to have been important sites for the exchange of knowledge and all sorts of print materials, as well as the presentation of lectures and demonstrations in natural philosophy. These could be of sufficient formality to compete or overlap with the provision of more formal schools or Gresham College.[92] Taverns were also ideal meeting places for clubs and societies, including the Mathematical Society that met to dine at the Bull's Head Tavern in Cheapside in the 1680s, and the Spitalfields Mathematical Society that began meeting at the Monmouth's Head tavern in Monmouth Street in 1717.[93]

Taverns also acted as sub-institutional sites for some guilds. The Mitre Tavern in Fenchurch Street was a temporary home for Trinity House when its building was

[89] Quoted in Kilburn-Toppin, *Crafting Identities*, 5.

[90] Ibid., 100.

[91] Anya Matthews, 'Honour, Ornament, and Frugality: The Reconstruction of London's Livery Halls after the Great Fire', in Hallett, Llewellyn and Myrone, *Court, Country, City*, 73–104 (pp. 73–4); and in reference to Mark Knights, 'A City Revolution: The Remodelling of the London Livery Companies in the 1680s', *English Historical Review* 112, no. 449 (1997): 1141–78 (pp. 1143–4).

[92] Stewart, 'Other Centres of Calculation'.

[93] Beeley, 'Practical Mathematicians', 244–5; Larry Stewart and Paul Weindling, 'Philosophical Threads: Natural Philosophy and Public Experiment among the Weavers of Spitalfields', *The British Journal for the History of Science* 28, no. 1 (1995): 37–62 (p. 40).

destroyed by the fire resulting from the 1715 gunpowder explosion, although their business was such that they soon rented a large house on nearby Mark Lane until the house at Water Lane was rebuilt. Taverns were, however, the regular meeting space for two craft guilds of early-seventeenth-century foundation, the Spectacle Makers' and Clockmakers' Companies. These were the two guilds with greatest interest in the optical, mathematical and philosophical instrument trade and, although minor players with little money at their disposal, they regulated training and products.[94] As well as offering a small amount of charitable support to superannuated freemen and their widows, they played a role in defining the nature of their trade, adapting this to suit the interests of their leading members and ensure institutional survival. In their seventeenth- and eighteenth-century form, they reveal what were the near-minimum requirements for the existence of an incorporated body.

Each, of course, had a charter, though the sum of the corporate rights that these granted was only slowly taken up, or not at all. That of the Spectacle Makers was granted in 1627 and of the Clockmakers in 1631, giving them power to oversee their trade within a three- and ten-mile radius of London, respectively. The Clockmakers' charter conferred the right to have a common seal but they only gained a grant of arms in 1672 and it was not until 1692 that their minutes reveal that they ordered a 'Common Seal of Silver about the breadth or circomference of a milled Half Crown piece, or Oval to be made'.[95] It was this that allowed them to act as a corporate body, in purchasing stock or property or making petitions to parliament.[96] The corporate existence of the Company resided in its chest, with three keys to be held by different officials – guild master, upper warden and renter warden. This contained the charter and seal as well as money, plate and the institutional memory of its minute books. Given the lack of a permanent meeting place, it was by necessity moved around, which could cause problems. In 1658 it was arranged that it should be moved to the house of a former Master, Robert Grinkin, but 'it would not enter into his house', presumably because of its size.[97]

Subsequently, arrangements were made for the use of company halls. In 1642 the Clockmakers gifted the Painter-Stainers a clock 'for which they were to have the use of their Hall for some Meetings as occasion should require', and in 1653 they also paid an additional three pounds 'for the standing' of the Company's Chest.[98] In 1654 they arranged to pay the Broderers' Company 4 pounds annually for use of their hall, plus another 20 shillings if they had their feast there, and later similar arrangements were made with the Blacksmiths, Goldsmiths, Founders, Cutlers, Mercers and Stationers.

94 Robischon, 'Scientific Instrument Makers in London', Ch. 5; M. A. Crawforth, 'Instrument Makers in the London Guilds', *Annals of Science* 44, no. 4 (1987): 319–77; Thomas H. Court and Moritz Von Rohr, 'Contributions to the History of the Worshipful Company of Spectaclemakers', *Transactions of the Optical Society* 31 (1930): 53–90; Samuel Elliott Atkins and William Henry Overall, *Some Account of the Worshipful Company of Clockmakers of the City of London* (London: Privately Printed, 1881).

95 Atkins and Overall, *Some Account*, 74.

96 They made investments in stock of the Bank of England in 1697, 1701 and 1702 and in the South Seas Company in 1720 and 1730: Atkins and Overall, *Some Account*, 214–15.

97 Court Minute Books, Clockmakers Company, March 1657/8, CLC/L/CD/B/001/MS2710/1, f. 94 (Guildhall Library).

98 Court Minutes September 1642, January 1653, CLC/L/CD/B/001/MS2710/1.

These were interspersed with periods of meeting at taverns, including the Half Moon in Cheapside and the George in Ironmonger Lane. From the 1760s they turned exclusively to taverns, meeting first at Paul's Head tavern in Cateaton Street, near Guildhall, and from 1800 at the King's Head in Poultry. Ultimately the Company set up a considerable presence there: by the early nineteenth century, they had installed a bookcase and collection of some 110 books, insured for 9 shillings annually, as well as 40 'Models and Specimens' and their essential corporate archive.[99]

The Spectacle Makers' Company, meanwhile, lacked even a formal coat of arms, though they used their three pairs of spectacles informally, and kept their books in 'a Bagg' between 1739 and 1763, when they finally acquired a chest.[100] They consistently met in taverns, generally keeping close, geographically, to where the trade's shops were gathered, around St Paul's Churchyard, the area north of St Pauls and around the Exchange.[101] Later in the seventeenth century they were regularly at The City Armes in Milk Street, and in the eighteenth century they settled in St Paul's Churchyard, at The Sun Tavern and Queen's Arms. In 1803–9 they were, like the Clockmakers, making use of the King's Head Tavern in Poultry.[102]

In the early nineteenth century both of these companies reasserted some of their corporate rights and, tellingly, clashed over who had jurisdiction over mathematical instrument makers. While this was about which company could gain payments from members of the trade, it was also about which could lay claim to representing a world of practical and theoretical endeavour now labelled 'science'. The specific clash was a challenge by the Clockmakers when the instrument maker and retailer Robert Brettell Bate chose, when pressed, to become free of the Spectacle Makers. Bate was a significant businessman who supplied hydrometers to the Board of Excise, weights and measures to the Exchequer and balances to the Bank of England and East India Company, as well as retailing a wide range of goods from his shop at 17 Poultry. The proximity of his well-stocked shop to the King's Head Tavern was probably not coincidental to his being the target of both companies. Both could point to a long tradition of mathematical instrument makers being part of the fellowship and apprenticeship training that they represented, but both also wished to associate themselves with the current rising esteem of science and claims about its national role. In 1810 the Spectacle Makers had adopted a new coat of arms, which added dividers, a globe and prisms to its three spectacles. This represented, they claimed, the fact that they were 'not only Manufacturers of Spectacles but <u>also</u> Constructors of the various Instruments by which practical Science is conducted & by which the Mathematician is assisted in his Labours'. They were essential to the work of the astronomer, geographer, surveyor, sailor, miner, teacher, mechanic and theoretician, and their forebears had

[99] CLC/L/CD/B/001/MS2710/001; Minute Book of the Library Committee, 1814–1836, CLC/L/CD/F/009/MS03967, entries for 29 August 1817, 2 January 1819, 17 December 1819 (Guildhall Library).

[100] The order to purchase a bag appears in the Court Minutes, 29 March 1739, CLC/L/SH/B/001/MS5213/003, and the donation of a chest referred to in the Account Book 1754 to 1765, CLC/L/SH/D/001/MS6027/001, Spectacle Makers Company (Guildhall Library). The informal coat of arms was used in William Maitland, *The History of London, from Its Foundation by the Romans to the Present Time* (London: Printed by Samuel Richardson, 1739), 611.

[101] Robischon, 'Scientific Instrument Makers in London', 178–83.

[102] Spectacle Makers Minute Books, MS 5213/2–4 (Guildhall Library).

> Assisted & Kept pace with the Progress of the Sciences, which have within the two last Centuries so greatly expanded themselves & in fine they have raised the fame of English Artists, & have set the pattern of real Scientific Construction combined with soundness of execution.[103]

The Clockmakers, meanwhile, could not only point to mathematical instruments being mentioned in their charter but also asserted their importance to city and nation, with their products and improvements producing £600,000 worth of trade and 120,000 timepieces annually, employing thousands and being 'incalculably beneficial' to 'a Maritime, Manufacturing, and Commercial Country'. The 'principles and ingenuity of Clockmakers', they claimed, 'enabled them to invent and to improve all the Machines of which in the silk, woollen and cotton trades labour has been diminished and Manufacturers have been supported'. They were essential 'to every branch of manufacture in which machinery is used'.[104]

While guild control was much weakened, and about to lapse entirely, these companies were concerned to reinvent and reassert their roles and significance. 'Science' and 'manufacture' were concepts they could draw on to give context and identity to their histories and future activity. These claims were no longer focused solely on the benefit for members of the trade in London, but their – and by extension the city's – significance to the prosperity of the nation. In this they were following the lead of much of the iconography of London's buildings. Representations of navigation and its supporting sciences of astronomy and geometry, often identified by relevant instruments, were associated with commerce, and goods from overseas proffered to the city. The public culture of natural and experimental philosophy in London had likewise become associated with success in producing or improving machinery and manufacturing techniques. The possibility of practical application and 'public' (usually corporate, civic or national) benefit had been held out by those hoping for individual advancement or aiming to promote the wider project of the new philosophy, shaping the avenues explored by practitioners and its wider meanings.

The success of such claims is evident by the end of our period. Men of science continued to assert that they could benefit and improve practical processes, but London's artisans, corporations and businesses could also find it useful to state that exercising their skills and pursuing their interests were contributions to science. The institutions that we explore in the following chapters all existed within a London that was from the beginning of our period a rich environment for generating and sharing skills and knowledge. Because of its scale, variety, connections, culture and governance, it also developed active public, intra- and inter-institutional knowledge cultures. It was a city that increasingly advertised itself within a rhetoric that linked London's trade success with skilled production, knowledge of the world and nature, and with mathematical and experimental approaches to learning and discovery.

[103] Papers & Letters referring to dispute between the Spectacle Maker's Company & the Clockmaker's Company, Clockmakers Company MS3941, 56–65 (p. 58) (Guildhall Library).

[104] Atkins and Overall, *Some Account*, 276, 283.

3

Technical knowledge, experiment and secrecy at the seventeenth-century Mint

Introduction

We usually think of the Tower of London as a notorious prison, or perhaps a royal storehouse for precious things. But in the early modern mind the Tower evoked a more varied range of associations. In his renowned early-seventeenth-century survey of London, John Stow wrote:

> This tower is a Citadell, to defend or commaund the Citie: a royall place for assemblies, and treaties. A Prison of estate, for the most daungerous offenders: the onely place of coynage for all England at this time: the armorie for warlike prouision: the Treasurie of the ornaments and Jewels of the crowne, and generall conseruer of the most Recordes of the kings Courts of iustice at Westminster.[1]

The impression we get from Stow's account – written when the White Tower, the largest and most ancient building in the Tower complex, was about five hundred years old – is of a highly significant, multifunctional urban site (Fig. 3.1).

Within its walls the early modern Tower of London encompassed, among other institutions, the Mint (where money was made, and precious metals tested and refined), the Ordnance Office (responsible for weapon procurement and storage) and the Royal Menagerie (a space for housing and viewing 'wild Savage Creatures').[2] These varied Tower institutions employed hundreds of artisans and technicians, who undertook practices of testing, measurement and material observation and production. Many of these Tower employees lived as well as worked on site, blurring boundaries between domestic, institutional and operational spaces. Moreover, the Tower acted as a kind of magnet for expertise, drawing in skilled practitioners – relating to the testing of metals,

1 John Stow, *A Survey of London. Reprinted from the Text of 1603*, vol. 1, ed. C. L. Kingsford (Oxford: Clarendon, 1908), 59.

2 Geoffrey Parnell, *English Heritage Book of the Tower of London* (London: B.T. Batsford/English Heritage, 1993), 54; John Strype, *A Survey of the Cities of London and Westminster* (London, 1720), I.xx.118.

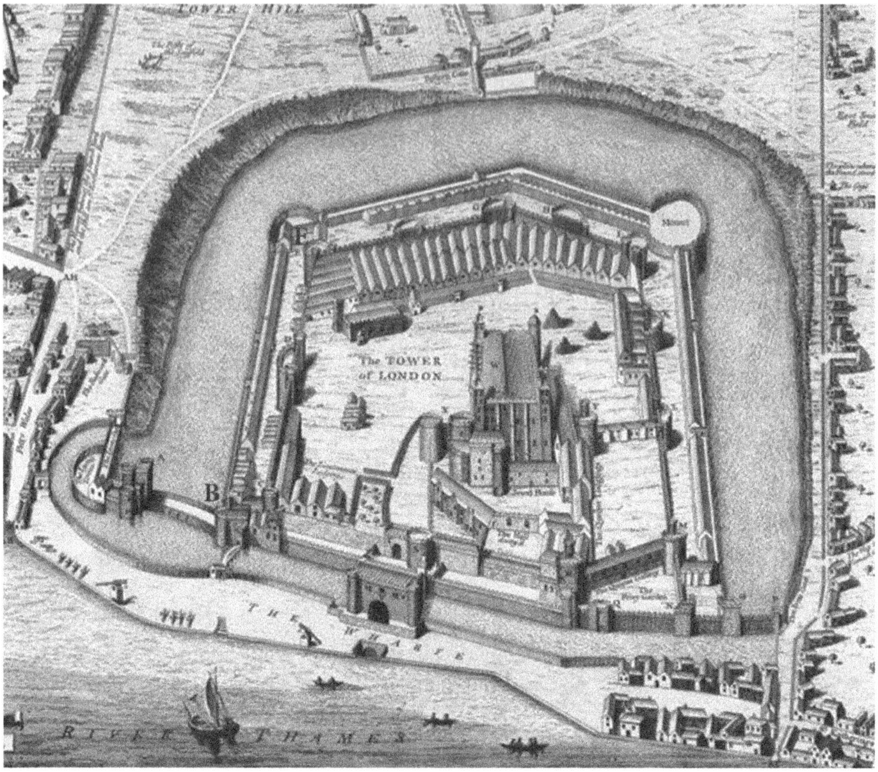

Figure 3.1 Detail from 'a true and exact draught of the Tower Liberties, surveyed in the year 1597' by Gulielmus Haiward and J. Gascoyne. Pictured here within the walls of the Tower are the Jewel House, the Lieutenant's lodgings, royal apartments, Mint and Ordnance buildings. Mint Street initially stretched between point B and E on the drawing (British Museum: Maps.K.Top.24.23.b). © British Library Board.

the manufacturing of coins and weapons, the welfare and reproduction of animals – from across London, Britain and Europe more broadly.[3]

The Tower was also a tourist destination for affluent and well-connected foreign travellers, dignitaries and merchants. By the late sixteenth century, Tower warders were paid by visitors for tours, and there were customary routes through the extensive institutional complex, which typically took in the Royal Mint, armoury, Jewel House and the exotic animals in the menagerie.[4] More humble local visitors from the surrounding city came to observe the lions at close quarters (perhaps as early as the thirteenth century).[5] Thomas Platter, the Swiss writer and humanist, wrote in exalting

3 For example, foreign engineers for the recoinage effort, of which more below, or for the production of explosives and casting of cannon, discussed in the following chapter.

4 Kristen Deiter, *The Tower of London in English Renaissance Drama: Icon of Opposition* (London: Routledge, 2008), 48, 53–4.

5 Ibid., 34.

tones of his 1599 visit to the 'very magnificent and well fortified' Tower of London. Having been assigned 'a guardsman, who was to act as guide round the sites' (and who required a 'gratuity' at each significant location), Platter and his companions visited the armoury, tapestry chamber, Mint, lieutenant's apartments and menagerie. In his account Platter focused particular attention upon his tour of the Royal Mint 'to watch all kinds of money being coined'. He was particularly impressed by the view of 'an oven, the clever invention of a German, in which the metal, encased, so as not to crack, in square iron receptacles, is annealed [subjected to fire], and no fire gets into the oven'.[6] This 'public' aspect of activities at the Tower, and the Mint specifically, has important implications for understanding the nature of knowledge production within this prominent institutional space.

This chapter takes as its focus the particular site of the Royal Mint within the wider built fabric of the Tower of London. We contend that the Mint was a significant urban space of technical knowledge and industrious labour, and, at certain moments, experimental practice. This is a new perspective, for the Mint has customarily been framed in terms of its institutional politics, bullion supply and output, specific levels of coin production and overall profits and losses.[7] To adopt Steven Shapin's turn of phrase, here we explore 'the siting of knowledge production'.[8] The Mint was an institution within which labourers, artisans, technical experts, institutional officials and gentlemen interacted. These varied social orders had a shared interest in metallurgy, and, more specifically, the production and testing of coin. Moreover, the Mint was a Royal institution and, from the early seventeenth century, favoured courtiers competed for the most lucrative roles of Mint master and warden; yet, its operation was dependent upon the technical knowledge and expertise of city workers. This was additionally a productive and experimental urban space which had a complicated relationship with the institutional and craft ideal of 'secrecy'. The operations of the Mint were meant to be secret, a form of 'intentionally concealed' technical knowledge, understood and observed by only a select group of officials and workers.[9] And yet, the everyday reality could be quite different. The site had seemingly porous boundaries; movement through the Mint was hard to control, and those with the right social connections might gain legitimate access to view operational practices and personnel. A handful of its workers also authored seemingly revelatory 'books of secrets'. By exploring the various material practices and operations of the seventeenth-century Mint, and the engagement of its practitioners with the thorny notion of secrecy, this chapter elucidates new outlines of its knowledge cultures.

6 Thomas Platter, *Thomas Platter's Travels in England 1599*, trans. Clare Williams (London: Jonathan Cape, 1937), 159–63.

7 C. E. Challis, *A New History of the Royal Mint* (Cambridge and New York: Cambridge University Press, 1992).

8 Steven Shapin, 'The House of Experiment in Seventeenth-Century England', *Isis* 79, no. 3 (1988): 373–404 (p. 373).

9 Pamela Long, *Openness, Secrecy, Authorship: Technical Arts and the Culture of Knowledge from Antiquity to the Renaissance* (Baltimore and London: John Hopkins University Press, 2001), 7, for formulation of secrecy as 'intentional concealment'.

The Royal Mint, technical knowledge and secrecy

From the thirteenth century, the Royal Mint was housed at the Tower of London. The Tower Mint was the primary institution in England for the production of silver and gold coin. Its officials and moneyers 'were incorporated as "one body perpetual" with the right to use their own common seal, to hold land, and to sue in any court'.[10] The built environment of the Mint consisted of a series of structures which gradually spread, by the end of the fifteenth century, to fill the narrow space between the inner and outer walls, or curtain walls, of the Tower. The most significant Mint buildings were situated along what was known as Mint Street, which ran northwards from Byward Tower up to Legge's Mount (Fig. 3.2).[11] In addition to workshops were stores, stables, coach houses, domestic residences and even gardens.[12]

Those employed at the Tower Mint engaged in a range of different activities and were experts in a variety of technical practices. Workers included weighers (who calculated the weight of metals brought into the Mint), melters (who liquefied the bullion), refiners, engravers, assayers and moneyers ('who are, some to sheer the Money, some to forge it; some to beat it broad, some to round it, and some to stamp or coin it').[13] Frustratingly, there are very few archival traces for the moneyers of the

Figure 3.2 Mint Street, looking towards Beauchamp Tower. © originalpickaxe/Wikimedia Commons.

[10] Challis, *A New History*, 184.
[11] Christopher E. Challis, *The Tudor Coinage* (Manchester: Manchester University Press, 1978), 1–3.
[12] Challis, *A New History*, 1-3; Craig, *Newton at the Mint*, 1.
[13] Strype, *A Survey*, I.xviii.96.

Tower who fashioned the blanks and struck the coins. It is likely that they formed a craft fellowship or guild from around the mid-fifteenth century, providing training for apprentices and mutual support in times of need.[14] A provost was responsible for discipline and paying wages. There were about twenty Tower moneyers in the early sixteenth century, fifty-nine in the early 1650s; twenty-four moneyers according to Mint records at the time of the Great Recoinage. Additional labour was hired at times of recoinage.[15] Moneyers' Hall was situated within the Tower site (opposite the Flint Tower) and used by the Moneyers' Company for socializing and administration (Fig. 3.3/Plate 2).[16] As was customary within early modern corporate culture, individual moneyers left funds in their wills for collective memorial dinners; Robert Fuller (d. 1606) bequeathed to his 'brethren the monyers in the Tower of London … 40s to be spent for a repast or drinking for them on the day of my buryall' (he also named moneyers as his executors).[17] Thomas Denham (d. 1605), who had served as provost of the moneyers (1576–1605), left the significant sum of £3 6s 8d for 'a

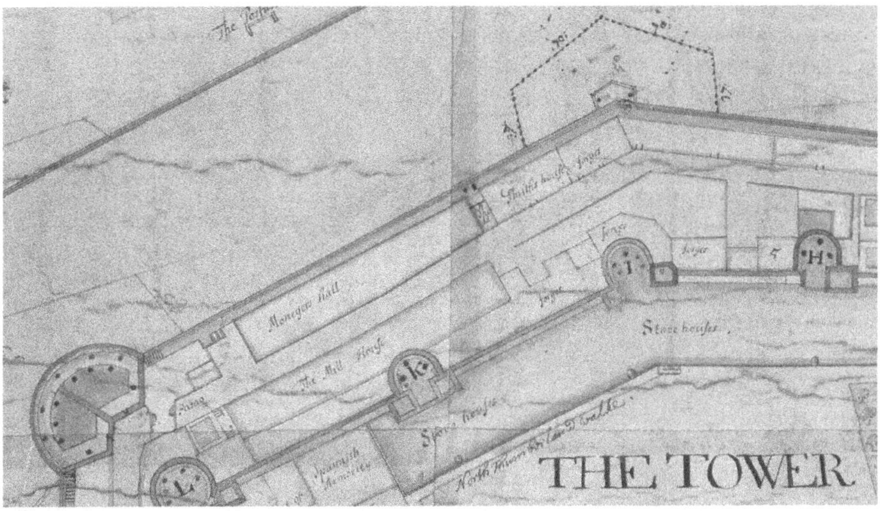

Figure 3.3 Detail from a 1682 plan of the Tower, showing Moneyers' Hall and the adjacent Smith's House and forges within the Mint (above the tower labelled 'I') (RA: TOL MP/TOWER/73/6-8). Courtesy of the Board of Trustees of the Royal Armouries.

See also colour version, reproduced in plate section 2.

14 J. Freeman, 'The Mistery of Coiners and the King's Moneyers of the Tower of London, *c.*1340-*c.*1530', *British Numismatic Journal* 70 (2000): 67–82; C. E. Challis, 'Mint Officials and Moneyers of the Tudor Period', *British Numismatic Journal* 45 (1975): 157–97.

15 Martin Allen, *Mints and Money in Medieval England* (Cambridge: Cambridge University Press, 2012), 93–4, 96; Christopher Challis, 'Hackney, Shoreditch and Moneyers in the Mint in Later-Medieval and Early-Modern London', *Hackney History* 16 (2010): 4–15 (p. 7).

16 Challis, *A New History*, 363.

17 PROB 11/107/6 (The National Archives).

supper amongst them'.[18] Pointedly, in the archival records of the Mint, seventeenth-century workers in coin spoke of their labour as 'industry' or 'industrious'.[19] By this they were probably emphasizing the ingenious and skilful elements of their practice.[20]

Responsibility for testing the precious metal content of ores, bullion, coinage and plate lay with assayers – a highly significant group of skilled artisans at the Tower site. All metals brought into the Mint were assayed (tested to determine their content and quality) before being melted down for coin production and, after minting, batches of coins were assayed to ensure they were of the correct weight and fineness. Maintaining rigorous 'standards and procedures' was not simply a matter of institutional reputation, however, for the very 'stability of the monetary system depended on them'.[21] As Sebastian Felten has written, 'the objective properties that determined a coin's intrinsic value were constructed by credentialled individuals operating in highly regulated spaces'. The credibility and reputation of the metallurgical testers mattered.[22] At the annual trial of the pyx, the integrity of the institution and the expertise of the master assayer were put to the test.[23] This highly ceremonial semi-public testing process at the court of Star Chamber at Westminster involved the assay of a sample of coins produced at the Mint by a jury of experienced goldsmiths, to ensure the coinage met the standards set by the crown.[24] Coins were compared against part of a gold trial plate, stored in the Pyx Chamber at Westminster Abbey (Fig. 3.4). Engaged in the separation and transformation of matter, assayers and the affairs of their workshops were a curiosity for those interested in the secrets of nature, material trials or experiments. As Patricia Fara writes, 'the processes of refining and testing precious metals involved the same chemical procedures and knowledge as isolating new elements, or producing drugs, or searching for the philosopher's stone'.[25]

It is probable that the varied workshop processes at the Mint were undertaken in distinct 'houses' (or certainly, at least, separate chambers). There were separate

[18] TNA, PROB 11/105/419.

[19] 'In discharge of the great trust of which place hath imployed his best *Industry* to do his majesty's faithful service', MINT 1/1, fo. 98 (The National Archives); 'in Satisfaction of His Labour and *Industry*', MINT 1/1, fo. 145.

[20] 'Characterized by or showing intelligent or skilful work; skilful, clever, ingenious, artful'. See 'industrious, adj'. *OED Online*. Oxford University Press, March 2022. Web. 1 June 2022.

[21] Allison Stielau, '"Adding to the Good Silver with Other Trickery": Purity and Contamination in Clement VII's Emergency Currency', in *Contamination and Purity in Early Modern Art and Architecture*, ed. Lauren Jacobi and Danile M. Zolli (Amsterdam: Amsterdam University Press, 2021), 157–82 (p. 159).

[22] Sebastian Felten, *Money in the Dutch Republic: Everyday Practice and Circuits of Exchange* (Cambridge: Cambridge University Press, 2022), 133–4.

[23] Jasmine Kilburn-Toppin, '"A Place of Great Trust to be Supplied by Men of Skill and Integrity": Assayers and Knowledge Cultures in Late Sixteenth- and Seventeenth-Century London', *British Journal for the History of Science* 52, no. 2 (2019): 197–223 (pp. 5, 15, 20).

[24] For a contemporary account of the trial of the pyx, see Goldsmiths' Hall Archive, MS C II.2.1, fos. 28^r, 30^r.

[25] Patricia Fara, *Life after Gravity: Isaac Newton's London Career* (Oxford: Oxford University Press, 2021), 164.

Figure 3.4 Gold trial plate of 1707 against which coins were tested. Trial plates had serrated edges as multiple portions were cut from the same sheet; together at the Trial they would fit together like pieces of a jigsaw. © The Royal Mint Museum.

keys for different doors to the blanching, melting and assay houses.[26] Archaeological excavations undertaken in the 1970s at Legge's Mount, the northwest corner of the Tower, uncovered brick buildings with furnaces, and the remains of crucibles (clay pots), bone ash cupels, parting and distillation vessels, and scrap metal – all crucial apparatus and materials for metallurgical workshop processes. This was almost certainly the location for the assaying of silver, and parting of silver and gold.[27] The pronounced separation of the different workshop practices of coin production and testing was no doubt partly a practical measure, concerning different types of equipment. But the spatial organization of the Royal Mint also held perhaps a deeper symbolic significance concerning security and trust. As with Italian mint architecture, 'the presence of distinct spaces within the mint for different types of activity reassured coin users that the materiality of coins had not been compromised, and thus inculcated confidence in the monetary system'.[28]

Within this industrious space there was a principle of secrecy concerning all minting activities: the production (and testing) of coin was meant to be secret, a concealed body of knowledge, entrusted to the Mint workers and particular royal representatives. Across Britain and Europe, this was an era in which 'coinage was regarded as a crucial state secret, and the ability to handle and keep secrets was therefore indispensable'.[29] From the perspective of the state this secret approach to coin production was closely

[26] Challis, *The Tudor Coinage*, 15.

[27] Parnell, *The Tower of London*, 59–60; Justine Bayley and Harriet White, 'Evidence for Workshop Practices at the Tudor Mint in the Tower of London', in *The Renaissance Workshop*, ed. David Saunders, Marika Spring and Andrew Meek (London: Archetype Publications, 2013), 138–43.

[28] Lauren Jacobi, *The Architecture of Banking in Renaissance Italy: Constructing the Spaces of Money* (Cambridge: Cambridge University Press, 2019), 74.

[29] Daniel Jütte, *The Age of Secrecy: Jews, Christians, and the Economy of Secrets, 1400–1800* (New Haven: Yale University Press, 2015), 104.

linked to fears of counterfeit currency. No doubt the secrecy imperative was also rooted within an institutional culture in which access to words, policies and documents of authority was closely monitored. Secrecy, Paul Griffiths contends, 'protected existing structures of authority'.[30] And, as has been observed by anthropologists, 'exclusion from knowledge' more generally 'tracks the topography of power'.[31] Across guild, civic and parochial government in London, elites 'preferred to meet behind closed doors, and the exclusive nature of proceedings helped to institutionalize inequality by restricting access'.[32] A similar association between access and authority likely operated within the Royal Mint. Furthermore, the arts of metalworkers were craft mysteries, bodies of technical knowledge which were not to be observed or repeated by the uninitiated. This was part of a broader late medieval and early modern urban culture of 'proprietary attitudes towards craft knowledge'.[33] And yet, the impetus towards secrecy at the Mint was also perhaps related to a more esoteric trend within metallurgy. Metallurgists shared many processes, tools, ingredients and recipes with alchemists and thus 'the boundaries between metallurgy, coinage, and alchemical knowledge were fluid'.[34] At the Tower Mint the drive for secrecy operated at a number of institutional, social, epistemological, political and cultural levels.

Recoinage and regulating space

Sensitivities on the part of English royal and institutional elites concerning the discussion of bullion, coin and procedures for their regulation were especially heightened at certain moments. These included the mid-sixteenth-century monetary debasements and when new techniques were being applied to the production of coinage, as in the 1660s. When the Parisian engineer Pierre Blondeau was employed at the Mint for the purpose of mechanizing the production of coinage for the newly crowned Charles II, he was sworn only 'to discover his Secrets in rounding the pieces before they are sized, and in making the edges of the moneys with letters and grayvings' to officers of the Mint and 'unto his Majesty if he shall please to do him the Honour of being a witness unto his Art and Inventions'.[35] More generally, with the arrival of Blondeau and the mechanization of the coinage, new Mint officers were required to swear on oath, not to 'reveale or discover the said new Inventions … unto any persons or persons whatsoever directly or indirectly in part or in whole'.[36] Crucially, Blondeau's coinage machines – which edged coins before they entered the screw press – vastly

30 Paul Griffiths, 'Secrecy and Authority in Late Sixteenth- and Seventeenth-Century London', *The Historical Journal* 40, no. 4 (1997): 925–51 (p. 927).

31 Lenore Manderson, Mark Davis, Chip Colwell and Tanja Ahlin, 'On Secrecy, Disclosure, the Public, and the Private in Anthropology', *Current Anthropology* 56, no. 12 (2015): 183–90 (p. 184).

32 Griffiths, 'Secrecy and Authority', 928.

33 Long, *Openness, Secrecy, Authorship*, 1.

34 Jütte, *The Age of Secrecy*, p. 104.

35 TNA, MINT 1/1, fo. 145. Here we can take 'sized' to mean: to regulate or control in relation to a fixed standard (now an obsolete meaning of the verb). See 'size, v.1'. *OED Online*. Oxford University Press, March 2022. Web. 1 June 2022.

36 BL Add. MS 18759, fo. 72r (British Library).

Figure 3.5 Striking coins with a screw press, from 'The Art of Coining', *Universal Magazine* (1750). University of Michigan/Google Books.

reduced the possibility of clipping coins around the edges (Fig. 3.5). Blondeau had previously visited the Interregnum Mint from 1649 with an 'innovative device' for 'an edge-marking process' for newly minted coins, but he was not formally employed by the Commonwealth.[37] The seeking out of French engineering and artisanal expertise at the Tower Mint was not a new phenomenon; indeed, 'a sequence of French specialists and enthusiasts had been promoting such a change [mechanisation] for a century: Eloi Mestrelle under Elizabeth, Nicolas Briot under Charles I, and most recently Pierre Blondeau'.[38]

The corporation of moneyers at the Mint consistently opposed external involvement in their Tower workshops. This opposition was rooted in a generalized resistance to foreign interference, and a deeper socio-economic concern. The moneyers claimed in print in the mid-seventeenth century that 'it is the livelihood and subsistence of above two hundred Families, when the Mint is full of work'. They spoke of 'twenty of our Fellows' who are 'fallen to so great decay, that both themselves and families are brought to great distress and poverty for want of imployment in the Mint'.[39] No doubt the mechanization scheme proposed by 'the French man' would result in further loss of work. For his part, Blondeau claimed that the silver and gold coins produced

[37] Challis, *A New History*, 329–30.

[38] B. J. Cook, 'Stampt with your own Image': The Numismatic Dimension of two Stuart Successions', in *Stuart Succession Literature: Moments and Transformations*, ed. Paulina Kewes and Andrew McRae (Oxford: Oxford University Press, 2018), 303–18 (p. 315).

[39] *The answer of the Corporation of Moniers in the mint, at the Tower of London, to two false and scandalous libels printed at London, and lately come forth without date* (London, 1653), 39.

by the Tower moneyers were not justly made. He also mocked the very idea of the moneyers' craft mystery, suggesting that their collective arcane knowledge was simply a self-interested profit-making scheme:

> the officers themselvs and the workmen of the Mint do cull or pick out the heaviest pieces to melt them again ... And that is the thing which among themselvs they call the mysterie or secret of the Mint, and which is the caus that, by themselvs and by their friends, they make use of all their joint power and credit.[40]

To Blondeau, the depleted state of the coinage was linked to this practice of 'Goldsmiths and others, who reciv the monie from the Mint' to 'pick out the heaviest of them that are left, to transport them beyond Seas; so that onely the light, the fals & the clipped monie remain's within the State'.[41]

Blondeau had his own ideas about the imperative of secrecy, closely linked to the desire to protect his invention. He expressed these preferences for concealed technology in the early 1650s in a series of bitter-printed exchanges, effectively position papers, with the corporation of moneyers at the Mint. The French engineer wrote of how 'the Invention needs not be made publick; But if it be the pleasure of the State, the Engines wherewith the brims are marked, may be kept secret among few men, who shall be sworn to keep it, and not to reveal it to anie'. (Moreover, he claimed that duplication was unlikely because the engines are 'so difficult to be made, and afore they are perfected ought to goe through so many hands'.) Blondeau stressed that since his arrival in England two years previously (September 1649), 'the Officers and Workmen of the Mint have made their utmost endeavours to finde out my Invention, which they could not attain unto, although the Master of the Mint hath had in his hands my Patterns above nine moneths together'. Besides, he claimed the moneyers at the Mint had entreated the royal committee 'that I should bee commanded to make a new tryal here, hoping they should be able to discover my secret [...] but they could not'.[42] Indeed, when it came to a competitive trial of 'skill' between Pierre Blondeau and engraver David Ramage, provost of the moneyers, the former produced a much higher quantity and quality of coin using the edge-marking process.[43] The London moneyers were presented by Blondeau as incapable of translating the new technology into words and diagrams.

40 *A most humble mem[o]random from Peter Blondeau, concerning the offers made to him by this Commonwealth, for the coyning of the monie, by a new invention, not yet practised in any state of the world, the which will prevent counterfeiting, casting, washing, and clipping of the same* (London, 1653), 4.

41 *A most humble mem[o]random from Peter Blondeau*, 3.

42 *The answer of the Corporation of Moniers in the Mint*, 6–7, 9–10. The administration of the oath of secrecy from the Board of the Mint and the workers handling the machine continued in London until 1811. See John Craig, *The Mint: A History of the London Mint from A.D. 287 to 1948* (Cambridge: Cambridge University Press, 1953), 162.

43 Ibid., 152; Challis, A *New History*, 330. It is likely that the London moneyers used Nicolas Briot's old equipment for the trial.

> Asked to draw and present their Proposition for coyning of the monie marked upon the thickness or edge, as that of the said Blondeau is .... they brought such a Proposition, that the said Committee having read it over and over, could not understand it, or the sens of it; and even those that brought it could not explain it.[44]

This example of challenges of artisanal interpretation has important implications for our understanding of technical secrets. It demonstrates how seeing plans, or even the technology itself, was not necessarily sufficient to replicate the process or machinery. As Paola Bertucci has suggested of technical secrets in eighteenth-century France, specific, local, embodied knowledge of the method or technology was fundamental to 'what constituted technical success – or what it meant that a machine "worked"'.[45]

In explicit acknowledgement of the broader concern of coining as a state secret, Blondeau asserted in his position paper that should his new way of minting be adopted at the Tower, he would 'not have the power to admit any one to the work of the Mint, without the consent and approbation of the State'; further, he would 'not have the keeping of the dices or stamps, but they shall be intrusted in the hands of such as the State shall admit to work'.[46] Incidentally, the aforementioned coiner Ramage, who probably cut dies for Commonwealth Farthings, was himself highly proprietorial over workshop equipment; he was accused of trying to 'supress all Tools for making Farthings but his own, ... that so he may have the sole Trade in his hands ... '.[47] We can see how secrecy for engineers and coiners vying for control at the Mint was in part a performance aimed at community building and exclusion.[48]

The institutional ideal of secrecy surrounding the production of coinage and testing of metals was reflected and reinforced through security measures on the Tower site.[49] Once through the Tower gates manned by yeomen of the guard, porters and warders, a prospective Mint visitor would encounter the Mint gates. A porter was employed to oversee the flow of people and goods in and out of the Mint gates and the office of receipt, the chamber to which customers (typically merchants and goldsmiths) brought their bullion.[50] That this official was also required to regulate the movement and potential indiscretions of institutional employees is reflected in a statement by the officers of the Mint in December 1664:

[44] *A most humble mem[o]random from Peter Blondeau,* 13. See also W. R. Hamilton, 'Blondeau's Proposal for Reforming the Coinage of England', *The Numismatic Chronicle (1838–1842)* 1 (1838–9): 165–80 (p. 170).

[45] Paola Bertucci, 'Enlightened Secrets: Silk, Intelligent Travel, and Industrial Espionage in Eighteenth-Century France', *Technology and Culture* 54, no. 4 (2013): 820–52 (p. 830).

[46] *The answer of the Corporation of Moniers in the mint,* 19.

[47] Leonard Forrer, *Biographical Dictionary of Medallists*, vol. V (London: Spink & Son, 1902), 21.

[48] Timothy McCall and Sean Roberts, 'Introduction: Revealing Early Modern Secrecy', in *Visual Cultures of Secrecy in Early Modern Europe*, ed. McCall and Roberts (Missouri: Truman State University Press, 2013), 1–23.

[49] Such close regulation was typical across Europe. In Rome 'the mint was ... a site of high security and strict controls'. See Stielau, 'Adding to the good silver with other trickery', 163.

[50] Challis, *A New History*, 268, 286. A payment to the porter to secure access was likely customary. See Martin Allen, 'Medieval Merchants and the English Mints and Exchanges', in Martin Allen and Matthew Davies, *Medieval Merchants and Money: Essays in Honour of James L. Bolton* (London: University of London Press; Institute of Historical Research, 2016), 197–212 (p. 208).

> that in Order to the preserving of the Reputation of his Majestys said Mint and the better securing the money and bullion of Gold and Silver from time to time delivered into his Majesty's Office of Receipt in the Mint, there ever was, and still is a Porter appointed to take accompt of all persons repairing thither and (with the Assistance of some of the Monyers) to search dayly in his lodge (in times of coyning) workmen at their going out and all suspicious persons as also to stop all such moneyers or workmen as should depart to leave their work before the hours appointed or allowed them.[51]

Effectively managing the movement of people and objects through the Tower complex proved to be an immense challenge, no doubt in large measure because of the multifunctionality of the site, and the social diversity of its working population. Correspondingly, particular groups and institutional workers were keen to assert their free movement through the building complex and associated privileged status. And thus, in December 1664, officers of the Mint were eager to remind royal governors that they had always

> had a particular and distinct Jurisdiction and Authority within themselves independent of the Constable or Lieutenant of the Tower and of the Master and Officers of the Ordnance with free Ingress, Egress, and Regress at all times for themselves and Servants and all persons bringing any Gold or Silver to the Mint without any disturbance, letting or gainsaying.[52]

Moreover, despite the porter at the Mint gates, the illicit movement of people and material things (specifically tools) was clearly taking place; the regularity with which prohibitions of movement were repeated in the Mint records is itself highly suggestive of the continuation of forbidden practices. Mint officials were especially worried about this prohibited traffic during intense periods of coin production; we see this concern manifested with particular force during the 1660s, coinciding with the mechanization of the new coinage. The central anxiety was evidently counterfeiting. In January 1661/2 in relation to the 'fabrick of money by way of press or screw [the mecahnization of the coinage] and the preventing of Abuses therein' it was decreed 'that no Graver or Gravers whatsoever shall henceforth grave or work any original master puncheons, matrices, stamps and dyes, or any irons for coyning either by way of the press or hammer in *any place but* in his majestys Mint in the Tower of London'. More explicitly, it was stressed that 'for the better preventing the Counterfeiting of the new current moneys of this Kingdom', moneyers were not permitted to 'keep or use any of the Presses, Cutters, or any other Tools or Engines employed in the new way of Coyning or in the making of the moneys with cutters or grayvings about the edges'. If 'such presses, rowlers and other tools and engines' were found outside the site of the Mint, they were to be seized,

[51] MINT 1/1, fos. 150-1.
[52] MINT 1/1, fo. 150.

secured and destroyed by its officers.[53] Regular inventories were also made of coining workshops in an attempt to keep track of their material contents.[54]

From Old Bailey records we hear of opportunistic institutional workers who attempted to smuggle coins out of workshops. One 'C-J' of the parish of Alhallows Barking, a workman at the Mint, was indicted 'for feloniously stealing 7s. 6d'. 'Being at work .... upon half Crowns, the Prisoner took his opportunity, and put his Hand into the Tray, and took the Money, and was going out with it, but being seen they called him back; he suspecting that they would search him, stoopt and laid them down.' The case was proved against C-J; he was found guilty and sentenced to branding.[55] Another Mint labourer received the same sentence for an ingenious attempt to smuggle coins from institutional workshops. Mint worker Richard Cannard deposed that 'carrying a Tray of Guineas' behind the prisoner and, turning his head 'observ'd the Prisoner to whip something into his Mouth ... searching the Prisoner found it in one of the Knees of his Breeches within the Linings; and it was a new one just come from the Press.'[56]

Aside from the distinct possibility of coins and tools being smuggled out of the Mint workshops, there was also the issue of managing the broader population of the Mint site. We noted earlier how the Mint had spread by this date to encompass the entire outer horseshoe-shaped complex between the walls of the Tower. In the early 1660s the officers of the Mint gave 'a particular accompt of all persons who have Intruded and now Inhabit within the Mint or precincts thereof and have no relation thereunto'.[57] These were individuals and families that had no specific employment at the Mint. Several years later, in 1664, there were further complaints of 'many incroachments made upon the Jurisdictions and Libertys of his Majesty's said Mint' which were said to cause the 'direputac[i]on therof' and to act as a 'great danger and prejudice of his Majesty's Officers and moneyers there by the keeping of common drinking houses and building of workhouses, sheds, and forges for wheelrights, carpenters and smiths etc.'[58] There was evidently a double concern at play here. First, the unease that these spaces and their activities and persons were largely unregulated; second, that these structures were taking up space which was now urgently needed by the Mint to accommodate new coining equipment. It was ordered by the royal council that 'all persons using or inhabiting any lodgings or houses situate or being between the two Mint Gates of the Mint' should 'remove themselves and their families'.[59] This was said to be 'for the better security of our said Mint, and for the accommodating of such other officers and workemen as hereafter shalbe employed in the making of Tooles and Engine, and in Coyning our Moneys by the Mill and Presse'.[60]

53 Ibid., 148.
54 MINT 1/3.
55 *Old Bailey Proceedings Online* (www.oldbaileyonline.org, version 8.0, 9 August 2021), December 1699, trial of C – J – (t16991213-32).
56 *Old Bailey Proceedings Online* (www.oldbaileyonline.org, version 8.0, 9 August 2021), October 1719, trial of (t17191014-4).
57 MINT 1/1, fo. 141.
58 Ibid., 151.
59 MINT 1/4, fo. 23. [Order repeated: fo. 35].
60 MINT1/4, fo. 39.

The distinct concern that the Mint site was overcrowded and thus unregulated and insecure continued into the next decade. In 1679 Sir Henry Slingsby, Master of the Mint, warned of the 'dangers and hazards which may happen by reason of the great concourse of people … inhabiting within the said Mint (though not belonging to it) especially to the smith of the Ordnance who making all sorts of Iron work there do dayly cause the crowding up of ye Mint w[i]th land and sea carriages, and the drawing number of carpenters, painters, and other workmen thither' (see Fig. 3.3/ Plate 2).[61] Here we see how the multifunctionality of the Tower site impacted upon the industrious and tightly regulated activities of the Mint. As a result of Slingsby's complaint the royal council 'viewed the several parts of the mint', 'examined the Rights and Boundarys therof' and found that the smith employed by the Ordnance Office was indeed 'dwelling and working in ye mint and making and fitting double headed shott, and all sorts of Iron work for that service'. This industry had resulted in 'a great and dayly concourse of people thither the crowding it up with land and sea carriages and the drawing a great number of Carpenters, Wheelrights, Painters and other workmen to your Majesty's Mint which *is thereby rendered as publick as the common streets* and lyable to many dangers losses and Inconveniencys'.[62] The allegedly 'public' nature of the Mint site in the 1670s was in clear contrast to the institutional ideal of a closely managed, concealed space.

The pressure for space within the walls of the Tower grew in the late seventeenth and early eighteenth centuries. As Mint Master at the turn of the eighteenth century, Isaac Newton requested that during artillery practice, 'the Gunner of the Tower do order the Guns in such a manner that upon firing they may do least harm to the glass windows of the Mint'.[63] This was certainly the thin edge of the wedge, and throughout his tenure, Newton waged a persistent battle 'defending the Mint's domain against the encroachment of the Board of Ordnance housed in the Tower'.[64] In his 1695/6 account of 'the state of the royall Mint in the Tower when the grand Coynage began' (the complete silver recoinage), Mint employee and theological writer Hopton Haynes made a number of barbed observations of how the Mint 'had been much incroach'd upon, before and since the Restauration of King Charles the 2d'. Tellingly, Haynes linked the encroachment by the Ordnance to the threatened security and secrecy of the Mint, and the corruption of the coinage:

> after the Fortifications of the Tower were repaird and a Garrison plac'd in it under the Command of the chief Governors for the time being, they have notwithstanding the speciall Privileges of the Mint marchd their Troops, posted their Centinells, and erected Barracks for lodging soldiers within the Liberty of the Mint, just by the work houses wherein all the whole mystery of the Coynage was transacted; a thing of very dangerous Consequence, and perhaps gave a very unlucky opportunity to

61 MINT 1/1, fo. 182.
62 Ibid., 183.
63 Craig, *Newton at the Mint*, 36.
64 Richard Westfall, *Never at Rest: A Biography of Isaac Newton* (Cambridge: Cambridge University Press, 1981), 556, 563, 614.

> diverse of the privat soldiers, who had been bred to any of the trades in smithery, to counterfeit and debase the monys.[65]

Haynes further claimed that in recent years the Mint had 'become a Common Thoroughfare. The garrison have taken possession of all its avenues; forreigners inhabit within the Liberty, and the old mint gates remain only to shew, how farr the bounds of the Mint extended in former times, and how jealous our Ancestors were of exposing the mystery of Coyning.'[66] As Alice Marples writes, 'it was common knowledge that money and metal were regularly embezzled by workers and officers alike and "a good number of blanks" had been found in the Tower ditch and in the Thames at low tide'.[67] Haynes explicitly admired the closely guarded secrecy of historic mint sites:

> The Romans took a particular care of the security and secrecy of their Mint: 'twas plac'd in the Temple of Saturn 'twas an inviolable sanctuary for the publick Treasure, and truly deserved the appellation of *Ærarium Sanctius*. And Charles the Great was so very sensible of the importance of secureing and keeping secret the coynage, and so tender of the publick faith in issueing Standard Monys, that he erected a Mint within his own Palace, under his own inspection.[68]

Suggestively, a contemporaneous survey commissioned by Newton of the *Mint in the Tower of London* (1701) by William Allingham reinforces this sense of the territorial encroachment by the Ordnance Office (Fig. 3.6/Plate 3). For instance, the space labelled 'H' is 'Barracks formerly ye Assay Masters house turned into Barracks', and 'O' represents the 'Smith of ye Ordinance formerly belonging to ye Smith of ye Mint'.[69]

Writing in June 1699 on the subject of 'the Mint privileges', Newton also reported an unwelcome barracks within the liberties of the Mint from the time of Charles II's Restoration, and a dwelling house for a clerk of the Ordnance. Echoing Slingsby's and Haynes's earlier complaints, Newton warned how

> the Mint-streett is now become, & made a great thorough-face, by ye Soldiers in Garrison, by strangers living in it, And by people [gentlemen visitors], whome ye Warders being to see ye Mint, & shew them ye tower: which ill practice very much exposes ye gold and silver monys, that must be carried from one Workhouse to another, to perfect ye Coynage.[70]

A further troubling development was the posting of sentries within the Mint on the order of the Lieutenant of the Tower, 'who together with ye Labourers, Smiths, & other

[65] Lansdowne MS 801, fo. 65v (British Library).
[66] Lansdowne MS 801, fo. 66r.
[67] Alice Marples, 'The Science of Money: Isaac Newton's Mastering of the Mint', *Notes and Records* 76, no. 3 (2022): 507–25 (p. 511).
[68] Lansdowne MS 801, fo. 66r.
[69] William Allingham's *Exact Survey of the Ground-Plot or Plan of His Majest[ies]s Office of Mint* (1701) (Royal Mint Museum).
[70] MINT 19/3, fos. 416r-v.

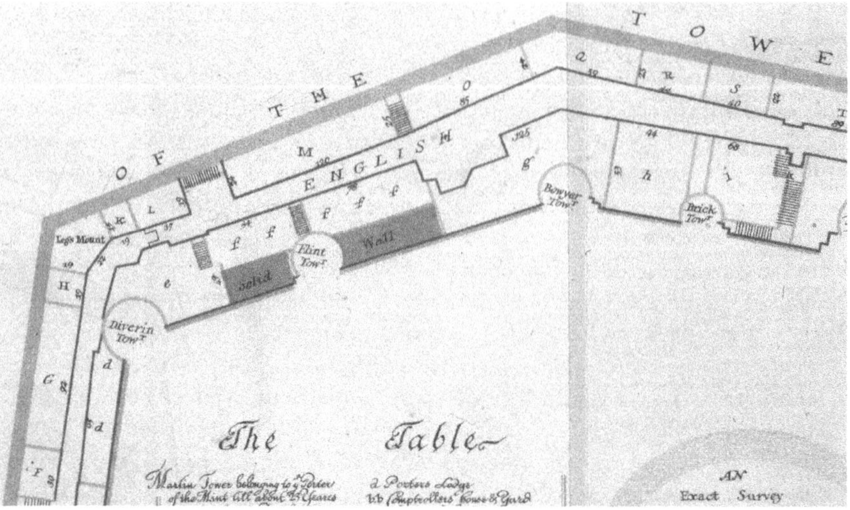

Figure 3.6 Detail from William Allingham's *Exact Survey of the Ground-Plot or Plan of His Majest[ie]s Office of Mint* (1701). At 'H', adjacent to Legges Mount, at the Northwest corner of the Tower, were 'Barracks formerly ye Assay Masters house' and at 'O' was the smith of the Ordnance, 'formerly belonging to ye Smith of ye Mint'. © The Royal Mint Museum.

See also colour version, reproduced in plate section 3.

Artificers belonging to ye Ordnance [...] resort to ye Mint, to ye great & apparent hazard of divulging ye mystery of ye Coyngage; by insinuating themselves into ye acquaintance of our mony[e]rs'.[71] Newton beseeched the royal council that 'the bounds & Limitts between ye Mint & ye Ordnance be exactly adjusted', and he asked for the expulsion of those not directly employed by the Mint. Further, he sought greater control over entry points to the Mint: 'strong gates [...] which may be shutt-up every night, & open'd by day, at such times as ye Warden of the Mint shall direct.'[72] Nearly twenty years later in around 1717, Newton was still complaining about encroachments on Mint property: 'we desire that no soldiers may be lodged in the Mint between the two gates thereof least it render the custody of the gold & silver unsafe.'[73]

Newton's long-standing grievance concerning the boundaries and privileges of the Mint site extended beyond the Office of Ordnance. In July 1697, Newton asserted in letters to the Treasury the customary exemption by royal warrant of the Mint and its employees from the jurisdiction of the Lieutenant of the Tower, Lord Lucas and his staff. Matters had come to a head in relation to the arrest of Mint labourer Philip Atherton by a constable of the Tower 'in our Melting house', and a serious altercation on the same day between a Tower guard and the son of a porter of the Mint which

[71] MINT 19/3, fo. 416v.
[72] MINT 19/3, fo. 417v.
[73] MINT 19/3, fo. 433r.

threatened to descend into a public riot. Newton reported how 'one of the Warders, to let out of the Mint some people w[hi]ch he had brought in [to arrest Atherton] seized the Gate of ye Mint & took ye Porters son by the throat; whence arose a fray between them w[hi]ch caused such a tumultuous concourse of people as rendred ye money unsafe w[hi]ch was then coming down the street of the Mint in Trays'.[74] Moreover, an extraordinary rumour that the exiled King James II was hiding on Mint territory in the house of former assistant engraver James Roettier had led Lieutenant Lucas to twice search the Mint without permission from Mint officers. The Tower guards were said to 'begin to be rather a grievance then security to us'.[75] Newton, as warden of the Mint, and Lucas, as governor of the Tower, rapidly collected conflicting testimonies in their jurisdictional battle over the Mint site. Newton requested that the Treasury consult historic charters so that 'the rights between the Tower & the Mint thence adjusted stated & limited so that both parties may know their Duties Privileges & Powers in respect of each other'.[76]

The ideal of a securely bounded site of metallurgical testing and production was exceedingly hard to uphold in practice. The complaints and commentaries from Mint officials from the second half of the seventeenth century suggest that the encroachment of men, machinery and working practices external to Mint output were not simply an inconvenience, but rather a threat to 'secureing and keeping secret the coynage'.[77]

Assay, experiment and secrets

Coin production, the focus of this chapter so far, was fundamental to the working identity of the Tower Mint, but so too was the testing of coinage. Trial of bullion, specie and newly minted coin was crucial to the integrity of the institution. In effect, these evaluations were a judgement upon the honour of the institutions and employees responsible for upholding the standard. As merchant and natural philosopher Clement Draper (*c.* 1542–1620) wrote from his cell in the King's Bench prison, Southwark, where he was imprisoned for debt, without assay: 'no man can justlye sell or buy or receave gold or silver or other metals', and no 'mynt master or goldsmythe or gold fyner or gold beatter can well exersyse their artts'.[78] The wardens of London's Goldsmiths' Company described the role of assayer as 'a place of great trust to be supplied by men of skill and integrity'.[79] Those who tested the coinage were expected to act with 'knowledges and

[74] MINT, 19/3, fo. 409r.

[75] MINT 19/3, fo. 409v. See also Westfall, *Never at Rest*, 562, Roettier 'still occupied the engraver's house in the Mint, though he no longer engraved'.

[76] MINT 19/3, fo. 410r.

[77] Lansdowne MS 801, fo. 66r.

[78] Sloane MS 1423 (British Library). And for a broader discussion of Clement Draper and his writings and experimental practices, see Harkness, *The Jewel House*, Ch. 5.

[79] Wardens accounts and court minutes (WA/CM), P2, fo. 15v. (Goldsmiths' Hall Archive, hereafter GHA).

discretions'.[80] The integrity of the coinage depended upon the trustworthiness of those involved in refining and metallurgical trials: 'metal in flux posed the opportunity for loss through spillage or the tampering with alloys for gain.'[81]

The skill of the assayer was rooted in technical aptitude and experiential knowledge. His evaluations of metals required the manipulation of precision tools (balances) and natural elements, specifically fire, and the interpretation of matter in states of flux. This complicated art was said by contemporary artisans to demand the full range of sensory and critical faculties. For fire assay, also called cupellation assay, the practitioner took a sample of precious metal from the article to be tested, which was weighed, and then melted down multiple times in a cupel (cup of bone ash) with lead until the base metals in the sample were absorbed into the cupel (Figs. 3.7 and 3.8). Base and precious

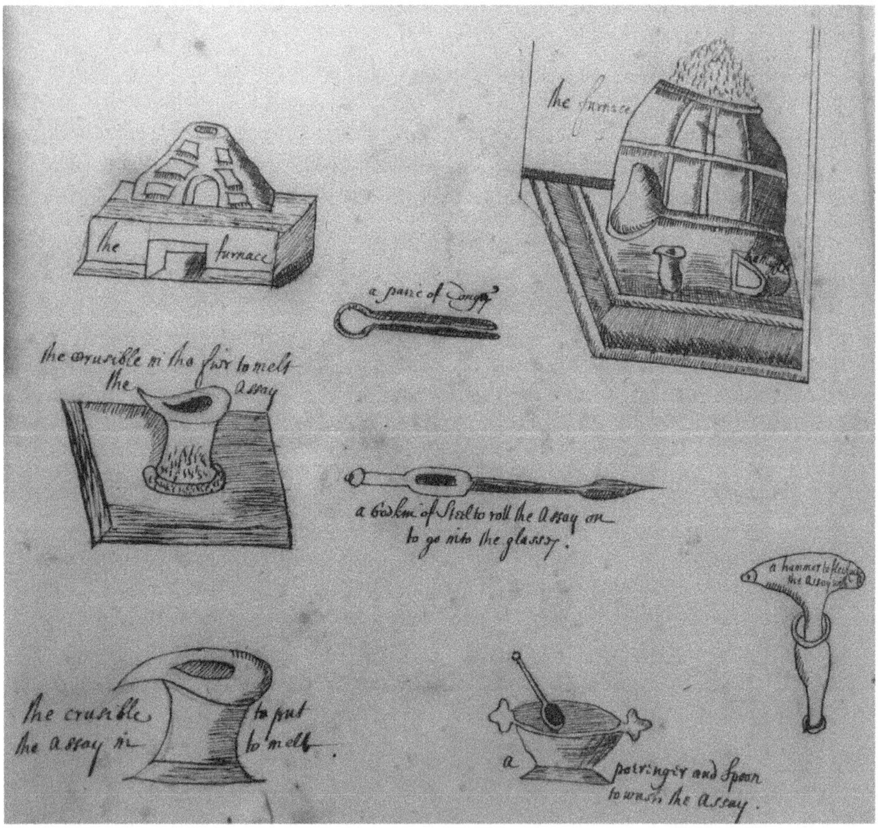

Figure 3.7 Detail from an illustration in *Mint and Moneta* of equipment and tools used for assay, including the furnace, the 'crusible in the fier to melt the assay' and a pair of tongs. The National Archives, ref. T48/92.

80 GHA, MS C II.2.1, fo. 27v.
81 Stielau, 'Adding to the Good Silver with Other Trickery', 164.

Figure 3.8 Cupellation furnace, 1696-1700, used to refine metals at the Royal Mint when Isaac Newton was Master. During cupellation precious metals are separated from impurities (SM: 1876-3). © Science Museum/Science and Society Picture Library.

metals thus separated; the fineness of the pure sample, or bead, was then weighed and calculated.[82] Newton wrote of the precise mensuration required of practitioners in weighing the testing sample before and after the assaying process, 'His scales turn w[i]th ye 128th part of a grain … They are fenced about w[i]th glass windows to keep them from ye motion of ye air & have in them little thin brass platters to take away the weights by w[i]thout handling the scales' (Fig. 3.9).[83] Moreover, broader skills of mathematical literacy might be required. Alloying calculations – working out proportions of alloy – were crucial in the production of legally fine silver and gold from raw materials of differing compositions.[84] Goldsmith and assayer John Reynolds, Mint and Goldsmiths' Company official, published several short texts containing reference tables and mathematical instructions for assaying and alloying (the mixing of metals).[85]

82 In the sixteenth century the customary weight of the sample taken for testing was either 15 or 30 grains. See John S. Forbes, *Hallmark: A History of the London Assay Office* (London: Unicorn, 1999), 21.

83 MINT, 19/1, fo. 10r. At the same time and reflecting a broader social condescension about the production of knowledge, Newton was dismissive of the assay master 'as a manual Artificer', see Kilburn-Toppin, 'A place of great trust', 27.

84 J. Williams, 'Mathematics and the Alloying of Coinage 1202-1700: Part II', *Annals of Science* 52, no. 3 (1995): 235–63.

85 Reynolds was 'either assistant to the assay master, or deputy assay-master nearly continuously from 6 October 1607 to 29 September 1665, as well as being clerk to the Tower for much of that period as well'. See Williams, 'Mathematics and the Alloying of Coinage', 254.

Figure 3.9 Detail from an illustration in *Mint and Moneta* of balances used for assay. The National Archives, ref. T48/92.

Most significantly, his 1651 tract, *A brief and easie way by table to cast up silver to the standerd of XI ounces ij penny-weight, and gold to the standerd of XXII carrots: with questions wrought by the golden-rule, also by decimall tables calculated by John Reynolds of the Mynt in the Tower,* includes instructions on how to calculate the amount of alloy in any given workshop context.[86] Assay by fire also involved an understanding, gained through lived experience, of many workshop variables, including furnace temperatures and the malleability of metals. Materials and elements might behave in unexpected ways. A fourteenth-century manuscript note from the Royal Mint recommended that in every instance of testing 'at least three impeccable assays should be made, lest through overheating or otherwise the silver should have spurted out from one of the assays and lest from draughts or a failure of the fire, the assay should have cooled, or by the fall of coals or in any other way the assay or silver should have been diminished.'[87]

The master assayer ideally had both innate and well-developed sensory perceptions that were honed through constant repetition of material experiments in the workshop. In order to make informed assessments of material properties and transformations

[86] John Reynolds (d. 1666) was assistant to the Tower assay master from 1607 and clerk to the master worker in the 1620s and 1630s, see Challis, 'Mint Officials and Moneyors of the Stuart Period', 157–97 (p. 172). For Reynolds' work as deputy assayer for the Goldsmiths' Company, see Kilburn-Toppin, 'A Place of Great Trust', 10–11.

[87] *The de moneta of Nicholas Oresme and English Mint Documents*, trans. Charles Johnson (London: Thomas Nelson and Sons, 1965), 81. This reference also highlights the extent to which 'one assay alone rarely settled the matter, as the results tended to diverge': see Felten, *Money in the Dutch Republic*, 119.

he was expected to employ his full range of senses, not simply sight.[88] The famed sixteenth-century metallurgist Agricola advised practitioners to pay attention to the odour emitted when assaying gold.[89] In a discussion of the evaluation of tin, Italian Vannoccio Biringuccio spoke of 'the whitest and hardest, or, when it is bent or some thin end of it is held tightly by the teeth, is heard to crackle as frozen water does'.[90] London goldsmith Hannibal Gamon wrote of the importance of hearing when testing precious metals with acid: 'to have surer knowledge therof laye your eare unto the saide glasse and yf it be full laden and charged w[i]th sylver it will sounde in this wise. bott, bott, bott'.[91]

We observed in Chapter 2 how London's artisanal workshops were sites in which the manipulation of matter, natural materials and instruments might be observed at close quarters.[92] The history of trades programme, enthusiastically taken up by fellows of the Royal Society between 1665 and 1680, pursued this Baconian vision of increasing understanding and improving the technologies of the mechanical arts. In his *History of the Royal Society*, Thomas Sprat listed the many planned improvements to 'the multiplying, and beautifying of Mechanick Arts … They intend the perfection of Graving, Statuary, Limning, Coining, and all the works of Smiths, in Iron, Steel, or Silver …. They purpose the trial of all manner of operations by fire'.[93] The experimental activities of the assayer's workshop were especially intriguing for London's gentlemen natural philosophers. As we have seen, proficiency in assay involved a complex blend of experiential and theoretical knowledge, and ultimately an ability to transform materials. Ideally its practitioners were gifted with extraordinary sense perception. Master assayers were also experts in managing and interpreting precision instruments. It was, moreover, an established practice in London's institutional assay workshops to record every trial which took place, including detail of metallic quantities, tools and materials – even tests that went badly wrong, through human error, a faulty furnace or unexpected material reaction, were recorded.[94] Experiential and instrumental knowledge, repeated trials, and the reporting and replication of experiments were

[88] For a further discussion of the sensory knowledge of artisans, see Pamela Smith, 'Vermillion, Mercury, Blood, and Lizards: Matter and Meaning in Metalwork', in *Materials and Expertise in Early Modern Europe: Between Market and Laboratory*, ed. Ursula Klein and E. C. Spary (Chicago and London: University of Chicago Press, 2010), 29–49 (pp. 32–3).

[89] Georgius Agricola, *De re metallica*, trans. Herbert and Lou Hoover (New York: Dover Publications, 1950), 243.

[90] *The Pirotechnia of Vannoccio Biringuccio*, trans. Cyril Stanley Smith and Martha Teach Gnudi (New York: The American Institute of Mining and Metallurgical Engineers, 1942), 211.

[91] GHA, MS C II.2.1, fo. 76v.

[92] Pamela Smith, 'Laboratories', in *The Cambridge History of Science. Volume 3, Early Modern Science*, ed. Katherine Park (Cambridge: Cambridge University Press, 2006), 290–305 (p. 296); Rob Iliffe, 'Material Doubts: Hooke, Artisan Culture and the Exchange of Information in 1670s London', *BJHS* 28, no. 3 (1995): 285–318; Bennett, 'Shopping for Instruments'.

[93] Thomas Sprat, The *History of the Royal Society of London* for *the Improving of Natural Knowledge* (London, 1667), 149.

[94] This connection or parallel – between the reporting of assay trials and the recording of experimental practice – has also been made in Cesare Pastorino, 'The Mine and the Furnace: Francis Bacon, Thomas Russell, and Early Stuart Mining Culture', *Early Science and Medicine* 14, no. 5 (2009): 630–60 (pp. 654–5).

of course central also to the methodology of the new experimental science.[95] It is particularly revealing that in early modern England the word 'assay' connoted both the trial of metals specifically, and 'experiment' more generally.[96]

The assay house at the Royal Mint was a specific site of curiosity. Gerard Malynes, merchant and writer on economics, reported in 1622 that 'coming to the Assay-house [within the Tower liberties], there we found diuers gentlemen desirous to see the manner of making of Assayes of Gold and Siluer, as also diuers Goldsmiths which brought some Ingots of Gold'.[97] Malynes claimed in print to have 'myne owne knowledge concerning Assayes, which I had observed and knowne above forty yeares, my father also having been a Mint-master'.[98] In February 1660/1 Fellows of the Royal Society undertook experiments 'on the weight of bodies increased in the fire' – heating cupels containing lead and copper – using furnaces at the Tower.[99] John Evelyn, diarist, writer and founder member of the Royal Society, recorded in July 1678 how he 'went to the Tower to try a Mettal at the Say-Masters, which [disappointingly] onely proved Sulphur'.[100] Surviving notebooks and recipes compiled by amateur gentlemen natural philosophers, such as Hugh Plat and Clement Draper, are also replete with techniques pertaining to the assayer's testing and separation of metals by fire, solution and touchstone.[101] Part of the appeal was certainly the contemporary perception that assaying was entangled with the broader sciences of 'chimistry' and 'alchimy' [*sic*].[102] Alchemists and assayers had mutual interests in the purification, testing and precise measurement of metals; practitioners of both employed precision balances.[103] It was these shared instruments, recipes and workshop techniques, and their experiential and experimental features, that proved mutually fascinating to eminent seventeenth-century gentlemen experimenters (including Robert Boyle and Isaac Newton) in their quest to uncover secrets of nature.[104]

95 Steven Shapin and Simon Schaffer, *Leviathan and the Air-Pump: Hobbes, Boyle, and the Experimental Life* (Princeton, NJ: Princeton University Press, 2011), Ch. 2; William Eamon, *Science and the Secrets of Nature: Books of Secrets in Medieval and Early Modern Culture* (Princeton, NJ: Princeton University Press, 1994), 289.

96 'assay, n.' *OED Online*. Oxford University Press, March 2022. Web. 2 June 2022.

97 Gerard de Malynes, *Consuetudo, vel, Lex mercatoria; or, The ancient-law merchant*, vol. 1 (London, 1622), 284.

98 Malynes, *Consuetudo, vel, Lex mercatoria*, 2nd edn (London, 1636), 188.

99 Thomas Birch, *The History of the Royal Society of London*, vol. 1 (London, 1756), 16. An assay furnace was specifically built for the Royal Society's use later that month (p. 17).

100 *The Diary of John Evelyn, Vol. 4. Kalendarium, 1673-1689*, ed. E.S. de Beer (Oxford: Clarendon Press, 1955), 138.

101 Sloane MSS 1423, 2210.

102 'The words "alchemy" and "chemistry" were used interchangeably to refer to the same body of activities throughout most of the seventeenth century, and only during the eighteenth century were distinctions similar to those in common modern usage rigidly drawn between the two.' See Lawrence M. Principe and William R. Newman, 'Some Problems with the Historiography of Alchemy', in *Secrets of Nature: Astrology and Alchemy in Early Modern Europe*, ed. William R. Newman and Anthony Grafton (Cambridge, MA and London: MIT Press, 2001), 385–431 (p. 386).

103 William R. Newman and Lawrence M. Principe, *Alchemy Tried in the Fire: Starkey, Boyle, and the Fate of Helmontian Chymistry* (Chicago and London: University of Chicago Press, 2002), 38–9, 41, 46, 49. See also William R. Newman, 'Alchemy, Assaying, and Experiment', in *Instruments and Experimentation in the History of Chemistry*, ed. Frederic L. Holmes and Trevor H. Levere (Cambridge, MA and London: MIT Press, 2000), 35–54.

104 Tara E. Nummedal, 'Words and Works in the History of Alchemy', *Isis* 102, no. 2 (2011): 330–7 (p. 331).

Frustratingly, there is very little surviving evidence for the day-to-day running and operations of the assay house at the seventeenth-century Mint.[105] However, a series of manuscripts authored by goldsmiths and assayers employed at the Royal Mint across the early modern period give us a fantastic insight into the culture of the craftsmen who worked within the Tower assay site. What comes across clearly from this corpus of writings – which were addressed to institutional authorities and fellow goldsmiths – is the significance of secrecy. Describing the process of assay, a seventeenth-century practitioner explained how institutional officials 'lock up the [workshop] door, with their several keys'.[106] A fuller description by an early-seventeenth-century goldsmith of a pyx trial for which he was an expert jury member stresses again the secretive dimension of the workshop practice. Hannibal Gamon wrote of how the trial was undertaken 'in one parte of ye Inner Chamber in ye Starre Chamber'.[107] The spatial dynamics of this trial reflected and reinforced its secretive nature. Each testing, or assay, of coin and bullion was meant to be kept secret in order that the process be – and crucially be perceived as being – impartial. Assay was also part of the goldsmith's craft or 'mystery', and as we have already observed, craft traditions were meant to be 'hidden from the understanding of those not experienced [or formally initiated] in the craft'.[108] Full knowledge could only be gained through apprenticeship and repeated practice, 'p[er]fection must be grounded upon Artificiall Exercise'.[109] There was probably a political aspect to this concealment too. The secretive nature of the trial of the pyx added to its theatrical significance as a ceremony of state. Inclusion suggested enhanced authority; exclusion its opposite.

Suggestively, as Master of the Mint Isaac Newton insisted in 1701 that Mint officers observe the pyx trial; he was not prepared to leave testing and witnessing to the goldsmith jury alone. The secrecy of the goldsmiths and exclusion of the Mint officers was, in Newton's eyes, likely to give the Goldsmiths' Company unwarranted influence over Mint operations.[110] In a memorandum on the subject of the trial, Newton stressed:

> the Iury are not to exclude the Officers of the Mint from the triall nor to order them to withdraw but to suffer them to take their places (as has been the custome) bot [*sic*] at the triall & at the dinner in the same order according to their Offices as if the whole buisiness were transacted in the Mint & to continue with them untill the veredict be given in.[111]

105 Challis, *A New History*, 295–6.

106 T48/92, fo. 23r.

107 GHA, MS C II.2.1, fo. 27v. The goldsmiths are undertaking a material trial, and their honour and expertise are on trial. See Simon Wortham, 'Sovereign Counterfeits: The Trial of the pyx', *Renaissance Quarterly* 49, no. 2 (1996): 334–59 (p. 336).

108 Pamela Smith, 'What Is a Secret? Secrets and Craft Knowledge in Early Modern Europe', in *Secrets and Knowledge in Medicine and Science, 1500–*1800, ed. Elaine Leong and Alisha Rankin (Farnham and Burlington, VT: Ashgate, 2011), 47–66 (pp. 49–50).

109 GHA, MS C II.2.1, fo. 4v.

110 Westfall, *Never at Rest*, 609: 'The goldsmiths insisted that they should try the coins in private. Newton insisted just as strongly on the right of officers of the Mint to be present … Ensuing silence on the subject implied that Newton won his point.'

111 MINT 19/1, fo. 264r.

Further, Newton suggested:

> a private Triall would not bind the [Mint] Master because not made according to his Contract with the King nor perhaps would it be advisable for the King to let the Mint by such private Trialls come into the hands of the company of Goldsmiths and by being at their mercy be secretly swayd and Governed by their interest.[112]

Institutional boundaries of secrecy were evidently political and might be negotiable (for those with sufficient persuasion and authority). Moreover, there is probably at play here the further epistemological dynamic of 'rules for trustworthy knowledge-making', especially witnessing of a trial which Newton would have been accustomed to as a fellow of the Royal Society (president from 1703).[113]

There was also a social dynamic at work which complicated the principle of secrecy associated with the assay house at the Royal Mint, and this relates to curious visitors. Warders of the Tower 'went to shew the Tower & Mint to 2 Gentlemen (as is Usuall)'.[114] The activities of refining and testing metals, and the newly mechanized coining process, were of considerable interest to those we might broadly define as gentlemen natural philosophers. There is evidence to suggest that those with the right social and institutional connections might gain access to Mint workshops associated with monetary production and metallurgical testing. On Tuesday 19 May 1663 Samuel Pepys (soon to become a fellow, and later president, of the Royal Society) thoroughly enjoyed a social and investigative visit to the Tower Mint, together with a naval associate. A personal friend of the Master, Henry Slingsby, also a fellow of the Royal Society, Pepys recorded in his diary how he was

> shown the method of making this new money, from the beginning to the end; which is so pretty that I did take a note of every part of it and set them down by themselfs [*sic*] for my remembrance hereafter. That being done, it was dinner time, and so the Comptroller would have us dine with him and his company [...] and after dinner went to the Essay- Office and there saw the manner of essaying of gold and Silver, and how silver melted down with gold doth part again being put into aqua fartis [*sic*], the silver turning into water and the gold lying whole in the very form it was put in, mixed of gold and silver; which is a miracle [...].[115]

Tellingly, Pepys described the process as both an 'experiment' (perhaps meaning a trial or test) and a 'mystery' (being both 'a hidden and secret thing', and more specifically a concealed body of collective craft knowhow).[116] It is worth repeating his notes in full, for they also demonstrate his very close observation of the process of assay:

[112] MINT 19/1, fos. 307–8.
[113] Marples, 'The Science of Money', 8.
[114] MINT, 19/3, fo. 397r.
[115] *The Diary of Samuel Pepys*, vol. 4, 1663 (1971), 143.
[116] 'mystery, n.1.' *OED Online*. Oxford University Press, March 2022. Web. 1 June 2022.

> If it be silver, they put such a quantity of that alone and wrap it up in lead; and then putting them into little earthen cupps made of Stuffe [*sic*] like tobacco pipes [the cupel], and put them into a burning hot Furnace; where after a while the whole body is melted and at last the lead in both is sunk into the body of the cup, which carries away all the copper or dross with it and left the pure gold and silver embodyed [*sic*] together [...] And to part the silver and the leade in the first *experiment*, they put the mixed body into a glass of boyling [*sic*] aqua fortis, which separates them by spitting out the silver into such small parts that you cannot tell what it becomes; but turns into the very water and leaves the gold at the bottom clear of itself [...] which is a great *mystery* [...].[117]

The culture of the assayers themselves, specifically the metallurgical manuscript culture emanating from the early modern Tower, also complicates the ideal of workshop secrecy. Through their authorship of books of secrets, a small but significant group of London assayers and goldsmiths created an imaginative workshop space which was rooted in, but also expanded beyond, the confines of the Tower assay house. Thomas Aunsham, for example, who served as deputy to the comptroller Sir Henry Wyatt and gave 'daily attendance' at the Mint from *c.* 1509 to *c.* 1520, authored a treatise of techniques, recipes and advice relating to assay. He counselled, for instance, on 'the most assured way of partinge of gould from siluer w[i]th aqua fortis and the least losse of the finding againe of ye silver after it is parted as farre as I Thomas Aunsham marchant can practis or knowe by any other minte masteres of strange Countryes'.[118] In so doing Aunsham presented himself as one of those 'masters of this science'.[119] This manuscript was copied and modified across the sixteenth and seventeenth centuries as later assayers added their own tried-and-tested recipes.[120] While these manuscripts could never fully reveal craft secrets – embodied workshop knowledge could only truly be understood through practice and experience – their revelatory nature muddies the boundaries between 'secret' and 'open' productive knowledges. Here the partial revelation of arcane knowledge might have heightened the social and institutional status of workshop practitioner-authors at the Royal Mint.

Conclusion

The Royal Mint was a significant space for material production, transformation and exchange in the early modern metropolis. Nestled within the increasingly crowded, multifunctional site of the Tower, the seventeenth-century Mint was a challenging place to manage and police. There existed an institutional ideal of secrecy, and of a clearly bounded site of monetary production and regulation. Yet the lived reality proved more complex, and Mint officials could not exercise complete authority over

[117] *The Diary of Samuel Pepys*, vol. 4, 1663 (1971), 144–5.
[118] Harley MS 38, fo. 262r (British Library).
[119] Harley MS 38, fo. 247r.
[120] For further discussion see Jasmine Kilburn-Toppin, 'Writing Knowledge, Forging Histories'.

the metallurgical knowledge reproduced and generated therein.[121] Understanding its complex knowledge culture requires paying close attention to the physical built environment, and the movement of people and things (including tools and coins) through the space. The socially varied, physically expansive and multifunctional nature of the Tower site impacted upon its institutional culture of secrecy.

This chapter has shown how motivations for secret knowledge and technological processes were also various, perhaps as diverse as the different groups employed at the Tower Mint.[122] At the highest official levels, secrecy surrounding monetary production was significant to protect the integrity of the coinage. It was also – as we saw with Newton's dispute with the goldsmiths – a means of restricting access and structuring the boundaries of authority. For engineer Blondeau, secrecy was significant in protecting his intellectual rights to a new form of technology for coin production; he hoped that 'the Engines wherewith the brims are marked, may be kept secret among few men, who shall be sworn to keep it, and not to reveal it to anie'. Among the moneyers at the Tower, a self-defined 'antient Corporation and Company, settled by charter for many hundred yeares' secrecy had perhaps yet another shade and consequence; it was connected to their deep-rooted craft sense of a collective mystery.[123] Within the assay house we have yet another 'community of knowers', who stressed the importance of secrecy and yet whose metallurgical understanding might be translated from tacit culture to the written word, likely expanding the audiences for their practices and technologies.[124] And at the Tower, complicating things further, we also have visitors, often gentlemen natural philosophers hunting secrets of knowledge that might expand the boundaries of experimental knowledge.

These visitors also interacted with other groups and individuals that used this complex site. In the next chapter we turn to another institution within the Tower, the presence of which so irked those charged with protecting the boundaries of the Mint and the secrets which it housed. This was the Office of Ordnance, and the jumble of trades, skills and activities that it required to provide arms, stores and buildings to the army and navy.

[121] In a similar vein (of workshops within institutions), Tara Nummedal found that 'however hermetically sealed alchemical patrons and practitioners imagined their laboratories to be, their location in the semi-public arena of princely courts mean that boundaries were always permeable': *Alchemy and Authority in the Holy Roman Empire* (Chicago and London: University of Chicago Press, 2007), 143.

[122] In a different context, Paola Bertucci has written of 'multifaced cultures of secrecy': 'Enlightened Secrets', 847.

[123] *The answer of the Corporation of Moniers in the mint, at the Tower of London*, 6–7, 23.

[124] Leong and Ranking, *Secrets and Knowledge*, 9, 'building communities of knowers was one of the most crucial functions of secrets'.

4

Manufacture, mathematics and experiment at the Ordnance Office and Arsenal

Introduction

As the previous chapter showed, the Tower of London was crowded with people and activities, as well as buildings, goods, vehicles and animals, both exotic and quotidian. Its multipurpose nature meant that, while there were disputes and attempts to control space, there were also fruitful overlaps. An industrious, if not industrial, site, the work of both the Mint and the Ordnance Office meant it was a place of manufacture and trial, as well as of storage and management. As the first part of this chapter will show, the proximity of people, and of skills, interests, objects and materials, could bring opportunities to those employed and living within the seventeenth-century Tower complex. The Ordnance's role in making and storing provisions for war or occupation meant both the regular practice of relevant skills and interest in new or standardized approaches that might reduce costs, increase efficiency and raise the status of the work involved. The middle two sections of the chapter thus explore Ordnance cultures of trial, innovation and training. While always linked to other garrisons, forts and castles, the geography of the Ordnance increasingly expanded to Thames-side sites of manufacture. The final part of the chapter looks to the Woolwich Arsenal in the eighteenth century as a site that, like the Tower but at greater scale and in the context of a more public culture of improvement, brought together craft skills, practical and theoretical knowledge, space, materials and manpower.

In the early sixteenth century, the role of the Master of the Ordnance was outlined as being responsible for receiving and safely storing munitions and other goods, overseeing their distribution and going to battlefields to manage the use of ordnance. He was also to ensure that the Master Gunner attended to his duties and 'see that there be attending on the Office of Ordnance certain artificers, as carpenters, wheelwrights, smiths, bowyers, fletchers, masons and such other necessary men, meet and convenient therefor'.[1] Over the following century, the appointment became more political and less hands-on, the scale of the Office increased and the key trades shifted as firearms overtook arrows. However, the close connection to artificers remained.

[1] Quoted in Archibald Forbes, *A History of the Army Ordnance Services* (3 vols, London: Medici Society, 1929), vol. 1, 35–6.

After the Restoration, the Earl of Clarendon characterized the Ordnance officers as 'conversant only with smiths and carpenters and other artificers and handicraftsmen'.[2] However, the physical work involved in making ordnance was something that could be celebrated as reflecting the power of weapons of war, tamed for the protection and expansion of the kingdom. The coat of arms of the Ordnance reflected this work by including Cyclopes supporters carrying smiths' forceps and a hammer to signify their role as assistants to Vulcan in his forge, crafting thunderbolts for Jupiter. These tools are now missing from the arms that were installed on the new Office of Ordnance building of 1672–3 (Fig. 4.1/Plate 4), but above them is an arm grasping a thunderbolt and below the motto, *sua tela tonanti*, loosely translated as 'to the warrior his arms', and more literally 'to the thunderer his weapons'.[3]

Figure 4.1 Coat of arms of the Ordnance at the Tower of London. Photograph by Richard Nevell, Wikimedia Commons.

See also colour version, reproduced in plate section 4.

[2] Quoted in H. C. Tomlinson, *Guns and Government: The Ordnance Office under the Later Stuarts* (London: Royal Historical Society, Swift Printers, 1979), 16. This was in contrast to the Treasury, which 'had much to do with the nobility and gentry of the kingdom' and was therefore, Clarendon argued, less suited to undergo the kind of regularization and reform recently instituted at the Ordnance.

[3] Geoffrey Parnell, 'The Buildings and Works of the Office of Ordnance at the Tower of London, 1660–1722' (2 vols, unpublished PhD thesis, University of London, 1996), vol. 1, 132–3, suggests this location for the late-seventeenth- or early-eighteenth-century coat of arms. Relocated more than once, it is now on the New Armouries, built 1663–4. The photograph in Parnell's thesis (vol. 2, Figure 7) shows the top of the forceps still in place. J. H. Leslie, 'The Arms of the Board of Ordnance', *Journal for the Society for Army Historical Research* 1, no. 5 (1922): 210–13. These arms were only officially granted in 1805.

However, brute power, like the explosive forces of gunpowder, had to be managed with appropriate knowledge and experience. Condescending attitudes such as that expressed by Clarendon undoubtedly rankled with both officials and artificers. Emphasizing the mathematical or philosophical principles connected to their craft or practice was one means of according to it greater status, and one which gunners, both in the context of artillery and fireworks, whether made for military purposes or display, had been promoting since the sixteenth century.[4] As Jim Bennett has shown, by the seventeenth century practical fields relevant to the Ordnance, such as surveying and gunnery, were widely understood to be underpinned by geometry and mathematics and there was increasing exchange between philosophers and practitioners.[5] Many scholars have taken issue with Rupert Hall's view that theoretical work on ballistics had little real-world impact, suggesting instead that there was 'a fruitful Archimedean synthesis of scientific theory and military practice'.[6] By centring on an institution and its site, rather than on social groups or disciplines, this chapter shows the extent to which people and activities overlapped and interacted.

The work of the Ordnance's officers and artificers was set out for the recently appointed Master General, George Legge, in 1683 as a series of 'Rules and Orders', thought to be a codification of existing (if idealized) practice.[7] They stated that the Master General should oversee the officers while the Lieutenant General should 'Direct and Oversee the Practice of the Master Gunner, Firemas[te]r, Mates, Fireworkers and Gunners' and 'See that all Officers, Ministers, Artifficers &ca perform their Dutys'. The Surveyor General was to see 'That all Labourers Artifficers &ca be Strictly kept to their Work' as well as to have oversight of all building work and the checking and proving of stores of arms and powder. The Principal (or Chief) Engineer, the Master Gunner and the Firemaster were each in charge of – and responsible for teaching, training and assessing – their workers: inferior engineers, gunners, fireworkers, petardiers, bombadiers and proofmasters. The Master Gunner should 'teach', 'Exercise' and 'Certifye the Abilitys of all Gunners', while the Chief Engineer was expected 'To be well skilled in the Mathematicks and Architecture &ca' – making drawings, plans and models as well as overseeing the building of fortifications and planning of sieges – and 'To provide good, able Engineers, and give an Account of their Service'. Inferior

4 See e.g. Steven A. Walton, 'Technologies of Pow(d)er: Military Mathematical Practitioners' Strategies and Self-Presentation', in *Mathematical Practitioners*, ed. Cormack, Walton and Shuster, 87–113.

5 J. A. Bennett, 'Geometry and Surveying in Early-Seventeenth-Century England', *Annals of Science* 48, no. 4 (1991): 345–54 and *idem*, 'The Mechanics' Philosophy and the Mechanical Philosophy'. See also Lesley B. Cormack, 'Mathematics for Sale: Mathematical Practitioners, Instrument Makers, and Communities of Scholars in Sixteenth-Century London', in *Mathematical Practitioners*, ed. Cormack, Walton and Shuster, 69–85.

6 Brett D. Steele and Tamera Dorland, 'Introduction', in *The Heirs of Archimedes: Science and the Art of War through the Age of Enlightenment*, ed. Steele and Dorland (Cambridge, MA and London: MIT Press, 2005), 1–33 (p. 9). See also e.g. Willmoth, 'Mathematical Sciences and Military Technology'; Catherine Ann France, 'Gunnery and the Struggle for the New Science (1537–1687)' (unpublished PhD dissertation, University of Leeds, 2014) and Werrett, *Fireworks*.

7 'Rules Orders and Instructions for the future Goverm[en]t of the Office of the Ordnance' (1683–1840), Records of the Ordnance Office and its Successors, WO 55/1790 (The National Archives); Tomlinson, *Guns and Government*, 16–7, 57–8.

Engineers were expected to follow orders, draw up designs and 'To Qualifye themselves for being Engineers as much as possible'.[8]

Like the Mint, the Ordnance Office operated as a group of officials, which came to be known as the Board of Ordnance, supported by lower-ranked officers and clerks, overseeing relations with and the work of those skilled in particular crafts and practices. These relations could be long-term arrangements, on per diem wages or annual salaries, or one-off contracts. Either way, payments were often in arrears.[9] Arrangements were usually made with individual men and occasionally women, such as Jane Hill the paintress, but could also be made with groups, as was the case with the Armourers' and Gunmakers' Companies.[10] The trades involved were the many required for supplying and maintaining stores of all sorts, for designing and constructing fortifications, storehouses and other buildings, for transporting and sheltering soldiers and stores on expedition, and for making, testing and using gunpowder and large and small ordnance. While, in the seventeenth century, gunpowder, guns and iron and brass ordnance were supplied by contractors based elsewhere, the Tower complex was home and/or workplace for smiths, gunners and proof masters, and often individuals who could simply profit from proximity. We find craft lineages associated with the Tower that supported the passing on of knowledge and skills between generations. While the political disruptions of the century meant change among the chief officeholders, there was much greater continuity among artisans and lower officials.

Smiths, gunners and mathematicians at the Tower

Ordnance contracts were rarely exclusive, with artificers being able, or needing, to take work from a range of sources, perhaps also combining regular payments with one-off commissions. For example, Thomas Hodgkins was smith to the Mint from 1644 to 1673 and was included in a 1664 list of Ordnance Gunners as 'M[aster]r Smith to the Office of the Ordnance in the Tower' by warrant dating from 1661, which paid 6d a day on top of his salaries as smith to both institutions.[11] This dual institutional link accounted for the presence of the smith's house and forges within the Mint at the Tower (see Fig. 3.3/Plate 2) and the later annoyance of Mint officials as the smith's Ordnance

[8] Index to WO 55/1790.

[9] For example, Ordnance artificers were in 'great difficulties' when the Ordnance applied to the Treasury for funds to pay them 1710/11 (see 19 September 1710 and their threat to apply to the House of Commons if not forthcoming on 7 April 1711), Minute Books, *Calendar of Treasury Books*, vol. 24, 51–68 and vol. 25, 37–50 (available at https://www.british-history.ac.uk/cal-treasury-books/). John R. Millburn, 'The Office of Ordnance and the Instrument-Making Trade in the Mid-Eighteenth Century', *Annals of Science* 45, no. 3 (1988): 221–93, indicates that payment of bills was often very tardy.

[10] Jane Hill was mentioned frequently in Ordnance bill books in the 1710s–30s for work painting interiors and exteriors at the Tower, Woolwich and Royal Observatory, Greenwich: see N. P. Barker, 'The Architecture of the English Board of Ordnance: 1660–1750' (PhD, University of Reading, 1985), vol. 2, passim. On Gunmakers and Armourers, see Forbes, *A History of the Army Ordnance Services*, vol. 1, 55–6 and Schwoerer, *Gun Culture in Early Modern England*, Chapters 1–2.

[11] Challis, *A New History*, 298; 'A List of the Gunners belonging to the Office of His Ma[jes]tes Ordnance as they were Monstred at the Tower Aprill 15th 1664', WO49/112, p. 5.

work increased. Edward Silvester (d. before 1719) briefly succeeded Hodgkins as smith to the Mint at £10 per annum but from Lady Day 1674 seems to have worked solely in 'the office of Master Smith of all His Majesty's iron works belonging to the Office of Ordnance within the Tower'.[12] However, he and his brothers, Thomas (d. 1710/11) and John (d.1722), all at some point based at the Tower, continued to do work for both institutions.[13]

Thomas was officially named as smith to the Mint in January 1697/8 and continued there until his death in 1711, by which time he received £50 per annum.[14] There he made dies and puncheons, including for recoinage in Scotland after the Act of Union, working with the Mint's engravers.[15] He and John were, however, together paid £270 'for smith's work ... for the Mint in the Tower' between 1 January 1697/8 and 31 January 1698/9.[16] They also received payments of various sorts from the Ordnance, in the 1690s supplying locks, bars, miners' tools, repairing guns and so on. One or other was the Mr Silvester who met commissions such as 'to make a Door to the Stove at the Proof Place' or to 'repair the bars and locks at the Wharf Gate' in the early years of the eighteenth century.[17] John was earlier based at the Ordnance's outpost at Portsmouth but returned to the Tower and was also able to buy considerable property in Yorkshire in the 1690s. He was presumably the Mr Silvester who cleared the house he had inhabited as smith and handed over the key in 1720.[18] The Silvesters thus represent a case of the flow of people and materials through the spaces of the Mint and Ordnance, within the Tower and beyond. It seems plausible that knowledge and practice pertaining to one sphere would have leaked easily to the other, perhaps helping to resolve unfamiliar problems.

Edward Silvester arrived in time to be involved in an unusual commission. That year, alongside what must have been more typical work on fixtures and fittings, he oversaw the production of the framework, including semicircles and gearing, of the 7-foot equatorial sextant destined for the new observatory in Greenwich (Fig. 4.2).[19] The observatory was an Ordnance outpost, with the building costs and salary of the astronomer, John Flamsteed, and his assistant on their books. Instruments, however, were paid for by Flamsteed or his patron, Jonas Moore, the Surveyor General, who had been the key lobbyist for the observatory's foundation. Moore was therefore

12 Entry Book: March 1674–5, *Calendar of Treasury Books*, vol. 4, 689–708; Challis, *A New History*, 352 (n. 306).

13 The family's genealogy has been explored at 'Linley & Jim Hooper's family history' website, though not without errors: http://www.linleyfh.com/p850.htm#i4892.

14 Minute Books: January 1698, *Calendar of Treasury Books*, vol. 13, 51–9; Warrant Book: November 1712, *Calendar of Treasury Books*, vol. 26, 507–22, indicates that payment had ceased on his death the year before.

15 Bill ... for equipment made for the Edinburgh Mint, 1707, MINT 19/1/178, transcribed by The Newton Project: http://www.newtonproject.ox.ac.uk/view/texts/normalized/MINT00394.

16 Entry Book, May 1690, *Calendar of Treasury Books*, vol. 9, 613–24.

17 For the 1690s see various entries in the Treasury Books. A John Silvester also appears as a falconer in the 1680s–90s. Index under Tower, 19 May 1705, Minutes Series 2, WO 47/22 and 19 February 1705/6, Minutes Series 2, WO 47/23.

18 Entry Book, May 1690, *Calendar of Treasury Books*, vol. 9, 613–24; Index under Tower, 13 May 1720, Minutes Series 2, WO 47/33.

19 Derek Howse, *Greenwich Observatory. Volume 3: The Buildings and Instruments* (London: Taylor and Francis, 1975), 75–9.

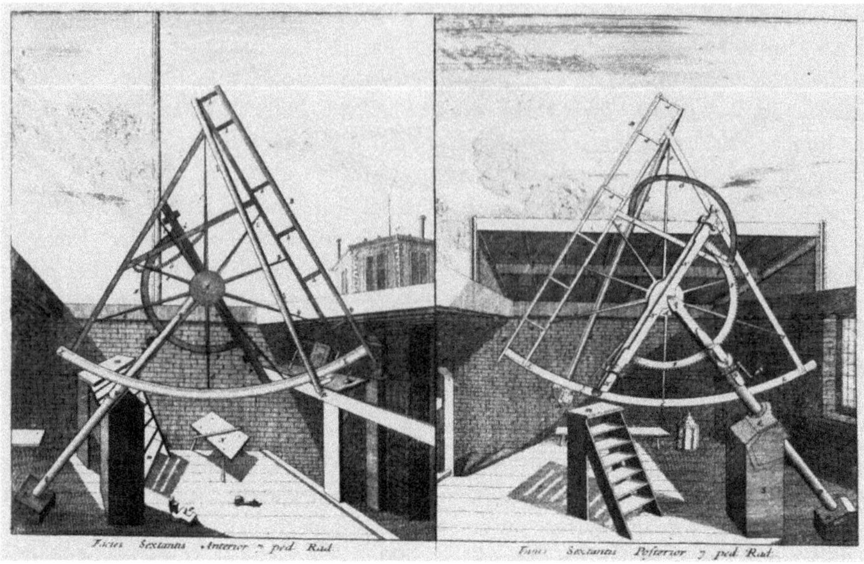

Figure 4.2 The 7-foot equatorial sector at Greenwich, engraved by Francis Place after a drawing by Ordnance draftsman, Robert Thacker, *c.* 1676, in John Flamsteed, *Historia Coelestis Britannica*, vol. 3 (1725). Bayerische Staatsbibliothek/Internet Archive.

to cover Silvester's payment of £120 for this instrument, although it was only fully settled after the former's death and appears in Ordnance records.[20] The sextant was designed by Flamsteed and made at the Tower while he was living there with Moore before the observatory was built. Thomas Tompion, a well-known clockmaker who was contracted by the Ordnance to supply and repair their clocks, produced the wheelwork, limb and indices.[21]

Flamsteed later complained that while the axles and semicircles were first modelled in wood, 'by the Tower smiths', in the event,

> The chief workman [presumably Thomas Hodgkins] dying before the semicircles were formed [in iron], I could not make his successor to understand how the bigger of them might, and was to be moved by a perpetual screw: but forced to suffer him to move it by wheel work. But, before the lesser semicircle, whereby the plane of the sextant was to be inclined, was ready to be applied, he understood me, and wrought it off so that it performed very well.[22]

The perpetual screw, which Robert Hooke claimed to have invented but Flamsteed said had derived from Tycho Brahe's description of his instruments, was clearly unfamiliar

20 Howse notes that Silvester (or Sylvester) received the final instalment in June 1683: Howse, *Greenwich Observatory*, 79.

21 Howse, *Greenwich Observatory*, 75; Parnell, 'The Buildings and Works of the Office of Ordnance', 91.

22 Francis Baily, *An Account of the Revd. John Flamsteed, the First Astronomer-Royal* (London: printed by order of the Lords Commissioners of the Admiralty, 1835), 39–41.

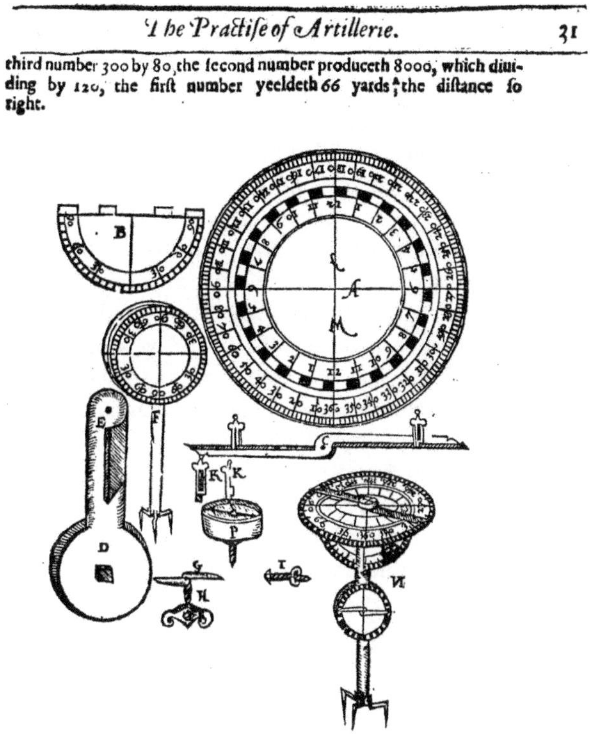

The Practise of Artillerie. 31

third number 300 by 80, the second number produceth 8000, which diuiding by 120, the first number yeeldeth 66 yards; the distance so right.

Figure 4.3 Robert Norton's cosmodelite in *The Gunner* (1628). Harvard University Library/ProQuest Early English Books Online.

to the new Ordnance smith, but Silvester quickly grasped what was required. Another of the instrument's unusual features, the toothed semicircles for moving and positioning the instrument, Flamsteed later reflected on in a letter, mentioning that one person who previously used them was 'Norton'. This was probably Robert Norton, a gunner and engineer at the Tower who, in 1628, published *The Gunner*, which included a description and illustration of his 'Cosmodelite', a kind of theodolite that had its main circle resting on a vertical semicircle (Fig. 4.3).[23] It is possible that, while he was living in the Tower, Flamsteed would have come across Norton's book. Even more likely is that Hodgkins, Tower gunner and smith, would have known and shared it with Moore and Flamsteed as they worked on the initial design and prototype.[24] There were close links between many individuals at the Tower and instrument making

[23] Eric G. Forbes, Lesley Murdin and Frances Willmoth (eds), *The Correspondence of John Flamsteed, First Astronomer Royal, Volume Two: 1682–1703* (Bristol and Philadelphia: Institute of Physics Publishing, 1997), 29, 32.

[24] Hodgkins probably died in 1674, and may have been consulted in April/May 1674 when Flamsteed stayed with Moore, when they also discussed instruments with Hooke at Garaway's coffee house: Willmoth, *Sir Jonas Moore*, 169.

and mathematical publishing; as Frances Willmoth has pointed out, the Ordnance was a hub for mathematical practice and its patronage well before the arrival of the experienced mathematician and surveyor Moore.[25]

Norton had received his training in engineering and gunnery from John Reynolds, the Ordnance's Master Gunner from 1623 to 1638, whose role included training and certifying gunners at the Artillery Garden outside Bishopsgate, north of the Tower.[26] This John Reynolds was no relation to the one we met in the previous chapter (and who we will encounter again below and in Chapter 5) but may have been the one described in 1582 as selling wooden scales and geometrical instruments at Tower Hill and, in 1590, as making and selling geometrical tables, with frames, rules, compasses and squares, at his 'dwelling right against the southeast end of Barking churchyard in tower streete', opposite Bulwark Gate, which led to the Tower's western entrance (see position AH on the left of Fig. 3.1).[27] The Bulwark itself was a key site. It had been encroached on by dwellings, sheds and yards belonging to private artisans from at least the end of the sixteenth century and had prompted concerns in the early seventeenth, not least because the activity was causing the moat to silt up.[28] Here, however, among timber, coal and wheelers' yards, we find a number of instrument makers, teachers of mathematics and printers or sellers of mathematical texts.[29] For example, the 1642 edition of Henry Bond's *The Boate Swaines Art* notes that the 'opening scale' it introduced 'is made in brasse by *Thomas Flowre* at the Dyall in the Bulwarke'.[30] Bond advertised himself as

25 Willmoth, Sir Jonas Moore, 138–57; see also Walton, 'Technologies of Pow(d)er'.

26 M. R. Glozier, 'Norton, Robert (d. 1635), Army Officer and Writer', *ODNB*; on the role of master gunners, see Steven A. Walton, 'The Art of Gunnery in Renaissance England' (unpublished PhD, University of Toronto, 1999), 64–6, and on the Master Gunner of England, or later of the Tower and Ordnance, 264–6, 415.

27 Edward Worsop, *A Discoverie of Sundrie Errors and Faults …* (London: 1582), Aiii verso; Cyrian Lucar, *A Treatise named Lucarsolace* (London: 1590), 10. Boris Jardine suggests caution in assuming they were the same individual: 'Henry Sutton's Collaboration with John Reynolds (Gauger, Assayer and Clerk at the Royal Mint)', *Bulletin of the Scientific Instrument Society* 130 (2016): 4–7 (p. 5). However, given Norton's interests and the other connections discussed here, it does not seem implausible. Worsop's book also mentions Humfrey Cole as a maker of instruments in metal, based 'neere unto the North dore of Paules'. Cole, widely seen as the key figure in establishing an instrument trade in London, was closely associated with the Tower, where he had a house and £20 p.a. as a die-sinker at the Mint from at least 1563 to 1578: B. J. Cook, 'Humphrey Cole at the Mint', in *Humphrey Cole: Mint, Measurement and Maps in Elizabethan England*, ed. Silke Ackermann (London: British Museum, 1998), 21–6 (p. 21). See also G. L'E. Turner, 'Queen Elizabeth's Instrument Makers' (ibid., 3–9) and Jardine, 'Instruments of Statecraft'.

28 Parnell, 'The Buildings and Works at the Office of Ordnance', 25.

29 Lesley Cormack included Reynolds in her survey of mathematical instrument makers 1550–1630 but her cutoff date means that Flowers and others are excluded and that she did not identify the area around the Tower as among her centres of the trade. The following also challenges her comment that 'we should not see instrument makers and booksellers as an interrelated group', because she did not find makers based in St Paul's Churchyard at this date: Cormack, 'Mathematics for Sale', 77–80.

30 D. J. Bryden, 'Evidence from Advertising for Mathematical Instrument Making in London, 1556–1714', *Annals of Science* 49, no. 4 (1992): 301–36 (p. 313), note 56. Also close, on Tower Hill, was John Bate, who made gunner's scales and spyglasses and wrote *Mysteries of Nature and Art* (1634), including a section on fireworks, and wrote a dedicatory verse in the *Pyrotechnia* (1635) of John Babbington, who described himself as 'one of his Majesty's Inferior Gunners'.

'teacher of Navigation, Surveighing, and other parts of the Mathematicks', also 'in the *Bulwark* neer the *Tower*', and he was still there when his 1663 *The Art of Apparelling and Fitting of Any Ship* was 'Printed for the Widow *Seyle* in the Bulwark by the Tower'.

The Gunners Glasse (1646), printed 'for Robert Boydel ... to be sold at his shop in the Bulwark neere the Tower', illustrates an instrument for the proof of gunpowder and states that 'These Instruments with others, are made by *W. Flower*, dwelling in the *Bulwark* neere the Tower.' Written by William Eldred in old age, after sixty years as a gunner (ultimately Master Gunner) at Dover Castle, there is an interesting mismatch between the instrument illustrated and the description of the process of proving powder 'as is used in the Tower', that may be a result of the book's production site.[31] While Eldred describes a method involving the senses – considering the colour and taste of the powder, urging the gunner to 'marke how it burns' and to observe the residue – the 'engine' or instrument made by Flower involved a fired sample of powder pushing

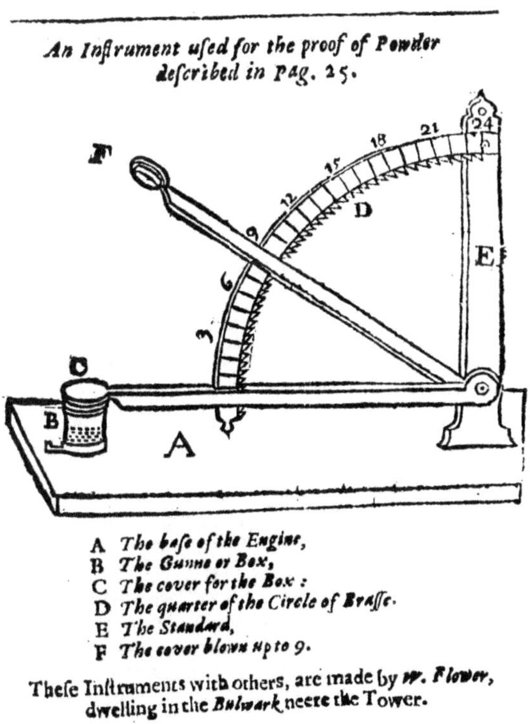

Figure 4.4 An instrument for the proof of powder, with advertisement for the business of W. Flower, in William Eldred's *The Gunners Glasse* (1646), BL: E331 (Wing 2nd ed., 1994, p. 91). © British Library Board /ProQuest Early English Books Online.

[31] William Eldred, *The Gunners Glasse* (London: 1646), 26, 91 (illustration).

an indicator up a scale (Fig. 4.4). Although other mechanical devices were described, qualitative methods similar to those of the assayers we met in the last chapter were more typical in manuals and in practice.[32] Perhaps, in encountering his bookseller at the Bulwark, Eldred learned of the neighbouring Flower's machine and added it to his text. Alternatively, and perhaps more likely, it was inserted by the publisher, as Boydell and Flower came to some mutually beneficial agreement in promoting each other's work to interested audiences within and surrounding the Tower.

Boydell had a number of other mathematical texts printed for him to sell at his shop 'at the Bulwark nere Tower Hill'. Directly linked to the Ordnance was the 1650 *Practical Geometry* by 'Captain Thomas Rudd, chief Engineer to his late Majesty', which declared itself '*A Worke very necessary for all Men, but* principally for *Surveyors* of *Land, Engineers*, and all other Students in the *Mathematicks*'. Its opening epistle to the reader called geometry '*the Princesse and Mother of all Sciences*', underpinning land survey, military architecture, engineering and gunnery. Rudd claimed that 'The utility and benefit of this most admirable *Art* is of such importance, that without the assistance thereof we can scarcely *live*, much lesse *preserve* our selves from danger, or hinder the *irruption* of a forreigne *Enemy*'.[33] Most of Boydell's mathematical texts were, from the 1650s, printed by Robert Leybourn who, in partnership with his kinsman William Leybourn, a surveyor and prolific mathematical author, specialized in mathematical texts until their Cripplegate workshop and stock were destroyed in the Great Fire.[34]

One Leybourn-printed text was John Wybard's *Tactometria* (1650), which mentioned a third Flower 'dwelling in the Bulwark, neer the Tower of *London*'. This advertisement told readers, who 'shall think good to make use of the Lines for Gauging, or for any other dimensions described in this Book: or of any other Instruments for mathematicall practice in general', that they could 'have the same very accurately made by Mr. *Christopher Flower* … besides divers others about this City'. Wybard's text, addressing 'all such as are either ingeniously studious of, or necessarily exercised and employed in the practice of the Art METRICALL', had significant connections with activity within the Tower.[35] This was a result of his collaborations with the other John Reynolds, of the Mint and Goldsmiths' Company.[36]

[32] On use of senses to ascertain the quality of products, see Patrick Wallis and Catherine Wright, 'Evidence, Artisan Experience, and Authority in Early Modern England', in Smith, Meyers, and Cook, *Ways of Making and Knowing*, 138–63. A typical set of sensory tests was described by Robert Norton, *The Gunner* (London: 1628), 145. Mechanical devices were described in Nathaneal Nye, *The Art of Gunnery* (London: 1647), 29–31 and William Bourne, *The Arte of Shooting in Great Ordnance* (London: 1587), 28–9. Quantitative methods were in use by the Ordnance by the 1740s, though not standardized even into the nineteenth century: Seymour H. Mauskopf, 'Chemistry in the Arsenal: State Regulation and Scientific Methodology of Gunpowder in Eighteenth-Century England and France', in Steele and Dorland, *The Heirs of Archimedes*, 293–330 (pp. 304–5).

[33] Thomas Rudd, *Practical Geometry* (London: 1650).

[34] Sarah Bendall, 'Leybourn, William (1626–1716), Mathematician and Land Surveyor', *ODNB*.

[35] John Wybard, *Tactometria. Seu, Tetagmenometria. Or, The Geometry of Regulars Practically Proposed* (London: 1650).

[36] See C. E. Challis, 'Presidential Address [John Reynolds and the Sterling Standard]', *British Numismatics Journal* 62 (1992): 235–46; Jardine, 'Henry Sutton's Collaboration'; Norman Biggs, 'More Seventeenth-Century Networks', *BSHM Bulletin: Journal of the British Society for the History of Mathematics* 32 (2017): 30–9 and *idem*, 'John Reynolds of the Mint: A Mathematician in the Service of King and Commonwealth', *Historia Mathematica* 48 (2019): 1–28.

Wybard described Reynolds as 'a man much noted by artists for his industry and ingenuity in the Mathematiques' and 'industrious' in 'mathematicall experiments'.[37] He recounts how Reynolds, 'very courteously entertaining me at his house in the Tower', engaged him in 'discourse about many experiments made by him in the mathematicks' and demonstrated some on how to determine the weight of bullets from their size – a properly Archimedean problem. This was done with an old, rusty bullet but 'he soon after, got another bullet, which was very sound and solid', which, given his proximity to the Ordnance, especially the house and workshops of the institutionally shared smith, surely would not have been difficult. They measured it with 'Callaper Compasses' and weighed it with both avoirdupois and troy weights and, by 'our experiment', worked out a rule for establishing the weight and centre of gravity of any cast-iron bullet or sphere.[38] It was no coincidence to his joining the Mint, and the range of his interests, that Reynolds's father, William, had been based in the Ordnance as Master Fletcher, and that the family was close to the then Master Smith, William Hopkins.[39]

Reynolds had a long-term interest in weights and measures, linked to the nationally significant role of Goldsmiths and Mint officers in testing and calculating precious metal content of coin, alloys and ores, and in standards of weights and measures.[40] He addressed the relationship of weight of metal and value of coin in his published texts and tables.[41] However, he pursued metrology well beyond professional necessity, in ways that reflect his long intimacy with the Tower. As well as accessing bullets, he investigated 'the weight of a cube-inch of Iron, which had been kept in the Treasury of the Tower of *London*, ever since King *Henry* the 7th his time or longer'.[42] Reynolds referred to the same weight in a later communication to the Royal Society, along with further measures comparing the weight to volume ratio of different substances, including metals, wood, liquids and gunpowder.[43] He also investigated the conversion between the (commercial) avoirdupois and (more specialist) troy systems and was experienced in gauging, having been involved in work relating to Excise.[44] Wybard recounts their investigations of standards at the Guildhall, noting that Reynolds considered 'those Measures in the Tower of *London*' to be 'the most ancient and true'.[45] These were presumably the 'true' standards incorporated into the 1655 wooden gauging rule made by Henry Sutton, of which Reynolds has recently been identified as the designer.[46] Likewise, Tower standards were the comparator for two surviving measures in museum collections: a wine-pottle inscribed in 1641 as 'TRYED BY JOHN

[37] Wybard, *Tactometria*, 216, 265. On Goodwyn see Biggs, 'John Reynolds of the Mint', 4.

[38] Wybard, *Tactometria*, 216–17.

[39] Biggs, 'John Reynolds of the Mint', 2. The son of William Hopkins, Thomas, became Reynolds's guardian on his father's death in 1597.

[40] Biggs, 'John Reynolds of the Mint', 8–11; A. D. C. Simpson and R. D. Connor, 'The Mass of the English Troy Pound in the Eighteenth Century', *Annals of Science* 61, no. 3 (2004): 321–49.

[41] Biggs, 'More Seventeenth-Century Networks', 31–2 and 'John Reynolds of the Mint', 17–24.

[42] Wybard, *Tactometria*, 226.

[43] Biggs, 'More Seventeenth-Century Networks', 34–5.

[44] Wybard, *Tactometria*, 218–19, 265–88.

[45] Wybard, *Tactometria*, 267.

[46] Jardine, 'Henry Sutton's Collaboration': the rule is now at the Whipple Museum.

RENALDS IN THE TOWER' and a yard 'EXAMINED BY ME IOHN REYNOLS AND SEALED AT THE TOWER OF LONDON APRILL 30 ANNO 1659'.[47]

Also at the Bulwark was Joseph Hone, who, like Hodgkins, was included in the 1664 list of Tower gunners, recorded as 'Mathematicall Instrum[en]t maker in the Bulwark'. Hone is referred to in the 1663 and 1664 editions of Bond's *The Boate Swains Art*, in place of Thomas Flower, as maker of the scale discussed in the book, perhaps indicating a succession in ownership of the Flowers' shop.[48] However, shortly after, the Ordnance began to remove the buildings that were cluttering and compromising the banks of the Tower's moat, including the decrepit Bulwark gate and wall.[49] Hone had to move, but only went as far as Tower Wharf, just metres south and east of the Bulwark. The Wharf was the first part of the Tower complex allocated to the Ordnance back in the fifteenth century and was still the focal point for stores to be received and dispatched. It was a site marked by the presence of large cannon and even larger cranes, visible in maps and frequently attended to and maintained by Ordnance labourers and artificers. Shortly before Hone moved, the wharf also became a place for 'the first landing and proveing of powder'.[50] In the 1670s we find Flamsteed referring to 'Mr Hones on the Wharfe' as a place to which letters could be directed, either to him, or from him to his patron, the Surveyor General Jonas Moore.[51]

Joseph Hone was certainly well placed, first in the Bulwark and then on the Wharf. His location – and its long connections to both mathematical practice and the interests of the Tower's gunners and engineers – is the answer to John Millburn's puzzlement that it was 'the relatively unknown' Hone who supplied instruments to the Ordnance. This connection is evidenced by the survival of fifteen bills in the Ordnance Bill Books between 1685 and 1702, including for gunners quadrants, gunners callipers, shot gauges, theodolites, plane tables, drawing instruments and rules.[52] Millburn was unable to judge Hone's work, as he knew of no surviving examples, but a theodolite, made for Sir John Colleton, the Royalist solider sent to settle lands in South Carolina, has since been identified, which indicates that Hone 'was a maker of exceptional quality'.[53] While this was probably a private commission, and for the Proprietors of Carolina rather than the army, it underscores Hone's early links to the military arts of surveying and

[47] On the pottle-measure (Museum of Archaeology and Anthropology, Cambridge, Z 2531), see Biggs, 'More Seventeenth-Century Networks', 34, and R. D. Connor, *The Weights and Measures of England* (London: H.M. Stationary Office, 1987), 155. The standard yard (Science Museum Inv: 1932–313) has stamps from the Tower and the Exchequer as well as one from the reign of William IV, when it was re-examined and inscribed 'Imperial Yard'.

[48] Bryden, 'Evidence from Advertising', 318, 313 (n. 56).

[49] Parnell, 'The Buildings and Works at the Office of Ordnance', 42, 44–5, 50.

[50] Parnell, 'The Buildings and Works at the Office of Ordnance', 13–14, 16, 24, 30. In the 1677 edition of Bond's book, Hone is located at Tower Wharf, although a 1704 edition, issued as *The Art of Apparelling*, places him at the Bulwark, presumably a legacy of earlier versions of the text.

[51] Eric G. Forbes, Lesley Murdin and Frances Willmoth (eds), *Correspondence of John Flamsteed, Volume 1: 1666–1682* (Bristol and Philadelphia: Institute of Physics Publishing, 1995), 331.

[52] John R. Millburn, 'Some English Military Instrument Makers of the Late 17th Century', *Bulletin of the Scientific Instrument Society* 68 (2001): 2–5 (pp. 2, 5).

[53] Jeffrey Lock, 'Restoration of a 17th Century Alta-Azimuth Theodolite', *Colonial Instruments*, 2018 (http://www.colonialinstruments.com/restoration-17th-century-alta-azimuth-theodolite/ [accessed 15 April 2020]).

fortification. His Ordnance bills show that he supplied military expeditions and also that he made some more unusual items associated with the trial of guns and mortars.[54] It was presumably through his connection to the Ordnance that he also supplied instruments to the Mathematical School at Christ's Hospital, with which Moore was closely associated.[55]

From Hone we can trace out an apprenticeship succession of instrument makers within the Broderers' Company who served the Ordnance and Mathematical School.[56] His apprentice John Rowley succeeded as the regular supplier of mathematical instruments to the Ordnance from 1704. A talented and much-admired instrument maker, working as the trade boomed in London, this was just one element among his many commissions. As well as supplying surveying and drawing instruments to the Ordnance, Rowley acted as 'hydraulic engineer', installing and maintaining engines to supply water and combat fire, and supplied scales, weights and measures. He received a commission to compare the Exchequer and Guildhall standards, after which he made standards for a range of Ordnance sites, from coastal forts to the Tower's eighteenth-century Drawing Room.[57] The last of these was to supersede those of both the Mint and the Exchequer in the investigations into standards carried out by the Royal Society in the 1730s.[58]

Ever more makers were available for the Ordnance to choose from, but from about 1723 they turned to Rowley's former apprentice William Deane, who became their mathematical instrument maker until just before his death in 1747. Over 100 bills for drawing, surveying and gunnery instruments surviving in the archive are testament to the increasingly typical, indeed essential, use of instruments by gunners for aiming as well as checking sizes of shot, shells and bores.[59] While Deane, like Rowley, is known for elaborate sundials, the Royal Armouries have a cased set of shot and calibre gauges made by him for Albert Borgard, Chief Firemaster from 1712 and later first commander of the newly formed Regiment of Royal Artillery.[60] Borgard was also said to have been responsible for designing a 1710 instrument by Rowley for levelling cannon, with sights and spirit level, now in the collection of the Royal Artillery Museum.[61] These

[54] Millburn, 'Some English Military Instrument Makers', 4–5.

[55] Clifton, *Directory*, 140.

[56] Crawforth, 'Instrument Makers in the London Guilds': Crawforth reproduces Rowley's apprenticeship indenture but misreads Hone as Howe (p. 321), meaning that the identification with a Spectacle Maker in Threadneedle Street called Joseph Howe is an error. It remains unclear when and why Hone became free of the Broderers' Company.

[57] John H. Appleby, 'A New Perspective on John Rowley, Virtuoso Master of Mechanics and Hydraulic Engineer', *Annals of Science* 53, no. 1 (1996): 1–27 (pp. 5, 13–14). On the Drawing Room see below and Geoffrey Parnell, 'The Ordnance Drawing Room 1716–52', *English Heritage Historical Review* 9, no. 1 (2014): 122–79 and Douglas W. Marshall, 'Military Maps of the Eighteenth-Century and the Tower of London Drawing Room', *Imago Mundi* 32 (1980): 21–44.

[58] Simpson and Connor, 'The Mass', 338.

[59] Millburn, 'The Office of Ordnance'; *idem*, 'William Deane and His Ordnance Bills', *Bulletin of the Scientific Instrument Society* 45 (1995): 12–18. In his textbook for the Royal Military Academy, John Muller emphasized the use of instruments but also noted that, because bores were not true, skill and judgement were required, and an instrument 'prevents the gunner from learning to judge by the eye': *A Treatise of Artillery* (London: Printed for John Millan, 1757), 229–30 (quotation p. 232).

[60] William Deane, Gauges, 1730 (Royal Armouries Collection: XIII.138).

[61] Millburn, 'The Office of Ordnance', 255; Royal Artillery Museum collection, INST 24/073.

makers' work for the Ordnance was a relatively small part of their income or output, although one that undoubtedly brought status and useful connections. Over a century, from the Flowers to Deane, this work was based on trade succession, stemming from close proximity to the Tower and interactions with officials and artisans based there.

Experiment at the Tower and beyond

The Ordnance's buildings were part of the increasingly popular tourist route at the Tower. While people paid to see old arms, armour and ordnance from at least the beginning of the seventeenth century, after the Restoration displays were more consciously aimed at visitors, with the Line of Kings – a display of historic royal armour on mounted mannequins – and decorative trophies of arms arranged in patterns and to mimic crests, animals and even a pipe organ, mounted in the Grand Storehouse (built 1688–91).[62] Nevertheless, the working site remained cheek-by-jowl with the attractions. The satirical *London Spy*, by Ned Ward, was not particularly impressed by the displays, from big cats to arms, or by the tales of the Warders keen to collect their fees, but at the door of the 'Horse Armoury', housing the Line of Kings, he noted the presence of

> two or three Smugfac'd *Vulcans*, who were as Amiable in Complexion, as if, to make themselves *Infernal Beaus*, they had Powder'd their Frizled Locks with Lamp-Black, and Beautified their Physiognomies with Kennel-Water, the Lines of their Faces being so pounc'd with Dirt'.[63]

The Tower remained a site of manufacture and, indeed, experiment. The latter term appears regularly in the surviving Ordnance minutes from the later seventeenth century, as it did in published writings on gunnery and pyrotechnics. The meaning at this date was usually closer to trial or, on occasion, a demonstration before witnesses.[64] However, as an institutional tradition, these test castings and firings link directly with later activity that would more readily be classed as empirical investigation.

A range of Ordnance personnel, contractors and sites could be involved in or prompt trials. In the 1680s, for example, there was a series of novel castings of iron and brass mortars that were centrally directed and closely overseen by Ordnance officers, although undertaken at contractors' foundries in the Weald of Sussex. Thus, in March 1688, Richard Leake, the Master Gunner, and Jacob Richards, the Third Engineer, were repaid their expenses for trips to the foundry owned by William Benge. Leake had been 'giving directions in the Casting sev[er]all Ordnance & Mortarpeeces' and Richards was 'there giving direction in Tryng An Experiment of preparing, Casting

62 Malcolm Mercer, 'Shaping the Ordnance Office Collections at the Tower of London: The Impact of Colonial Expansion, Diplomacy, and Donation in the Early Nineteenth Century', *Museum History Journal* 9, no. 2 (2016): 153–67 (p. 154); Parnell, 'The Buildings and Works at the Office of Ordnance', 65–7, 73–82.

63 Ward, *The London-Spy Compleat*, 302, 313.

64 Werrett, *Fireworks*, 56.

and Transmutateing the Iron ore, the Effects of w[hi]ch. Mr. Benge hath reported to the Hon[ora]ble. Board'.[65] However, in January 1696/7, the trials took place at the Tower, when Benge was directed to send fire-shells 'to be Experimented' there.[66] On another occasion, it was the smith, who two decades earlier had worked on Flamsteed's sextant that took the initiative. On December 1695, it was ordered that the contractor, Mr Wightman, 'have a Warrant to cast with an Iron Chamber a Morterpeice 13 or 10 inches, according to the Direccon of Mr Edward Silvester, who proposes to fire it 100 times in a very short space, and believes is more than Brass will endure'.[67] Wightman was also called on to 'immediately cast and Deliver into his Majesty's Stores such a Brass Gunn as the Marquis of Carmarthen shall direct for an Experiment in the new ship lately built at Chatham according to his projection'.[68]

Not surprisingly, places with greater open space and less risk of fire than the Tower were often used for trials and demonstrations. One significant seventeenth-century site was Faux-Hall, or Vauxhall, on the south bank of the Thames, upriver from Westminster. Here, from 1629, the courtier and mechanical experimentalist Edward Somerset, second Marquis of Worcester, had established a workshop and a forge on Crown land.[69] The site's relationship with the Ordnance is difficult to unpick, as the 'operatory' there appears to be an example of mixed private, royal and government patronage. However, Worcester employed the Dutch engineer, Caspar Kalthoff, who had previously been given space to work at the Tower, in investigations that were of clear military interest.[70] In 1655, after imprisonment in the Tower, and with Parliament threatening to take over what had been a considerable facility, Worcester described how his experiments and inventions might again 'with my directions' be undertaken 'by the unparallel'd Workman both for trust and skill, *Caspar Kalthoff's* hand'. He said Kalthoff 'hath been these five and thirty years as in a school under me imployed, and still at my disposal, in a place by my great expences made fit for publick service'.[71] However, a 1645 inventory of the premises listed the contents as 'his Ma[jes]t[ie]s Goods Engines and Materialls'. Likewise, another workman, William Lambert, petitioning Charles II after the site had been returned to the Crown, described himself as 'founder for guns and water work, or brass work, to the late King, at Foxhall'.[72] Kalthoff returned to Vauxhall

65 Quoted in Ruth Rhynas Brown, 'Notes on the Wealdon Furnaces in the Records of the Board of Ordnance, 1660–1700', *Wealdon Iron: Bulletin of the Wealdon Iron Research Group* 2, no. 13 (1993): 20–30 (p. 23) [28 March 1688, WO 51/36, 67v–68].

66 18 January 1695/6 and 11 February 1685/6, Board of Ordnance Journal of Proceedings, WO 47/18.

67 31 December 1695, WO 47/18. This was probably John Wightman, mould-maker and employee of Matthew Bagley, the Ordnance's supplier of brass guns until he was killed by an accident at his foundry in Moorfields in 1716: O.F.G. Hogg, *The Royal Arsenal: Its Background, Origin and Subsequent History* (London: Oxford University Press, 1963), 2 vols, 249, 246–7.

68 25 January 1685/6, WO 47/18.

69 Arthur MacGregor, '"A Magazin of all Manner of Inventions": Museums in the Quest for "Salomon's House" in Seventeenth-Century England', *Journal of the History of Collections* 1, no. 2 (1989): 207–12 (p. 208).

70 Anthony F. C. Wallace, *The Social Context of Innovation: Bureaucrats, Families, and Heroes in the Early Industrial Revolution, As Foreseen in Bacon's New Atlantis* (Lincoln and London: University of Nebraska Press, 2003), 38.

71 Marquis of Worcester (Edward Somerset), *A Century of the Names and Scantlings of Such Inventions as at Present I Can Call to Mind to Have Tried and Perfected …* (London: 1663) [n.p.].

72 Quoted in Macgregor, 'A Magazin', 208.

after the Restoration, and, until his death in 1667 or 1668, he was 'employed in the making of Guns and divers Engines and works for the King's service', with Lambert and a joiner, William Joulden, who described himself as having charge of the engines and models, 'under Gaspar so long as he continued at Fauxhall'.[73] The forges, workshops and mills, plus the 'Models, Utensils, or other necessaries for practical Inventions' they contained, were understood by all as a facility for, or with the potential to be turned to, state use.[74] Although the 'Vauxhall Operatory' never became a 'college of artisans', as Samuel Hartlib and Robert Boyle hoped in the 1650s, it is evident that it was long an industrious site and, at periods, one of innovation, trial, demonstration and experiment.

Another important site was the Artillery Garden, a facility sometimes unhappily shared between the Tower's gunners and the group that became the Honourable Artillery Company. This was where the Master Gunner lived and where gunners were regularly trained and exercised in the use of artillery and great ordnance. It was also used for proving guns made by the trade.[75] In 1669, the ever-curious Samuel Pepys recorded going there, at the invitation of Anthony Deane, who had recently been appointed master shipwright at Portsmouth. This was to see what Pepys variously called a trial, proof and 'great and first experiment' of a cannon designed by Deane, known as Punchinello 'from the shortness and bigness' of its shape.[76] They arrived too late, for 'the trial had been made' but there was 'extraordinary report of the proof of his gun'. They prevailed on William Legge, the Lieutenant General of the Ordnance before and after the Commonwealth, 'to stay and give us a sight of her performance'. This he did,

> and there, in short, against a gun more then as long and as heavy again, and charged with as much powder again, she carried the same bullet as strong to the mark, and nearer and above the mark at a point-blank than theirs, and is more easily managed, and more easily managed and recoyls no more then that – which is a thing so extraordinary as to be admired for the happiness of his invention [...].

Pepys reported that Legge did justice to the performance of the gun but that it was to 'the great regret of the old gunners and officers of the Ordinance that were there'.[77] Perhaps they objected to an innovation by an outsider to their trades, or perhaps they were aware of practical difficulties not apparent to Pepys, Deane and Legge. It does not appear that the Ordnance took up the Punchinello, although a gun of 'extraordinary

[73] Excerpt from the Books of the Duchy of Cornwall Office in Henry Dircks, *The Life, Times and Scientific Labours of the Second Marquis of Worcester* (London: Bernard Quaritch, 1865), 575 (Project Gutenberg Ebook: https://www.gutenberg.org/files/51903/51903-h/51903-h.htm); quoted in MacGregor, 'A Magazin', 208.

[74] MacGregor, 'A Magazin', 207–9 (quotation p. 209).

[75] Steven A. Walton, 'The Tower Gunners and the Artillery Company in the Artillery Garden before 1630', *Journal of the Ordnance Society* 18 (2006): 53–66; Hogg, *The Royal Arsenal*, vol. 1, 69–70.

[76] *The Diary of Samuel Pepys*, vol. 9, 1668–9 (1976), 528; Brian Lavery, 'Deane, Sir Anthony (*c.* 1638–1720?), shipbuilder', *ODNB*, 528.

[77] *The Diary of Samuel Pepys*, vol. 9, 1668–9 (1976), 528.

proportions', dated 1672 and marked with the Ordnance's broad arrow, was kept at the Tower and has tentatively been identified as of Deane's design.[78]

The firing of experimental guns, or long-range shooting practice, could only be carried out at the Artillery Garden while there were open fields on at least one side. By 1681, as the city expanded, the Ordnance was seeking to sell this land to property developers, if 'Some other convenient place could be provided for the proof of small guns and the exercise of the fee'd gunners at the Tower'.[79] While some proofs of powder and small guns continued at the Tower, other stores and activity were transferred downriver to Greenwich, where by the mid-seventeenth century the chief powder magazine was located, to the open ground of neighbouring Blackheath and, further east, to the expanding Ordnance site at Woolwich.[80] John Evelyn reported on 16 March 1687 that he 'saw a trial of these devilish, murdering, mischief-doing engines called Bombs, shot out of the morter-piece on Black-heath'. He was impressed: 'The distance that they are cast, the destruction they make where they fall, is prodigious.'[81] The diarists Evelyn and Pepys have left us with the richest accounts, but the presence of fellows of the Royal Society was by no means required. As well as the experimental castings discussed above, orders to the instrument maker Joseph Hone reveal institutional interest in investigating or standardizing performance. He was asked to provide, for example, a brass instrument to 'try the bores of mortar pieces', a 'large quadrant of brass with sites &c', which was 'for the use of mortar-pieces', and 13 'rammer heads of brass, for exact proving the natures of guns at the proofs'.[82] Elsewhere, John Baxter, the Laboratory's Storekeeper at Woolwich, was instructed to arrange 'the experim[en] t of his 2 small mortars to be made before the Principall Officers' and 'To make an Experim[en]t. of his Square Mortar at the next Proof', which was to be tried alongside a 'warlike Engine to throw a Granado' designed by a Mr Barre.[83]

While the gunners and fireworkers did not need experimental philosophers to carrying out such trials, the philosophers required the artificers if they sought practical knowledge of ballistics. Prompted by academic discussion about the experimental exploration of Galileo's parabolic theory by the London weaver and mathematician Robert Anderson, a group of Fellows of the Royal Society carried out their own investigations at Blackheath. These included the president, Viscount Brouncker, secretary, Henry Oldenburg, and Curator of Experiments, Robert Hooke, together with Jonas Moore and Thomas Chichele – and unknown gunners – of the Ordnance.[84] The London mathematician John Collins informed James Gregory, in Edinburgh, that

[78] Royal Armouries XIX.34, https://collections.royalarmouries.org/object/rac-object-24415.html.

[79] Quoted in Hogg, *The Royal Arsenal*, vol. 1, 72.

[80] Tomlinson, *Guns and Government*, 119–22; Hogg, *The Royal Arsenal*, vol. 1, 105–11.

[81] John Evelyn, *Memoirs, Illustrative of the Life and Writings of John Evelyn, Esq. F.R.S* …., ed. William Bray, 2nd ed. (2 vols, London: Henry Colburn, 1819), vol. 1, 638.

[82] Quoted in Millburn, 'Some English Military Instrument Makers', 4. Millburn suggests the brass rammers 'were probably precisely-dimensioned plug gauges for checking bore diameters'.

[83] Index under Capt. Baxter, 11 January and 7 February 1705/6; Index under Tower, 2 February 1705/6, Minutes – Series 2 WO 47/23.

[84] A. Rupert Hall, 'Gunnery, Science and the Royal Society', in *The Uses of Science in the Age of Newton*, ed. John G. Burke (Berkeley and Los Angeles: University of California Press, 1983), 111–41 (p. 123); France, 'Gunnery and the Struggle', 265–7.

Anderson's books had 'caused much bustle' and prompted this group 'to go many times a shooting with mortar pieces at Blackheath'.[85] Hooke in his diary recorded briefly the trips to Blackheath where they 'Tryd Experiment of bullet' to establish its path and, later, heard from Moore about 'his observation of Granadoes tried the day before at Blackheath'.[86] He also designed and built, at the prompting of the Society, 'an engine' or 'instrument for determining the force of gunpowder by weight'.[87] Here the strength was measured by placing a weight at different points on a beam, rather than using the indicator arm and scale suggested by Flower. A trial of it on 9 December 1663 was unsuccessful, and, although he was told to find ways to remedy the problems, it seems to have been dropped, although experiments on powders and guns certainly continued.[88]

As Simon Werrett has shown, London natural philosophers interested in fireworks and ballistics developed particularly 'close connections with the practitioners of artifice and spectacle'.[89] He suggests that for the rocket experiments of William Whiston and Humphrey Ditton, who proposed their use as a position-finding system on land or sea, the skills of Ordnance fireworkers and gunners were essential, not just to conduct the experiments but 'as guarantors of scientific accuracy'. It was, as Whiston and Ditton said, 'the business of their art to proportion the Match or Fusee to any particular time when [the gunpowder] shall give fire'.[90] The trial rockets and fireballs were fired from mortars on Shooter's Hill – a name in use from at least the thirteenth century for what is one of the highest points of London's surroundings, conveniently close to Greenwich, Blackheath and Woolwich. By this date, what had become known as the fireworks laboratory at Greenwich had moved to Woolwich as the Royal Laboratory. Here, two years later, the Royal Brass Foundry began to be built, as the Ordnance took on the manufacture of large guns itself. The importance of the site – for storage, manufacture, residence and training – and the increasing scale of operations, is traceable in the number and grandeur of the buildings of this period.[91] Royal and national ambition was also shown by the importing of overseas talent in the form of artillerymen, engineers, fireworkers and founders, although the selection and training of a more local workforce was also, as it had always been, a requirement and concern of the officers of the Board.[92]

85 Hall, 'Gunnery, Science and the Royal Society', 123.

86 Quoted in France, 'Gunnery and the Struggle', 266.

87 Thomas Birch, *The History of the Royal Society of London for Improving of Natural Knowledge* (London: A. Milar in the Strand, 1756), 295, 302–3, 335. Hooke's original paper, with figure, dated 9 September 1663, among the Classified Papers at the Royal Society, CLP/20/22.

88 Birch, *The History of the Royal Society*, 342.

89 Werrett, *Fireworks*, 75.

90 Werrett, *Fireworks*, 97.

91 Peter Guillery (ed.), *Survey of London Vol. 48: Woolwich* (London, 2012), https://www.ucl.ac.uk/bartlett/architecture/research/survey-london/woolwich, 134–45.

92 For example, Martin Beckman, who was employed by the Ordnance from 1661 and became Chief Engineer and Comptroller of Fireworks, was Swedish; the Dutch Bernard De Gomme was Chief Engineer from 1661 and later Surveyor General; the Danish Albert Borgard was in English service from 1693, becoming Chief Firemaster in 1712; the Swiss Andrew Schalch was Master Founder from 1718.

Training: Laboratory and classroom

From 1699 John Baxter was both Storekeeper to the Laboratory and Mate to the Chief Firemaster.[93] As well as managing stores, he played a central role in overseeing firewokers and other artificers, and in communicating with the officers of the Board in relation to their employment. His proximity to the men at work meant that he was asked to judge their ability, as for example in January 1705/6 when he was requested 'To send up a List of w[ha]t Bombardiers he thinks qualify'd to go to Catalonia' for an attempt to take Barcelona during the War of Spanish Succession.[94] Most of the training for fireworkers was done in the Laboratory and much of the emphasis was on good order. In the Rules set out in 1683, they were urged 'To be very Carefull in the Laboratory', which should 'be kept Cleane'. This appears to be exemplified in a later series of watercolours, attributed to Gamaliel Massiot, a drawing master at the RMA, showing laboratory workers in the 1750s (Fig. 4.5/Plate 5). These are exceptionally well-ordered spaces, with men at work wearing numbered 'frocks' over their clothes.[95]

Figure 4.5 Military and civilian workers at the Royal Laboratory moulding and driving fuses, attributed to Gamaliel Massiot, *c.* 1750 (NMM: PAI0745). © National Maritime Museum, Greenwich, London.

See also colour version, reproduced in plate section 5.

[93] Hogg, *The Royal Arsenal*, vol. 1, 225, 230.

[94] Index under Baxter, 5 January 1705/6, WO 47/23.

[95] 'Rules Orders and Instructions' WO 55/1790. There are eleven watercolours in the collection of the National Maritime Museum (PAG9664, PAH4071-72, PAI0744-46, PAI7701-03, PAJ2303 and PAJ2312). The numbers on the back of the 'frocks' are visible in PAI0744 and PAI770. The drawings are discussed in Celina Fox, *The Arts of Industry in the Age of Enlightenment* (New Haven: Yale University Press for The Paul Mellon Centre for Studies in British Art, 2009), 364–7.

Just as the Chief Gunner and Chief Engineer had pedagogical roles, it was the Firemaster's duty – which clearly often fell to his Mate – to 'Learne and informe such inferior Fireworkers or Scollars as shall be required by the Master Generall or Principall Officers of the Ordnance to Teach and instruct in any part or parts of the whole Art off Fireworkers for War or Pleasure'. They were to

> Exercise ones or twice a Weeke those that he hath under his Comand with a great and small Mortar Peece to Cast Stones Bombs and Granadoes filled with Sand at Severall Distances and to keep them Continually att Worke in the Laboratory to make unfilled Paper Rockets to prepare Saltpetre Brimstone and Powder for Service to make all manner of Covering of Filled Fire Balls with Lynes or Cords to shew them how to Fill Fitt and Apply a Petard for Severall Occasions and of Severall sorts.

It was indicated that they should be 'Able to give satisfaction of theire Proficiencyes' before the Master General or other officers 'appointed to Assist att his or theire Examination'.[96]

There is little evidence regarding Ordnance artificers' examinations. In 1762, Charles Frederick, then Comptroller of the Laboratory, referred to the 'Trial' that an individual who wished to be appointed as an armourer at the Tower would expect to undergo. Although this man had 'been bred to the trade of a Gunsmith', the trial was essential 'as none are Admitted into that Branch, but such as are very able Workmen'.[97] A late-eighteenth- to early-nineteenth-century manuscript (Fig. 4.6/Plate 6) that lists and illustrates the methods, ingredients and utensils for making shot, rockets, cannonade, powder, cartridges and more appears to be the work of a trainee fireworker, or possibly a cadet of the Royal Military Academy (RMA). Either way, he probably learned the craft of the Laboratory much as his predecessors had done, even if the range of ordnance had changed over time.[98] The importance of bodily discipline was reiterated, with an entry on 'Orders by the Chief Fire Master of the Royal Laboratory 1 March 1799', which stated that 'The People who may be employed in Driving of Composition of all Descriptions [i.e. different recipes for explosives] are desired to keep their Bodyes in an Erect Position so as to be Clear of the Mould' so as to 'preserve them from the Fatal Consequences attendant upon Accidents'.[99] This instruction is clearly visualized, presumably for the benefit of such learners, in one of the Massiot watercolours (Fig. 4.5).

[96] WO 55/1790, p. 634.

[97] Copy of a letter from Charles Frederick to Hans Stanley, 9 February 1762, 'Letters by and to Sir Charles Frederick', 62–3, MssCol 1075 (New York Public Library).

[98] Process Manual, SUPP 5/24 (The National Archives). The manuscript includes a 'method for examining the different nature of Lieut Col. Shrapnel Shells', which would date this part to after Henry Shrapnel's 1804 promotion and arrival at Woolwich as Assistant Inspector of Artillery. An inscription and additions to this volume suggest that it was still being used well into the nineteenth century.

[99] SUPP 5/44, [p. 21].

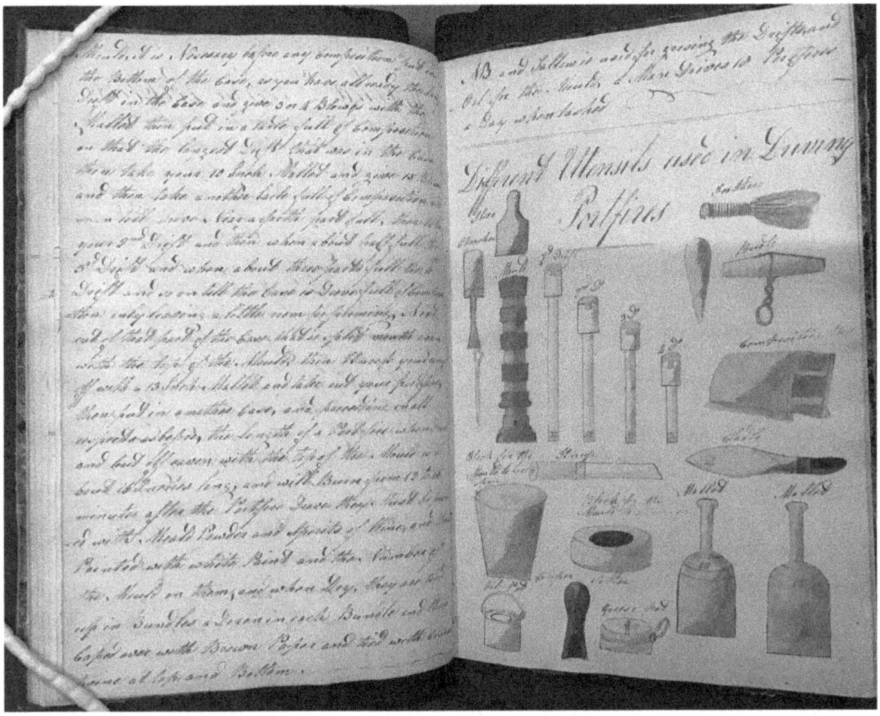

Figure 4.6 Page showing drawing of the utensils for driving portfires, 'Process Manual' *c.* 1800. The National Archives, SUPP 5/24.

See also colour version, reproduced in plate section 6.

Even in the late seventeenth and early eighteenth centuries, some Ordnance workers were given additional, mathematical training. Baxter, who was asked to recommend bombardiers, may be the same 'John Baxter, Bombardier' who was listed by John Flamsteed as one of his pupils at the Observatory in Greenwich in May 1689.[100] The link, and an ongoing connection, is also suggested by the letter of another former pupil, Captain Henry Stanyan, on 24 July 1706, in which he asked Flamsteed to send his service to, among others, Captain Baxter.[101] Flamsteed's pupils included a number of Ordnance clerks or storekeepers. Although taking on these pupils – which Flamsteed was wont to complain about – has been understood as a service to his patron, Moore, it is evident that his connection to the Ordnance was more widely made use of. Long after Moore's death, the Ordnance minutes record that a Mr Pratt was to be recommended to Flamsteed 'for an Engineer' in June 1706. Flamsteed duly listed a Thomas Prat as at the observatory, 'in order to learn mathematics'.[102]

[100] John Flamsteed, 'A list of my pupils Names & Employments … ', RGO 1/15, f. 165v–166v (Cambridge University Library).

[101] Flamsteed, *Correspondence*, vol. 3, 342.

[102] This, as was sometimes the case in the list, was recorded in Latin. See Flamsteed, *Correspondence*, vol. 3, 331.

The formal provision of mathematical education of some Ordnance employees and of the military companies of gunners (Royal Regiment of Artillery) and engineers (Corps of Royal Engineers), formed at Woolwich in 1716, became an increasing concern in the eighteenth century. Key spaces for this training were the Drawing Room, founded at the Tower in 1715, and the Academy, initially but abortively established at Tower Place in Woolwich in 1720 and then successfully in 1741 as the RMA. The Drawing Room resulted from a 1714 'Memorial relating to the present State of the Military branch of the Ordnance in order to the new remodelling of the same' by the reforming Surveyor General Michael Richards. He suggested better care be taken of 'plans, projects, Profills, Estimates, &ca of all Fortifications', making one individual responsible for overseeing their production and storage, who would be assisted by a draughtsman and 'all the Engineers not on Duty'. The draughtsman in question was Robert Whitehand, who had been paid £100 per annum since late in 1711 'Constantly to attend the Office'.[103]

In 1715–16 the eastern annex to the White Tower was altered and fitted out for the storage of drawings and plans, the copying of charts and the production of designs and estimates of buildings and fortifications. As with other Ordnance spaces, teaching was a key part of the activity, as is indicated by Whitehand's 1724 will. This directed the executors to

> deliver all and every my Draughts Designs Sketches Mathematical Instruments and Rulers the several Draughts in fframes and Schemes of Opticks in my Chamber excepted and excluded to the Surveyor General [John Armstrong, also Chief Engineer] humbly requesting him that he will be pleased to dispose thereof to and for the Use benefit and Improvement of the young persons instructed in the Drawing Room in the Tower for the Use of the Office of Ordinance as the said Surveyor Generall in his discretion shall think most meet and convenient [...].[104]

Over time, and under the Chief Draughtsmen Clement Lemprière and John Peter Desmaretz, the work and the staff in the office increased. While some made a career there, others used it as a stepping stone to military advancement. It has been suggested that 'the Drawing Room came to function as a source of training for officers', including selected cadets from the RMA, and effectively became a school in 1776 when the Master General, George Townshend, assigned a mathematical master. The first of these was Reuben Burrow, who also took his students out on outdoor training exercises in practical surveying.[105]

At Woolwich, mathematical teaching was at an early date seen as desirable for cadets and officers of the Artillery and Engineers, alongside their practical training. In 1720 it was admitted at a meeting of the Board that a school should already have

[103] Quoted in Parnell, 'The Ordnance Drawing Room', 128.

[104] Will of Robert Whitehand, Captain of Tower of London, Middlesex, PROB 11/604/61 (TNA), partially quoted in Parnell, 'The Ordnance Drawing Room', 129.

[105] Marshall, 'Military Maps of the Eighteenth-Century', 23–5. Marshall's article focuses on the Drawing Room's role in cartography while Parnell's emphasizes architecture and corrects some of Marshall's errors.

been established but that funds had been lacking. Richards, the key mover behind the Drawing Room, then indicated that he had succeeded in securing a £3,000 investment in the South Sea Company, which it was hoped would provide for an academy. Richards was ordered to provide drawings and estimates 'for another Great Room to be built at Woolwich for an Academy equall to the Great Room already built [at Tower Place, used as a Board Room], with a house for the mathematical master', as well as of 'a house for the exercise of the long gun and mortar at Blackheath in lieu of the old one now past repair'. A mathematical master was to be appointed at a salary of at least £100 a year.[106]

The academy room was built in 1721 and one of the fireworkers, Burnett Godfrey, was to teach mathematics and be paid by the students themselves: the 'cadet gunners are to be stopped £4 a year each toward gratifying the said Mr Godfrey or such other officers who shall instruct them in gunnery, fortification etc'.[107] It is not clear how much teaching took place, or how long Godfrey undertook – or attempted to undertake – this task, though he received a commission as a Second Lieutenant in the Artillery in 1728.[108] Presumably hampered by the almost immediate collapse in value of South Sea stock, it appears that there was little more attention paid to mathematical teaching at Woolwich before 1741, the year that the RMA was founded. We do know, however, that William Deane supplied some demonstration equipment to the Ordnance in the 1720s that was similar to that supplied directly to the Academy some forty years later by George Adams, who had succeeded as instrument maker to the Ordnance. These included a set of pulleys in a frame and 'a new machine for Gunnery which shews how any ball or shell mounts in the air, and what curve it makes according to any degree of elevation'.[109]

The royal warrant of 30 April 1741 allowed £500 a year (soon after increased to £1,000) for the institution and endowment of a school 'for Instructing the raw and inexperienced People belonging to the Military Branch of this Office, in the several parts of Mathematicks necessary to qualify them for the Service of the Artillery, and the business of Engineers'. It ordered that the existing room be fitted up, 'an able and Skilful Master and Assistants' be appointed, regulations written and 'an Aparatus of Instruments, Books, and other necessarys' be provided.[110] The 'Rules and Orders of the Royal Academy' emphasized that the provision of education 'both in Theory and Practice' was required to produce 'good Officers of Artillery, and perfect Engineers'. A Chief Master on £200 per annum and a Second Master on half the amount were to be appointed and, between them, operate the school for six hours each day, save Sundays and holidays.[111] A curriculum was outlined for 'Teaching the Theory' – mathematics, fortification, sieges, the 'Doctrine of Projectiles', knowledge of types of ordnance

[106] Hogg, *The Royal Arsenal*, vol. 1, 345–6 (quotation at p. 346).
[107] Hogg, *The Royal Arsenal*, vol. 1, 346.
[108] Hogg, *The Royal Arsenal*, vol. 1, 347.
[109] Millburn, 'William Dean', 17. The second bill, dated 1729, also included 'A New Contrived Ballance to find the Specific Gravity of several Sorts of Metals', a matter which had been the focus of Reynold's experiments back in the 1650s.
[110] Hogg, *The Royal Arsenal*, vol. 1, 347–8.
[111] Royal Military Academy, Woolwich, *Rules and Orders for the Royal Academy at Woolwich* (London: printed by J. Peele and W. Wilkins, 1741), 4–5.

and engines, and compositions of metals, gunpowder and fireworks – and 'Teaching the Practice' – setting up and firing of artillery, raising batteries, entrenchment and mining, making of floats and bridges, the making of gunpowder and fireworks, casting, working and proofing of ordnance, the surveying and disposing of stores, and setting up a train of artillery. Perhaps indicating that more was expected than pedagogical exercises, it was added that 'whenever … [the Chief Master] shall have Occasion to make Experiments of any Sort relating to the Art of War, the Commanding Officer of the Artillery shall assist and provide him with Men, and the Storekeeper of the Ordnance at *Woolwich* with all other Things necessary for the same'.[112]

A drawing master was added to the establishment by 1743, and drawing instruments, as well as those for surveying and gunnery, were regularly supplied over the following years. As Celina Fox has noted, at the Ordnance drawing was closely tied to mathematical practice and empirical observation.[113] It was embedded and institutionalized via the Drawing Room and the RMA, and later produced both the topographical drawing of the Sandby brothers and the maps of the Ordnance Survey. She has suggested that the series of drawings of the Laboratory, including the two reproduced above, may have been produced to support the cadets' training, noting their 'great clarity and graphic immediacy'. It is striking how recognizable the processes and equipment are when compared to diagrams and written descriptions.[114] The trained observing eye and drawing hand were as important as mathematical and practical exercises. Knowing the tools, using the instruments, learning the recipes for powder, undertaking repeated training and exercises and replicating them on paper were all part of attempts – not always successful, especially in the unruly early days of the Academy – to standardize not just the work of the Ordnance and its regiments but also their personnel.[115]

Woolwich Warren: Combining theory and practice

The work of the mathematician Benjamin Robins, published in *New Principles of Gunnery* (1742), applied Newtonian theory to and drew on experiments regarding the motion of projectiles. It provided questions and approaches that were taken up at Woolwich, linking the practical business of firing ordnance with the mathematical teaching of the RMA. Its influence on practitioners has, however, largely been dated to the late eighteenth and early nineteenth centuries.[116] Charles Hutton, professor of mathematics at the RMA from 1773, made use of the Robins ballistic pendulum to

[112] *Rules and Orders*, 14.

[113] Fox, *The Arts of Industry*, 362–78.

[114] Ibid., 364.

[115] On the difficulties with discipline at the early RMA see Hogg, *The Royal Arsenal*, vol. 1, 354–60 and F. G. Guggisberg, '*The Shop;' the Story of the Royal Military Academy* (London, Paris, New York and Melbourne: Cassell and Company, 1900), 4–8.

[116] Brett D. Steele, 'Muskets and Pendulums: Benjamin Robins, Leonhard Euler, and the Ballistics Revolution', *Technology and Culture* 35, no. 2 (1994): 348–82; Seymour H. Mauskopf, 'Gunpowder and the Chemical Revolution', *Osiris* 4 (1988): 93–118.

carry out his first 'course of experiments on fired gunpowder and cannon balls' with 'some able officers of the Royal Regiment of Artillery' in Woolwich in 1775. These experiments were published in the Royal Society's *Philosophical Transactions* in 1778 and won him the Copley Medal. His experiments continued in the 1780s, under the direction of Major Blomefield, the Inspector of Artillery, and on the orders of the Duke of Richmond, the Master General, because, said Hutton, of his concern 'for the good of the public service'. In the 1780s, too, began the experiments of William Congreve, who largely took control of the Royal Laboratory from 1782, on powder production and testing.[117] The Ordnance also sought and adopted the advice of Richard Watson, professor of chemistry at Cambridge, on the production of charcoal, which Congreve showed increased the ballistic force of the gunpowder produced.[118] However, we argue that this approach, combining theory and practice, had a longer history at Woolwich than is usually allowed, and that there is continuity with activity at other and earlier Ordnance sites.

Hutton's accounts of his 'New Experiments in Artillery' in the 1780s are clear about the nature of the resources at Woolwich, both in supporting and shaping his approach. He was supplied with 'Five very fine brass one-pounder guns' that 'were cast and prepared, in Woolwich Warren, for these experiments, and bored as true as possible'.[119] He also had barrels of 'government powder' procured from 'the best maker' and cannon balls 'cast on purpose, very round' and ready access to mathematical instruments.[120] He noted that 'our situation at Woolwich, close by the river Thames', surrounded by open land, allowed 'a good length of range', sites that could be accurately surveyed and locations where assistants could be placed to observe the landing of projectiles in water. He was, throughout, supported by Ordnance workmen and others, including the first class of Gentleman Cadets at the Academy, 'namely Messieurs Bartlett, Rowley, De Butts, Bryce, Wm. Fenwick, Pilkington, Edridge, and Watkins, who have gone through the science of fluxions, and have applied it to several important considerations in natural philosophy'. These students were, he said, of exemplary conduct – arising as 'the pure effect of true philosophical principles' – and willing and cheerful, carrying out their voluntary task even when soaked by rain.[121] All of this, of course, supported Hutton's claims to experimental and pedagogical virtue, and painted an image of the Warren as a veritable Eden for experimental philosophy, despite the frustration of what remained an imperfect science.[122]

[117] Seymour H. Mauskopf, 'The Crisis of English Gunpowder in the Eighteenth Century', in *Materials and Expertise in Early Modern Europe: Between Market and Laboratory*, ed. Ursula Klein and E. C. Spary (Chicago and London: University of Chicago Press, 2010), 288–320 (pp. 309–11) and 'Chemistry in the Arsenal'. Charles Hutton, *Tracts Mathematical and Philosophical* (London: 1786), 99, 104.

[118] Colin Russell, 'Richard Watson: gaiters and gunpowder', in Mary D. Archer and Christopher D. Haley, *The 1702 Chair of Chemistry at Cambridge: Transformation and Change* (Cambridge: Cambridge University Press, 2005), 57–83; Mauskopf, 'Chemistry in the Arsenal', 303.

[119] Hutton, *Tracts*, 108.

[120] Ibid., 137.

[121] Hutton, *Tracts*, 217, 219, 220.

[122] Benjamin Wardhaugh, *Gunpowder & Geometry: The Life of Charles Hutton, Pitt Boy, Mathematician and Scientific Rebel* (London: William Collins, 2019), 135–41.

It has been suggested it was 'Hutton's ethos for experimentation and military improvement' that had an impact 'on the cadets and staff with whom he was in contact' and set the scene for the work of Shrapnel and Congreve, making the Warren 'a semi-public place' for teachers and students 'to see experiments in progress'.[123] While it was a key site for science by the end of the eighteenth century, and products of the Woolwich system were among the scientific servicemen that were central to the reform of science in the nineteenth century, it is too much to credit it to Hutton alone.[124] In fact, Hutton's and Congreve's investigations were more of a continuum with earlier activity than the sudden arrival of a scientific mindset. Robins's book was a by-product of the Ordnance's interest in combining theory and practice in its cadets' training – it was, indeed, intended as 'a public Course of Fortification and Gunnery' and part of his unsuccessful application to be master at the RMA in 1741.[125] His work was at least partly introduced into the curriculum by Muller before it was more fully incorporated under Hutton.[126] Likewise, Hutton's experimental site was the product of the availability of space, materials, practical skills and theoretical knowledge at Woolwich that continued and elaborated the long history of trials undertaken at the Tower, Vauxhall, Bishopsgate and Blackheath. From the 1740s, rather than the 1770s, Woolwich was home and workplace to mathematically trained and theoretically adept individuals linked to the Academy, as well as Ordnance officials closely engaged with both practical work and London's entrepreneurial natural and experimental philosophers and experimentalists.

Two plans, of Woolwich and of the Warren, dating from 1749 show the RMA as associated not only with surveying, drawing and artillery practice but also with teaching mathematics and experimentation. Based on surveys of 1748, these were produced by John Barker, who was then based at the RMA, though whether this was as a visitor, cadet, private pupil or in some other capacity is unclear. It has been suggested that the humorous putti in the vignette on his survey of the Warren indicate that he was a pupil: two are studiously engaged with theoretical lessons in mathematics and fortification, offered by a figure representing Geometry, while two others, demonstrating the practice of artillery, appear to be intent on shooting down the Academy building (Fig. 4.7).[127] In his plan of Woolwich, Barker included a 'view of the Royal Academy, Laboratory, &c' with what appears to be the frame of Robins's ballistic pendulum set up in the foreground, with a cannon pointing towards it for trial or proof (Fig. 4.8). This surveyor may be the John Barker who accompanied Robins to

[123] Wardhaugh, *Gunpowder & Geometry*, 141.

[124] See e.g. Simon Werrett, 'Congreve's rational rockets', *Notes & Records of the Royal Society* 63, no. 1 (2009): 35–56 and Jenny Bulstrode, 'The Promiscuous Exercises of the Woolwich Bois Boys' (paper presented at All Souls College, Oxford, 17 December 2015); David Philip Miller, 'The Revival of the Physical Sciences in Britain, 1815–1840', *Osiris* 2 (1986): 107–34.

[125] Bejamin Robins, *New Principles of Gunnery* (London: printed for J. Nourse, 1742), iii. Robins's appointment seems to have been blocked for political reasons: Steele, 'Muskets and Pendulums'.

[126] Steele, 'Muskets and Pendulums', 373–4. Gucciardini notes the presence of 'good mathematicians' at Woolwich from 1741, although he suggests that teaching did not reach a particularly high standard before the end of the century: Niccolo Gucciardini, *The Development of Newtonian Calculus in Britain, 1700–1800* (Cambridge: Cambridge University Press, 1989), 109–10.

[127] Peter Barber, 'A Rum Lot of Maps', *Maps and Views Blog* (British Library, 2015), https://blogs.bl.uk/magnificentmaps/2015/03/a-rum-lot-of-maps.html.

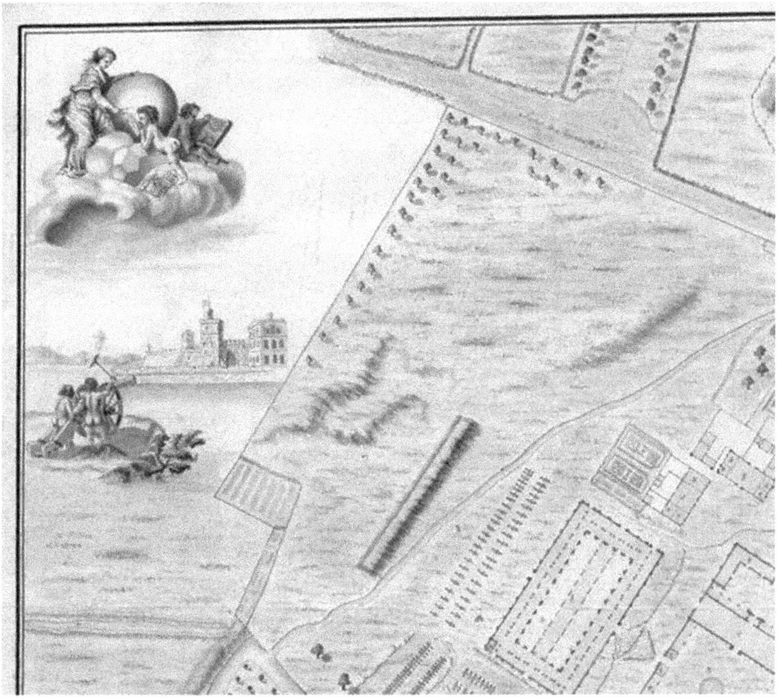

Figure 4.7 Detail from 'An Exact Survey of the Warren at Woolwich, drawn by John Barker at the R. Academy, An. 1749' (BL: Maps K. Top. 17.22). © British Library Board.

Fort St David as assistant when the latter was appointed Engineer General to the East India Company in 1749.[128]

Muller's *Treatise of Artillery* also testifies to a number of experiments that took place at Woolwich in the mid-eighteenth century. He tested some of Robins's assumptions, for example regarding the 'elastic force' of gunpowder in relation to its quantity and density, and referred to other experiments undertaken in 1736 by John Armstrong, intended to establish the most effective length for a 24-pounder gun. Armstrong, who had been both Chief Engineer and Surveyor General, was a Fellow of the Royal Society who had pushed for the founding of the RMA and was commended for his 'Experimental, as well as Contemplative Knowledge'.[129] There were also trials of the strength of lightweight guns and 'several experiments … with two small Mortars' that Muller made with Thomas Desaguliers, the Chief Firemaster.[130] Desaguliers was son

[128] H. D. Love, *Vestiges of Old Madras, 1640–1800*, 4 vols (London: John Murray, 1913), vol. 2, 408–9.

[129] Muller, *A Treatise*, 20; Paul Latcham, 'Armstrong, John (1674–1742), Military Engineer', *ODNB*. This judgement appears in a dedication to Armstrong in Thomas Weston's translation of Galileo. Weston, who had an Academy at Greenwich, had been on the Ordnance's books as 'labourer' at the Royal Observatory: Dedication by John Weston in Galileo Galilei, *Mathematical Discourses Concerning the Two New Sciences Relating to Mechanical and Local Motion*, trans. Thomas Weston (London: 1730).

[130] Muller, *A Treatise*, 11, 41–9, 7 (quotation).

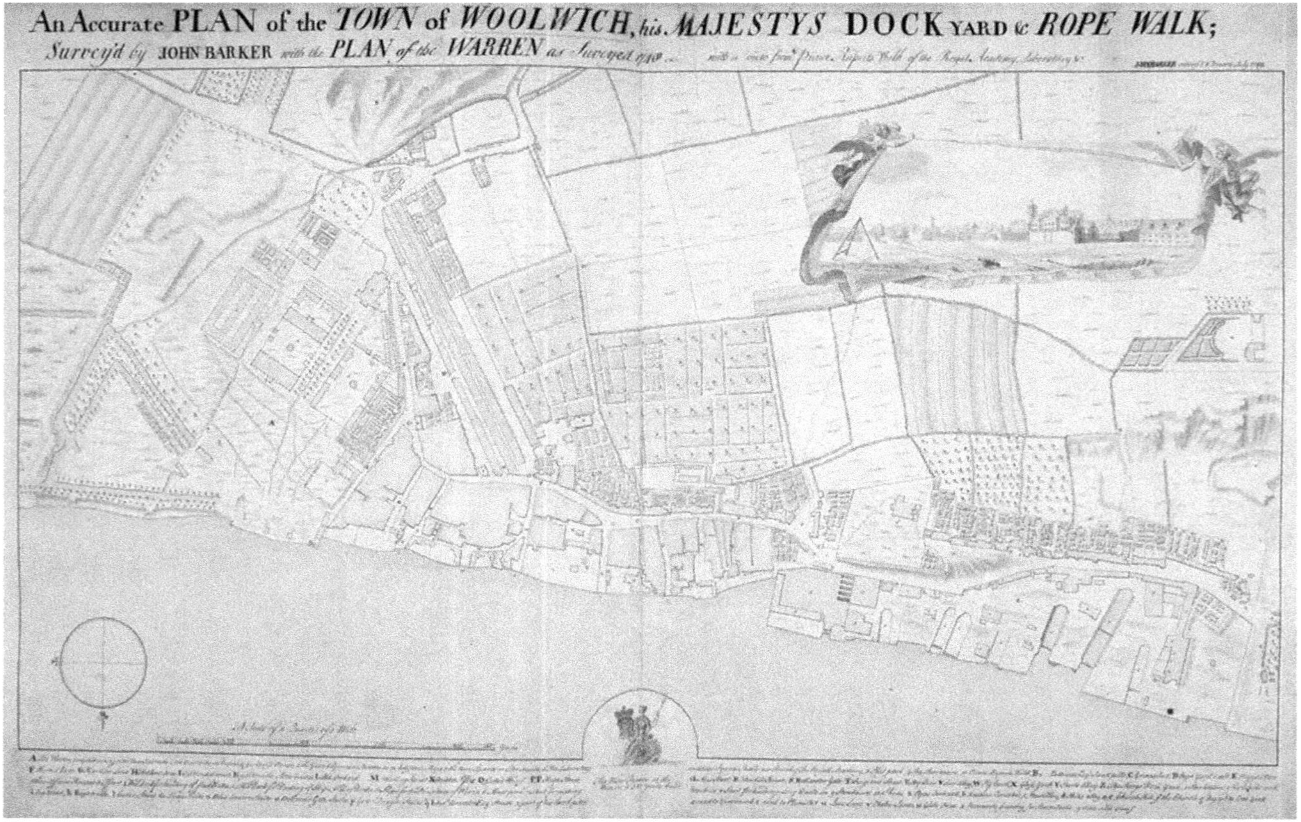

Figure 4.8 A plan of the Town of Woolwich, including the Warren (bottom left) and Dockyard, by John Barker in 1748 (BL: Maps K.Top.17.21). © British Library Board.

of John Theophilus Desaguliers, Larry Stewart's chief exemplar of London's public world of Newtonian science. He had entered the Royal Artillery in 1740, became Chief Firemaster in 1748 and has been described as 'the first scientific maker of cannon and the first regular investigator into the powers of gunnery in the English army'.[131] Over his long career, which included action, he worked frequently with Charles Frederick, who was both Comptroller of the Laboratory (1746–82) and Surveyor General (1750–82). Educated at Oxford and Middle Temple, Frederick was an MP and a Fellow of the Royal Society and Society of Antiquaries.[132] He did not, however, refrain from taking a hands-on approach. This was the case with the fireworks celebrating the treaty of Aix-la-Chapelle that, with mixed success, he and Desaguliers oversaw with the assistance of Italian pyrotechnicians. Horace Walpole described him during the preparations as 'bronzed over with a patina of gunpowder'.[133]

The interest of Desaguliers and Frederick in experimental and standardizing work at the Royal Laboratory is indicated by the rise in the number of instruments supplied for use there by George Adams in the 1750s.[134] Woolwich was also a key site for trying new or improved designs, which might be suggested by a range of actors. In 1752 Lieutenant Jacob Tovey informed the Board of Ordnance of his investigations 'to find out some more efficient method for taking the Centres of Guns and Mortars than has formerly been practised' and claimed to have found one 'more Exact as well as Easier'. He, 'having had the Instrument made, and tried at the Publick Practice', could report that 'Col. Belford [of the Regiment of Artillery] together with the Officers and Masters of the Academy agree it is the best of the kind they ever saw or heard of' and hoped for some encouragement or repayment of expenses. It was 'Ordered That he send the Instrument up to the Tower for the Board's Consideration' and, after considering Muller's recommendation, Tovey was paid thirty guineas and the instrument taken into the Tower stores. They ordered more instruments of this design, from Thomas Rybright, rather than Adams their usual supplier, who perhaps had been the maker that Tovey had worked with privately.[135]

Another experimental instrument, made the same year, resulted from a request to the Board by Major Albert Borgard Michelson of the Royal Artillery, a nephew of the Chief Firemaster Albert Borgard. In October 1750 he had asked for a quadrant for use at Woolwich, 'there being none fit for making Observations on Experiments with any Certainty'. The Board therefore ordered that 'a Warrant be made out for a new Quadrant according to the Surveyor-General's [Charles Frederick's] directions'. This order, of a brass quadrant 'for making Experiments at Woolwich', went to Benjamin Cole, who

[131] Stewart, *The Rise of Public Science*; Stephens, H. M. and Jonathan Spain, 'Desaguliers, Thomas (1721–1780), Army Officer and Military Engineer', *ODNB*.

[132] D. W. Bailey, 'Frederick, Sir Charles (1709–1785), Politician and Antiquary', *ODNB*.

[133] See Werrett, *Fireworks*, 148–54 (quotation p. 151).

[134] Millburn, 'The Office of Ordnance', 237–40. Frederick also personally received a 'Large Case of Instruments with 14 Ivory Scales' in 1753.

[135] Millburn, 'The Office of Ordnance', 250–2. Presumably Thomas Ribright, the optical instrument maker.

had taken over Deane's shop and purchased some of his gunnery instruments. Cole delivered the instrument straight to Woolwich early in 1751, which delayed payment in a system where all such supplies were meant to pass through the Tower before issue. It was only in March 1752 that he was paid £7 17s 6d for his rendering of Frederick's innovative design: a 'Brass Quadrant for Elevating Cannon with two Spirit Levels & Wheelwork withinside and Dial Plate & hand on the Back with Two Counter Ballance weights and Brass Plummet in a Wainscott Case'.[136]

Frederick and his wife Lucy were described by her sister-in-law Frances Boscawen: he was, she said, 'the busiest, she the idlest, mortal living'. During the build-up to the 1749 fireworks he was at the office built for him in Green Park from 8 am to 4 pm, while 'The rest of the day he gives audiences and worries his spirits and his person till 'tis reduced to a shadow'.[137] The official archive indicates that he remained a busy man and that he had an anxious concern for the effectiveness of his own work, and that of others, as a duty owed to the public. His concerns ranged widely across the Ordnance's various departments and sites, and he wrote of his 'Laborious Employment' which had 'left me but few leisure hours to retain the remembrance of my former amusing Studies, and none to improve my knowledge in them'. However, he retained the view that certain kinds of 'Science' were essential 'either to be correct in one's own Designs, or to Judge the Merits of Others'.[138] That said, it was the experience and daily practice of artisans that was actually required. Thus, when one William Dine was 'recommended to me as a very skilful Joiner & Carpenter', he was sent 'to be Tried in the Laboratory' and Captain John Mollman, Assistant Firemaster, was asked to 'Report to me his Abilities'.[139] Frederick also asked that Mollman and Desaguliers would try and report on Stephen George, 'recommended to me as an exceeding good Turner'.[140]

Frederick saw himself as punctilious about selecting on merit. As the train of artillery was assembled for shipping to Portugal to counter a Franco-Spanish invasion in 1762, he assured his correspondent, James Pattison, that 'the Civil Staff', which included Fireworkers and other artificers as well as clerks and storekeepers, 'will I think, ably and Diligently perform their duties; I have taken as much pains as I can to Select them'.[141] Pattison, the Artillery officer in command of the expedition, was a close friend, formerly the Chief Firemaster and from 1764 the Lieutenant-Governor of the RMA where he has been credited with instituting zealous and effective reform.[142] Elsewhere Frederick reported problems within the Drawing Room to the Master General, Lord Granby. One of the young draughtsmen had been struck by George Haines, who was

[136] Millburn, 'The Office of Ordnance', 282, 283, 287; Royal Artillery Museum, INST 24/099. Millburn suggests that this design was influenced by the patented quadrant design of John Barston, who around this time may have been engaged to repair the turret clock at Woolwich.

[137] A. N. Newman, 'Frederick, Charles (1709–85), of Hammersmith, Mdx', *History of Parliament* (1970) https://www.historyofparliamentonline.org/volume/1715-1754/member/frederick-charles-1709-85.

[138] Frederick to Joshua Kirby, 10 May 1766, MssCol 1075, [n.p.].

[139] Frederick to Mollman, 30 January 1762, Laboratory In-Letter Book 1761–1763, SUPP 5/25.

[140] Frederick to Mollman and Desaguliers, 11 August 1762, SUPP 5/25.

[141] Frederick to Pattison, 31 May 1762, MssCol 1075, p. 43.

[142] Guggisberg, *The Shop*, 9–10.

deputizing for the Chief Draughtsman, and both were reprimanded.[143] The trainee had made 'little progress' in the two years he had been there but in general, since Desmaretz had gone overseas, 'the Boys in the Drawing Room are grown so riotous and make so small a Progress in their Profession that excepting a few they are of little or no use to the Publick'. Frederick hoped to increase their sense of duty but feared they were the wrong sort: 'their Parents get them recommended for the sake of the 1/6 p[er] diem, without considering they would much sooner make a figure in Life by making a Suit of Cloaths, than ever they will do by drawing of Plans.'[144]

That Frederick had direct knowledge of work at the Woolwich Laboratory is indicated in his response to a request from John Cleveland, Secretary to the Admiralty, for incendiaries. He suggested that, as he was headed to Woolwich that morning, he would 'endeavour to Contrive some Composition that shall answer the purpose, and will inform Mr. Cleveland this night of his success, and the utmost secrecy shall be preserved'.[145] While this sounds like a hastily undertaken trial-and-error approach, Frederick was key to more systematic attempts to rationalize knowledge, procurement and standardization of gunpowder. Brenda Buchanan has judged that he had 'expertise with gunpowder making' and had a 'professional approach' that allowed him to rise to the challenge of supplies for the Seven Years War.[146] Powder was contracted from private makers, although in 1759 the Ordnance also acquired Faversham Mills as the question of supply and quality became critical.[147] A letter from Frederick to Bristol suppliers, whose powder had failed a proof in 1761, gave detailed advice about the ingredients, quantities and process gained from experience at Faversham.[148] He urged them to 'keep *one mill* at work trying Experiments (for a little while) on different Compositions different Times of grinding etcr' in order to find 'the right method as others have done'.[149]

Henry Strachey, then probably a clerk in the office of the Secretary at War, relayed Frederick's advice and endorsed it by stating that he had 'studied his Art more than any Man in England, and made more Experiments'.[150] Reframed as personal dedication, public duty and book learning, Frederick's experimental approach and close interactions with materials and artisans have, however, continuity with the ways in which his predecessors at the Ordnance worked. It also reveals the institutional setting within which Hutton's experimental and improving 'ethos' could thrive.

143 Marshall, 'Military Maps': 37.

144 Frederick to Lord Granby, 4 October, MssCol 1075, [n.p.]. Marshall quotes George Townshend, Master General from 1772, who he credits with turning the Drawing Room into a school, on preferring officers to those 'of a low rank in life': Marshall (1980), 29.

145 Frederick to Cleveland, 7 July 1762, MssCol 1075, 58.

146 Brenda J. Buchanan, '"The Art and Mystery of Making Gunpowder": The English Experience in the Seventeenth and Eighteenth Centuries', in Steele and Dorland, *The Heirs of Archimedes*, 233–74 (pp. 259–61).

147 Mauskopf, 'The Crisis', 294.

148 Buchanan, 'The Art and Mystery', 261–2.

149 Quoted in Buchanan, 'The Art and Mystery', 262.

150 Quoted in Buchanan, 'The Art and Mystery', 265; E. I. Carlyle and S. J. Skedd, 'Strachey, Sir Henry, First Baronet (1736–1810), Politician', *ODNB*.

Conclusion

At the Ordnance craft skill long interacted with practical mathematics and experimental activities. As an institution, it was often in debt or overstretched in responding to military action, while many of the artisans it supported had reason to maintain tradition over innovation. However, as this chapter has shown, the busy sites and surroundings of the Tower and Woolwich allowed for fruitful interactions of knowledge and skill in relation to a range of activities, materials and objects. These were multifunctional spaces within which hybrid activities were undertaken, combining the skills of individuals across different trades and social classes, and hands-on practice with book and classroom learning. The 'working worlds' of gunnery and surveying threw up problems and opportunities that prompted action from and interaction between artisans and scholars. These can be traced through the provision and use of mathematical instruments and the continuous, but increasingly self-conscious, culture of trial and experiment. Scholars and fellows of the Royal Society appear as curious visitors and as clients requesting access to resources at the Tower and Woolwich, but also as Officers of the Ordnance who engaged practically and pragmatically with the pressing demands of the moment and worked productively with artisans based within and beyond the institution.

The interactions described between smiths and mathematicians indicate the generative nature of this kind of hybrid and multi-institutional site. The metrology of John Reynolds was a product of both his office in the Mint and his more scholarly interests; it also benefitted from his access to the spaces occupied by Ordnance artisans and objects. Gunners and surveyors of the Ordnance relied on and helped support the businesses of neighbouring mathematical practitioners. Although skills that the Ordnance required were at times imported from overseas, we have also shown that the Tower and its environs produced craft lineages that can be traced throughout the seventeenth and into the eighteenth century. Family and business succession were key to the passing on of knowledge and skills, and personal relationships as well as geographical proximity created the potential for collaboration. Although formalized training and teaching, in the Drawing Room at the Tower and at the Laboratory and RMA Woolwich, became increasingly important in the second half of the eighteenth century, such relationships and the productive overlapping of spaces and activities remained significant throughout our period.

5

Artisanal and mercantile knowledge production and exchange at livery company halls

Introduction

By the turn of the seventeenth century there were approximately sixty livery company halls within the City walls. These were the headquarters of London's craft and mercantile guilds, responsible for regulation and training within the crafts and trades. Many guilds had late-medieval roots and foundations, and some mercantile companies had held property on the same site since the fourteenth century. Within these multifunctional institutional buildings, guilds governed and disciplined members, enrolled apprentices, socialized and commemorated founders and benefactors, administered large estates, managed substantial archives and material collections (of plate, textiles and other treasures), distributed charity and regulated standards of material production. Company halls were also spaces within which materials, tools and substances were tested, masterpieces were assessed, and training and education took place.[1] At Apothecaries' Hall in Blackfriars there was a laboratory under the hall chamber for the production of chemical medicines from the early 1670s and a 'designated retail space' for the same.[2] Goldsmiths' Hall housed a metallurgical workshop of international significance. Anatomies and lectures were undertaken at Barber Surgeons' Hall, home to London's first purpose-built anatomy theatre. In other words, company halls could be dynamic sites of knowledge production and exchange.

Guild halls were typically impressive courtyard houses with substantial gatehouses that fronted the street. Some were located at the centre of the trade cluster, others situated simply where suitable properties and donors had become available. In the fifteenth and early sixteenth centuries, the institutional complex usually included hall chamber, parlour, courtyard, garden or recreational space, and sometimes a chapel and almshouses. From *c.* 1540, livery halls were expanded outwards, and upwards, as new specialized rooms and grand staircases and galleries were added. Novel spaces included court chambers, dining rooms, counting houses, kitchens, butteries, pantries, larders, penthouses and domestic lodgings (for clerk and beadle). To manage

1 See Kilburn-Toppin, *Crafting Identities*, Chapter 4.

2 Anna Simmons, 'Trade, Knowledge and Networks: the Activities of the Society of Apothecaries and its Members in London, *c.*1670–*c.*1800', *BJHS* 52, no. 2 (2019): 273–96 (p. 278).

and safeguard growing collections of material cultures (such as textiles, furniture and cooking equipment) and archives, treasuries and armouries were constructed. Parlours were extended and richly furnished. Internal decoration became more ornate from the mid-sixteenth century, and exterior walls and gatehouses were enlarged and embellished.[3] The Great Fire of 1666 had a devastating impact on the institutional built fabric, with forty-four out of sixty livery halls destroyed.[4] However, reconstruction was quick: 'three-quarters [of the city's guild headquarters] were structurally complete by 1673.'[5] The rapid impetus for rebuilding on pre-Fire plots reflected the centrality of such sites to the status and antiquity of corporations and, more broadly, the view that even as control of craft and trade by guilds diminished they remained essential to the civic, religious, political, social and commercial life of the city.[6]

Movement within company halls was regulated by guild officials, such as the beadle, who often lived on site. A range of people might have entered the hall complex through the gatehouse into the courtyard – including journeymen, established artisans, merchants, gentlemen, foreign dignitaries and sometimes members of the royal court. All guildsmen had access at certain times of the ceremonial year to the hall chamber, a semi-public space with distinct 'high' and 'low' ends, where feasts took place and ordinances were read to the assembled company on election days. At Barber Surgeons' Hall this was also where lectures and dissections took place until the mid-1630s. Access to high-status rooms on the first and second floors of guild halls, reached by magnificent staircases, such as parlour and court room, was much more restricted, and reserved largely to men of the livery, or assistants – the guild governing elite. Limited too was entry to gardens, or bowling greens, which often lay adjacent to the parlour. Admittance to chambers for testing and material display might also be closely regulated. A goldsmith complained in 1572 that guild matters were being 'determined in the parlour "in hugger-mugger" [meaning secret or concealed] fashion'.[7] At Barber Surgeons' Hall none but the master and assistants 'shall or do presume at any time hereafter to come into the parlour of the said fellowship'.[8] On court days the Company Beadle was ordered 'to look and see that no manner of person or persons do presume or come in into the parlour door … nor none do walk in the Court before the parlour window'.[9] Regulation of access, visibility and the soundscape of the parlour or court chamber was about controlling privileged information, and reinforcing civic hierarchies.[10]

This chapter takes as its focus two significant seventeenth-century company halls: Goldsmiths' Hall at the northern end of Foster Lane and Barber Surgeons' Hall on

3 Kilburn-Toppin, *Crafting Identities*, 112–37.

4 Thomas Reddaway, *The Rebuilding of London after the Great Fire* (London: Arnold, 1951), 26.

5 Anya Lucas and Henry Russell, *The Livery Halls of the City of London* (London: Merrell in association with the Worshipful Company of Chartered Architects, 2018), 242.

6 Matthews, 'Honour, Ornament, and Frugality', 73–104 (pp. 73–4); and in reference to Knights, 'A City Revolution', 1141–78 (pp. 1143–4).

7 Walter Sherburne Prideaux, *Memorials of the Goldsmiths' Company*, 2 vols (London: Eyre and Spottiswoode, 1896–97), vol. 1, 75.

8 The Barbers' Company Archive, London, B/1/2, fo. 23r.

9 B/1/2, fo. 17r.

10 Griffiths, 'Secrecy and authority'.

Monkwell Street. The Goldsmiths' Company received its first royal charter in 1327 and was one of the so-called Great Twelve livery companies, the most affluent and politically influential mercantile (and craft) guilds. Its prestige stemmed from the precious metals at the heart of the trade. The membership of the Goldsmiths' guild was heterogeneous, ranging from journeyman craftsmen, yet to establish their own workshops and trading spaces, to men of the highest civic status, who also served as lenders to the royal court and held substantial positions at the Royal Mint.[11] The Barber Surgeons' Company was incorporated in 1540, with roots in the medieval Barbers' and Surgeons' guilds.[12] The manual nature of the craft and its fellows' routine contact with diseased bodies and corpses impacted upon the status of the guild. At its Company court it attempted to regulate the work of unlicensed practitioners, as well as the labour of its guildsmen. The Barber Surgeons also existed in tension with the College of Physicians (founded in 1518), whose university-educated membership held 'the task and intention of controlling all practitioners of physic in the capital'.[13]

There were, however, commonalities in the experiences and responsibilities of the Barber Surgeons' and the Goldsmiths' Company. Both were actively involved in civic governance, festivity and craft regulation. Perhaps more than any other City company of comparative longevity and status, the Goldsmiths and Barber Surgeons also held halls which were demonstrably metropolitan centres of knowledge production and dissemination about the natural world: concerning metals and human bodies, respectively. A later foundation, Apothecaries' Hall, with its chemical laboratory, would be another apposite case study, but it has already been the subject of significant scholarship.[14] The Blackfriars laboratory was a centre for drug production, serving not just London apothecaries, but Britain and its empire by the eighteenth century. This site also had an educative purpose: laboratory 'subscribers and apprentices were allowed access to the laboratory to "inform themselves by what they shall see in the knowledge of Chimicall part of Pharmacye"'.[15] In this chapter, such practices of education and knowledge distribution will be explored through a close examination

11 S. P. Harper, 'Royal Servants and City Fathers: The Double Lives of London Goldsmiths at the Court of Henry VII', in *Medieval Merchants and Money: Essays in Honour of James L. Bolton*, ed. Martin Allen and Matthew Davies (University of London Press: Institute of Historical Research, 2016), 177–93; David Mitchell, *Silversmiths in Elizabethan and Stuart London: Their Lives and Their Marks* (Woodbridge: The Boydell Press/The Goldsmith's Company, 2017), 11–13.

12 *The Annals of the Barber-Surgeons of London; Compiled from Their Records and Other Sources, by Sidney Young* (London: Blades, East & Blades, 1890), 78.

13 Margaret Pelling; with Frances White, *Medical Conflicts in Early Modern London: Patronage, Physicians, and Irregular Practitioners, 1550–1640* (Oxford: Clarendon Press; New York: Oxford University Press, 2003), 1. See also Celeste Chamberland, 'Between the Hall and the Market: William Clowes and Surgical Self-Fashioning in Elizabethan London', *The Sixteenth Century Journal* 41, no. 1 (2010): 69–89 (p. 71), on 'the surgeons' struggle to establish boundaries and construct an identity in an occupation that neither fit tidily within the practical setting of artisanal labor nor was suitably compatible with the private, professional milieu of the university-trained physicians'.

14 See the scholarship of Anna Simmons, especially 'Medicines, Monopolies and Mortars: the Chemical Laboratory and Pharmaceutical Trade at the Society of Apothecaries in the Eighteenth Century', *Ambix* 53, no. 3 (2006): 221–36; 'Stills, Status, Stocks and Science: The Laboratories at Apothecaries' Hall in the Nineteenth Century', *Ambix* 61, no. 2 (2014): 141–61; 'Trade, Knowledge and Networks'.

15 Simmons, 'Medicines, Monopolies and Mortars', 226.

of the Assay Office at Goldsmiths' Hall, and the Anatomy Theatre at Barber Surgeons' Hall, demonstrating how corporate cultures and institutional guild spaces shaped urban knowledge cultures in distinctive ways.

The Assay Office at Goldsmiths' Hall

The site of Goldsmiths' Hall, located in the parish of St John Zachery and the ward of Aldersgate, was acquired by nineteen goldsmiths on behalf of the Company in 1339. 'No other London livery company can claim a longer or earlier tenure.'[16] The guild headquarters lay in close proximity to Cheapside and Lombard Street, the chief locations for the homes, workspaces and retail sites of London's goldsmiths.[17] The guild acquired a mercantile courtyard house which was adapted to corporate needs over time. By the fifteenth century the building had a hall chamber (with bay window), parlour, chapel, armoury, assay house and courtyard.[18] John Stow wrote of how '*Bartholomew Read*, Goldsmith, Mayor in the yeare 1502. kept such a feast in this hall … which as they say, were more than an hundreth persons of great estate. For the messes and dishes of meates to them serued, the paled [enclosed] Parke in the same hall, furnished with frutefull trees, beastes of venery [game] … Westminster hall would hardly haue suffised'.[19] In the 1630s, concerned about the dilapidated state of the hall, the guild demolished their timber house and constructed a Palladian-style brick and stone-clad mansion, designed by master mason Nicholas Stone (likely in close dialogue with Inigo Jones, Surveyor of the King's Works).[20] The courtyard arrangement was retained. Elaborate new staircases – 'more large and convenient for the companyes use' – connected the courtyard to the high-status first-floor rooms, including court room and parlour.[21] In architectural prestige and interior decoration Goldsmiths' Hall now rivalled the most impressive livery company halls, including Mercers' Hall on Cheapside, with a battlemented stone façade of five bays, over ninety-eight-feet long.[22] Several decades later Goldsmiths' Hall was badly damaged by the Great Fire, though the external walls remained largely intact.[23] Before the flames took hold, the Company's plate and archive were rapidly removed from the treasury at Foster Lane 'for their p[re] sent p[re]servac[i]on' by a quick thinking warden, and taken to a house at Edmonton known to Charles Brattle, the guild's assayer.[24] The treasury was a small room with a

[16] T. F. Reddaway and Lorna Walker, *The Early History of the Goldsmiths Company, 1327–1509* (London: Arnold, 1975), 29–30.

[17] Prideaux, *Memorials of the Goldsmiths' Company*, vol. 1, xviii.

[18] Prideaux, *Memorials of the Goldsmiths' Company*, vol. 1, xxviii; 28.

[19] Stow, *A Survey of London*, vol. 1, 305–6.

[20] Kilburn-Toppin, *Crafting Identities*, 81–2.

[21] Goldsmiths' Hall Archive, London (GHA), Wardens Accounts and Court Minutes (WA/CM), T, fo. 31r.

[22] Jean Imray, *The Mercers' Hall* (London: The London Topographical Society and the Mercers' Company, 1991), 19; John Schofield, *Medieval London Houses* (New Haven and London: Yale University Press for the Paul Mellon Centre for Studies in British Art, 1994), 174.

[23] John Newman, 'Nicholas Stone's Goldsmiths' Hall: Design and Practice in the 1630s', *Architectural History* 14 (1971): 30–9+138–41 (p. 30).

[24] GHA, Court Book 5 (1665–9), fo. 62r.

single locked door, positioned between the high end of the Goldsmiths' hall chamber and the parlour, on the Maiden Lane side of the building (Fig. 5.1/Plate 7). The keys to this room were kept strictly in the custody of the upper warden.[25]

Goldsmiths' Hall was the focus for the Company's governance, socializing and administration, but perhaps most fundamentally it was the centre for the guild's testing and regulation of silver and gold. It was through 'the light of the assay' that artisans and merchants, makers and consumers, might know 'what virtues or evilness' existed within plate, coin and bullion.[26] Alongside the workshop in the Tower Mint, this was the most important metallurgical testing site in London. The Foster Lane Assay Office could act as a check on the activities of the Mint through the trial of the pyx, undertaken by expert juries of goldsmiths, selected by the assistants of the Company.[27] The Goldsmiths' Company had the responsibility of testing the precious metal wares and ingots of all London goldsmiths to ensure compliance with the standard. The guild also checked all balances and weights employed in workplaces.[28] During the fourteenth and fifteenth centuries, assay (by touchstone testing) and touch

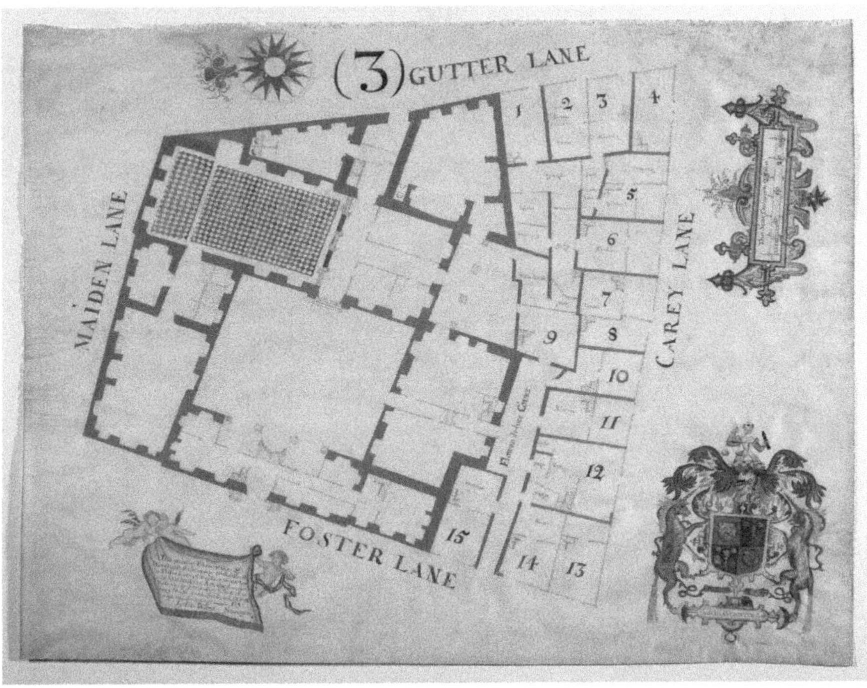

Figure 5.1 John Ward, Goldsmiths' Hall: ground-storey plan, 1692. Courtesy of the Goldsmiths' Company, London.

See also colour version, reproduced in plate section 7.

25 Kilburn-Toppin, *Crafting Identities*, 128.
26 *The Pirotechnia of Vannoccio Biringuccio*, 136.
27 Prideaux, *Memorials of the Goldsmiths' Company*, vol. 1, xxv.
28 Forbes, *Hallmark*, 31.

(referring to the wardens' marking of plate of the requisite standard with the mark of the leopard's head) were undertaken by the wardens in retail and work spaces.[29] Faulty wares (below standard) were broken up by wardens in the shop or at Goldsmiths' Hall. From December 1478 a centralized system based at Goldsmiths' Hall was adopted with a full-time salaried assayer, known as the Common Assayer, employed at the Assay Office.[30] Artisans brought their wares to Goldsmiths' Hall to be struck with a maker's mark and then tested in the assay workshop.[31]

When it came to appointing the Common Assayer at Goldsmiths' Hall, competitive trials were conducted for multiple candidates.[32] In 1619 there were four contenders for the post. Each man was required to undertake a practical examination across two days, overseen by the Company's refiner and three members of the Court of Assistants. These senior guildsmen composed a report of their observations which was read to the Court of Assistants who then voted for their preferred candidate.[33] In these examinations we see the significance of the collective dimension of guild decision-making and knowledge cultures. The communal aspect came to the fore also in the performance of the trial of the pyx, during which, as we have seen, a jury of around fifteen 'ancient and skilfullest goldsmythes' made their assessments of the quality of work undertaken at the Mint. This trial might even take place at Goldsmiths' Hall, should the furnaces prove inadequate at the court of Star Chamber at Westminster.[34] The collective nature of guild cultures is evidenced also in the routine search of workshops and retail spaces for faulty or below-standard wares. Searches were undertaken by small groups of guild authorities (often three or four men) and these citizens would collaboratively use their full range of senses and technical skills in making assessments about material quality.[35]

The Oath of the Assayer stressed the high degree of trust which the position held:

> truly assay all such gold and silver as shall be brought to you to assay. And also you shall melt all pieces of gold and silver delivered to you truly and impartially, without any deceit … and never take advantage of the owner … And you shall not accept gold or silver to be touched, wrought or sold that is below the standard ordained by a statute.[36]

This was the ideal, but trials and personalities could occasionally be contentious. One such quarrelsome character was John Reynolds, whom we have already encountered through his metallurgical work at the Tower. In addition to his roles at the Mint, Reynolds was involved in the Assay Office at Goldsmiths' Hall from 1607. In May 1629 a group of thirteen working goldsmiths presented a petition to the Company, suggesting

29 Forbes, *Hallmark*, 24.
30 Reddaway and Walker, *The Early History*, 164–5.
31 Mitchell, *Silversmiths in Elizabethan and Stuart London*, 174–5.
32 Reddaway and Walker, *The Early History*, 170.
33 Forbes, *Hallmark*, 104.
34 Goldsmiths' Hall Archive, London, MS C II.2.1, fos. 27r, 28v. Following the Coinage Act of 1870 the pyx trial was held annually at Goldsmiths' Hall. See Prideaux, *Memorials of the Goldsmiths' Company*, vol. 1, xxv.
35 Wallis and Wright, 'Evidence, Artisan Experience, and Authority in Early Modern England', 138–63.
36 Reddaway and Walker, *The Early History*, 215.

that Reynolds was not 'fit to judge or refuse our plate'. It was argued that Reynolds as Common Assayer had become 'partial in his office' and was 'causing the wardens to break some far better [plate of the standard] of such workemen disaffected by him'. Far from being balanced and focused, Reynolds was said to have a 'fury or rather madness' and 'in his rage hath misused many'. The petitioners also suggested that Reynolds was leaving much of the assaying work to his servants (presumably because he was busy at the Mint).[37] For his part, Reynolds claimed that his oath of office was 'a bond of Integrity laid upon his conscience his sufficiency of skill and knowledge required for that place, being not inferior to any his predecessors'.[38] Eventually Reynolds was dismissed from the post of Common Assayer, and in response he petitioned the King over unfair treatment; the Goldsmiths' Company sent a counter-petition to the Privy Council.[39] The dispute between the working goldsmiths and Reynolds suggests how, aside from technical abilities and skills, the workshop activities at Goldsmiths' Hall might be political and personal.[40]

A list of 'charges of the Say House Yearly', probably composed by Common Assayer Thomas Keeling in the late 1570s, gives us a deeper understanding of the equipment, fuel and labour employed at the institutional workshop for cupellation assay. As many as 14,000 cupels are referenced (which gives a sense of the sheer scale of the operation), cupel moulds, large quantities of coal and lead, '12 mufflers for the furnace', and pot earth for 'mending and repairing' the same. The assayer required ink and two books for keeping record of the trials – 'for entering the plate and says' – and '20 lbs candles'. Keeling's costs also included three meals a day for two 'men servants' and their clothing, including winter and summer coats.[41] Two servants were employed a century later when annual costs were again listed for the Assay Office. The workshop now required 600 cupels a week (approximately 31,000 per year).[42] A visual representation of the Assay Office was produced around this time as the frontispiece to William Badcock's *Touchstone for Gold and Silver Wares* (London, 1677) (Fig. 5.2). Overall, encompassing the goldsmith's shop (above) and Goldsmiths' Hall workshops (below), the image speaks of the fundamental importance of assay to the regulation of the craft and trade, and the trust of the consumer. Focusing on the Assay Office workshop interior, we see at the front of the image the refining furnace and men forging plate. In the workshop behind we see the assay furnace, and a servant putting assays into the fire. The seated figure at the table with balances is 'The Assay-Master making Assays'. The man at the far-right striking plate is 'the Warden marking the Plate on the Anvil'. Following a petition from workmen in 1660 – 'divers members of the Goldsmithery' – it was ordered that 'the Assayer shall make fire every working day [excepting Sunday] for the Assayes and touch of the plate brought to the hall'. Plate had to be delivered to the Goldsmiths' Weigher at the Weigh Office before 9 am.[43] By 1666 the Common Assayer

[37] GHA, WA/CM, Q1, fos. 112r-v.
[38] GHA, WA/CM, Q1, fo. 113r.
[39] Forbes, *Hallmark*, 116.
[40] For a more extended discussion of the working goldsmiths and Reynolds episode, see Jasmine Kilburn-Toppin, 'A place of great trust', 10–11.
[41] TNA, E 101/304/18, fo. 58r.
[42] GHA, Court Book 3 (1660–63), fos. 72r, 73r.
[43] Ibid.

Figure 5.2 William Badcock, *A Touchstone for Gold and Silver Wares* (London, 1677). Frontispiece showing the interior of a goldsmith's shop (top) and the assay workshop at Goldsmiths' Hall (bottom). Wellcome Collection: Wing B382.

was paid a salary of £70 per annum. This was the most lucrative office in the Company by some margin and must have reflected, from the perspective of the guild governors, the relative skill and industry of the post. For comparison, the Clerk was paid £42; the Beadle and Weigher £16.[44] Along with the Clerk and Beadle, the Assayer and his family were provided with accommodation at Goldsmiths' Hall and were expected to reside there.[45]

From court minutes and seventeenth-century plans of Goldsmiths Hall we know that the Assay Office had a separate side entrance to the main Hall site on Foster Lane, so artisans were not bringing plate through the main (ceremonial) gates on a daily basis (see Fig. 5.1: the entranceway between chambers '14' and '15' on the ground-storey plan of Goldsmiths' Hall. The chamber labelled '9' is the Common Assayer's house). Appropriately – considering the furnaces and labour – the institutional assay workshop was also adjacent to kitchens (on what was effectively the ground floor 'service side' of the livery hall). However, as at the Mint site in the Tower of London, managing the space of the assay house within the guild headquarters could be a challenge for Company governors. Here too there were issues of secrecy and accessibility. For example, assay worker John Kirk was disciplined by guild governors for selling secrets of the workshop – 'the feate of assayes making' – to interested gentlemen at the royal court. It was unlawful and in breach of his oath 'to open that or any other secret of his mystery to any man that is not free therof' [of the guild]. Kirk was unrepentant, saying 'he had taught others' and 'would do it again for money'.[46] The doors and even sightlines into the institutional assay workshop had to be carefully monitored too; as when the assistants ordered that no man walk on the elevated terrace above the Assay Office while the assayer and touch wardens were working 'and doe sitt and debate about the affaires of this society'.[47] At Goldsmiths' Hall secrecy was both a craft ideal – the concealed knowledge of the trade ('secret of this mystery') – and a tool of restrictive and hierarchical civic governance (controlled access to discussions about 'the affaires of this society').

Our understanding of how the labour at the assay workshop at Foster Lane was perceived by contemporary craftsmen and merchants is deepened through a manuscript presented to the Goldsmiths' Company at their Hall in the early seventeenth century. *The Gouldesmythes' Storehowse. Wherein is layde up many hidden secrets of that Ingenious Misterie* (1604) was probably a collaborative effort between father and son, Hannibal Gamon senior and Hannibal Gamon the younger, both guildsmen. The latter was an Oxford graduate, and the named author; the former a prominent goldsmith with an active workshop on Cheapside, and many apprentices enrolled over the course of his working life.[48] The manuscript was likely intended for a readership

44 GHA, Court Book 5 (1665–9), fo. 68v.
45 Kilburn-Toppin, *Crafting Identities*, 186.
46 Prideaux, *Memorials of the Goldsmiths' Company*, vol. 1, 62.
47 GHA, WA/CM, O2, fo. 194.
48 This combination of university and workshop-based knowledge, and gleaned from within the same family, was typical for metallurgical treatises. See Pamela Long, 'The Openness of Knowledge: An Ideal and Its Context in Sixteenth-Century Writings on Metallurgy', *Technology and Culture* 32, no. 2 (1991): 318–55 (pp. 334–5, 355).

of goldsmiths and institutional officials. Gamon suggests that the work is 'meete for a Gouldsmythe, Fyner [refiner], or Maister of a Mynte to understand, and haue perfit knowledge therin'.[49] Composed of three sections or books, the text includes many ideas and direct passages from contemporary and ancient sources on metallurgy (including Aristotle, Pliny the Elder and Agricola), and explores assaying, refining, monetary circulation, alchemical experiments and principles.[50] The second book is a lapidary. One of the overarching themes is the nature of the assayer's craft, more specifically how true understanding or knowledge might be acquired.

In *The Gouldesmythes' Storehowse*, the work of the Assay Office is explicitly framed as a knowledgeable and skilled practice; it is a 'scyence or mysterye'.[51] Gamon writes of degrees of knowledge about precious metals and money. Whereas 'everye ma[n] knowethe' the difference between bullion and money, 'most men doe not knowe' whether a coin or plate 'be fine golde, or fine sylver, or commixed w[i]th other metals'. Fundamentally, 'tryall at the Assaye' is required 'for the perfit knowledge of Golde & Syluer: It Requyrethe a p[er]fit Assay m[a]n, whose p[er]fection must be grounded vpon Artificiall Exercise; for these things doe rather consist in doinge, then in Resoninge, for they are not eselie redused to matter of Argument, unlesse Exercise be ioyned w[i]th speche'.[52] Actions, observation and verbal guidance (rather like the process of apprenticeship) were crucial for true comprehension. So too, as we observed in the Tower Mint chapter, were finely trained sensory responses. Reporting assays 'askethe a good Judgment, gotten rather by yeares and experience, then by speculation and dispute … that besydes his grownded experience in this scyence or mysterye should haue A perfit eie to vewe, and as stedye a hande to waye for other mens sense cannot serve him'.[53] Throughout the manuscript, Hannibal Gamon stresses both the importance of learned experience, or experiential knowledge, gained through years of observation and practice in the workshop, and the theoretical knowledge of the assayer's craft, formalized in a treatise like the *Storehowse*. When writing of the various specialisms encompassed by the broad umbrella of the Goldsmiths' Company, including workers in gold, silver and refining, Gamon notes how the 'Marchant goldesmythe, otherwise termed the Buyer and Seller', 'must be a ma[n] that must haue skill and knowledge, in all these aforesaid seuerall knowledges … All w[hi]ch cannot in manye yeares be attained vnto onlye by Tradition; vnles learning, which is gotten by Readinge seuerall Authors, be ioyned thereto'.[54] By emphasizing the growing body of propositional knowledge about metallurgy – to which the *Storehowse* itself contributed – the goldsmith Gamons underscored the theoretical dimension of assay practice, and concurrently the status of the art.[55]

49 GHA, MS C II.2.1, fo. 65r.

50 For a more extensive discussion of this manuscript and its themes, see Kilburn-Toppin, *Crafting Identities*, 47–60.

51 MS C II.2.1, fo. 5v.

52 MS C II.2.1, fo. 4v.

53 MS C II.2.1, fo. 5v.

54 MS C II.2.1, fo. 32r.

55 On writing about craft practice: Long, *Openness, Secrecy, Authorship*; Pamela Smith, 'Why Write a Book? From Lived Experience to the Written Word in Early Modern Europe', *Bulletin of the German Historical Institute* 47 (2010): 25–50.

The anatomy theatre at Barber Surgeons' Hall

In the early modern metropolis Barber Surgeons' Hall was a highly significant site for surgical education, regulation and dissemination of anatomical knowledge. It also figured strongly in the contemporary imagination as *the* metropolitan location for the dissection of human cadavers. And yet, though much has been written on London's competitive medical marketplace, corporate spaces, specifically Barber Surgeons' Hall, have not figured substantially in the discussion of skilled practice and identity formation.[56] Here we focus on the Hall at the northwest edge of the city, fronting Monkwell Street, the corporate home of the Barbers from the 1440s.[57] In the sixteenth century, as was typical, Company buildings were arranged around a courtyard (with a well and pump), 'paved with fair broad stone'.[58] The main point of access to Barber Surgeons' Hall was the courtyard gates on Monkwell Street that were monitored by the Company Porter ('in a decent and comely manner with a staff in his hand' on feast days).[59] A hall chamber, parlour, kitchen, buttery and a 'garden plot' are named in Company court minutes.[60] More specialized and prestigious spaces for governance, socializing and dissection were added to the hall complex in the 1630s.

Young men were presented by master surgeons at Barber Surgeons' Hall to be ceremonially bound as apprentices. Apprentice surgeons were required to be fully literate and 'to be clean in person and limb and meet for the exercising of the same mystery, science or art'.[61] Apprentices developed their skills and proficiency in the households of senior Company men for seven years. The apprenticeship requirement was in contrast to the College of Physicians, whose members were university-educated men.[62] As in any early modern craft, apprenticeship did not constitute a formal training programme but was rather an immersion in the master's household and broader spaces of surgical practice, through which motivated apprentices 'came to embody skills by participating in situated production' and 'repetitive practice'.[63] It was at Barber

[56] The literature on London's medical marketplace is vast; some of the more influential works include: Harold Cook, *The Decline of the Old Medical Regime in Stuart London* (Ithaca and London: Cornell University Press, 1986); Pelling, *Medical Conflicts*; Mark Jenner and Patrick Wallis (eds), *Medicine and the Market in England and Its Colonies* (New York: Palgrave Macmillan, 2007).

[57] Schofield, *Medieval London Houses*, 47; Jessie Dobson and R. Milnes Walker, *Barbers and Barber-Surgeons of London: A History of the Barbers' and Barber-Surgeons' Companies* (Oxford: Blackwell Scientific for the Worshipful Company of Barbers, 1979), 77.

[58] Paved in the 1580s. The Barbers' Company Archive, London, B/1/2, fo. 42r.

[59] B/1/2, fo. 71. There was probably an inner and outer gate. The 'outer streete gate' is mentioned in company archives in 1638. See *The Annals of the Barber-Surgeons*, 453.

[60] The company clerk was responsible for keeping the garden 'clean, swept, and weeded'. B 1/1/1, fo. 19v.

[61] B/1/2, fos. 18, 24 [an order from the 1560s]. In London the literacy requirement was only applied to apprentice surgeons, goldsmiths and apothecaries. See Patrick Wallis, 'Between Apprenticeship and Skill: Acquiring Knowledge Outside the Academy in Early Modern England', *Science in Context* 32 (2019): 155–70 (p. 160). See also Pelling, *The Common Lot*, 128, 'The ordinances of 1530 for the London Barber-Surgeons' Company … included the provision that all prospective apprentices were to be inspected to see that they were "clean without continual diseases or grievous infirmities whereby the king's liege people might take hurt", on pain of a fine.'

[62] Andrew Wear, *Knowledge and Practice in English Medicine, 1550–1680* (Cambridge: Cambridge University Press, 2016), 217.

[63] Wallis, 'Between Apprenticeship and Skill', 167.

Surgeons' Hall, however, that the expertise and knowledge of aspiring surgeons were put to the test. On completion of their training, apprentices who hoped to pursue 'the mystery of surgery' in London and become free masters were examined at the Hall by a small group of expert surgeons.[64] The Company court minutes suggest that there was both a theoretical and an applied element to this examination. Candidates were expected to demonstrate that 'he knoweth what is surgery and also what an anatomy is and how many parts it is, or what the four elements and the twelve signs [of the zodiac] be', and 'to practise and to put in use such knowledge as he hath'.[65] This examination was understood as being essential to set a benchmark for skilled practice, 'the worthy science of surgery'. It also ensured that unlicensed 'uncunning' persons did not blemish the collective reputation of the guild.[66] If successful the candidate with 'skill exp[er] ience and knowledge in the Arte or Science of Chirurgery' would pay a membership fee, join the ranks of the guild and be issued with a certificate to practice.[67]

In a 1570s publication concerning the cure of syphilis, guildsman William Clowes explicitly set out the corporate contribution to his skill as a surgeon, particularly activities focused upon Barber Surgeons' Hall. Clowes wrote how 'what so ever I have and am able to doe, in this our art of chirurgirie, I have receaved the greatest part thereof by the benefite of the lectures, anatomies, co[n]ferences, and practices, of the notable and most skillfull chirurgio[n]s that heretofore have bene, and are at this present the worthy members of this house'.[68] Indeed, for London's surgeons, education and training did not end upon completion of the apprenticeship and examination. Journeymen and master surgeons were also required to attend regular lectures on the theoretical and practical dimensions of their craft at Barber Surgeons' Hall, often on a weekly basis. This continuing institutional education, focused on the site at Monkwell Street, was totally unique among London's livery companies. Lectures were given by especially appointed Readers, rewarded with a fee of £10 per annum. These Readers were usually physicians, occasionally surgeons or, indeed, in the case of Alexander Reid (appointed lecturer in 1632), a physician also licensed as a surgeon. We might see this lecture programme, explicitly delivered in English, not Latin, 'as a viable alternative to the university'.[69] The emphasis on the theoretical dimensions of the surgeon's art was likely a claim for heightened status.[70] Attendees were tutored via 'oral instruction and visual demonstration, and generally focused on vernacular

[64] B/1/1, fo. 29v. The Barbers ordained this 'assay and examination' from 1376 (see *Annals of the Barber-Surgeons*, 35–6). Usually there were four examiners.

[65] B/1/1, fos. 29v, 30v.

[66] B/1/1, fo. 30v.

[67] *The Annals of the Barber-Surgeons*, 343; Celeste Chamberland, 'Honor, Brotherhood and the Corporate Ethos of London's Barber-Surgeons' Company', *Journal of the History of Medicine and Allied Sciences* 64, no. 3 (2009): 300–32 (p. 306).

[68] A short and profitable treatise touching the cure of the disease called (Morbus Gallicus) by unctions, set forth by William Clowes, of London, chirurgion (London, 1579), sig. A3r. Despite these complimentary words here, Clowes's relationship with the Barber Surgeons' Company was often quite fraught and conflictual; see Chamberland, 'Between the Hall and the Market', 69–89.

[69] Celeste Chamberland, 'From Apprentice to Master: Social Disciplining and Surgical Education in Early Modern London, 1570–1640', *History of Education Quarterly* 53, no. 1 (2013): 21–44 (pp. 33–4).

[70] Wallis, 'Between Apprenticeship and Skill', 162.

translations of ancient medical texts, anatomical commentary, or discussion of specific ailments'.[71] Attendance at Anatomy lectures was compulsory for all surgeons within the guild and the purpose was explicitly pedagogical (fines were issued to absentees).[72] These lectures were only open to freemen of the guild, and the wider public, including women and unlicensed practitioners, were excluded. The prestige of the surgeon Readers specifically was materially reinforced in the hall from the early 1630s, through having their names 'sett up in a table and hung up in the lecture house publiqlye [*sic*]'.[73] Against the backdrop of the competitive medical marketplace – between surgeons and physicians, surgeons and apothecaries, and most commonly, licensed surgeons and unlawful practitioners – the institutional education at Barber Surgeons' Hall also reinforced the boundaries between licensed and illicit surgical practice.[74]

In addition to the weekly lectures were the 'public' anatomies. These anatomical demonstrations took place quarterly – the guild had a legal right to acquire four bodies each year from Tyburn, London's chief execution site – and each lasted three days. The 'public' referred to the 'public body' of the executed felon.[75] The physician reader would give six lectures to the assembled audience, morning and afternoon. Each day was spent on a discrete system – visceral, muscular and osteological. Until the 1630s 'public' anatomies took place in the hall chamber, the largest and most accessible communal space in Barber Surgeons' Hall. There is evidence from court minutes that there were also 'private' anatomies taking place in other rooms of Barber Surgeons' Hall. Licensed surgeons might 'make and work the same [dead bodies] for their knowledge and more cunning'.[76] There was no limit on private demonstrations, provided the freeman could pay for the body, its transport to Monkwell Street and a decent burial. The skin of the cadaver was expressly not permitted to be 'carried or conveyed away' from the Hall.[77] Surgeons would invite their colleagues, friends and apprentices as a private audience to these dissections. Dissection of cadavers in private homes in London was expressly forbidden; such demonstrations should only take place 'within the common hall of the said mystery'; the Barber Surgeons policed this and prosecuted with substantial fines.[78]

A painting of surgeon and Oxford-educated physician John Banister (1532/3–1599?) giving the visceral lecture at Barber Surgeons' Hall depicts a dissection in action (Fig. 5.3/Plate 8).[79] We see Banister – who had been appointed Reader for public anatomies that year, at the age of forty-eight – with his left hand on the cadaver,

71 Chamberland, 'From Apprentice to Master', 35.
72 *The Annals of the Barber-Surgeons*, 313.
73 B/1/5, fo. 147.
74 Chamberland, 'From Apprentice to Master', 22.
75 Ibid., 35.
76 B/1/2, fo. 3r.
77 B/1/2, fo. 37r. The court minutes specifically mention the (illicit) practice of tanning the human skin 'like leather'. Perhaps this was linked to skin's 'magical and protective powers'. See Katherine Park, 'The Criminal and the Saintly Body: Autopsy and Dissection in Renaissance Italy', *Renaissance Quarterly* 47, no. 1 (1994): 1–33 (p. 26).
78 B/1/2, fo. 3r; *The Annals of the Barber-Surgeons*, 362.
79 John Banister studied medicine at Oxford and gained a licence to practice in 1573. In 1593 Banister received a licence from the College of Physicians of London to practise medicine and surgery. See A. Griffin, 'Banister [*formerly* Banester], John, (1532/3-1599?), Surgeon', *ODNB*.

Figure 5.3 John Banister delivering a visceral lecture at Barber Surgeons' Hall, *c.* 1580 (MS Hunter364). University of Glasgow Library.

See also colour version, reproduced in plate section 8.

directing attention with his right hand to a skeleton as a teaching aid. The skeleton 'is supported and crowned with the colours of the Barber Surgeons' Arms'.[80] Immediately to Banister's right-hand side is a lectern holding Realdo Colombo's *De re anatomica*, opened at a section on the visceral system.[81] Behind Banister is an attentive audience of young surgeons dressed in the appropriate attire of black gowns and caps and white linen over-sleeves. At the head of the corpse is the full-bearded mature figure of William Bovey, master of the guild. The two figures holding scalpel and probe and touching the dissected body or table are the masters of the anatomy. The two men at the front of the image, close to the cadaver and the surgical tools, are the stewards of the anatomy. Masters and stewards of the anatomy were accountable for the smooth running of public dissections at the Hall. These roles were appointed annually by the guild and those who served as steward would be masters the following year – having observed proceedings and thus 'better known how to work'.[82] Banister published a considerable number of medical treatises in English from the mid-1570s to the late 1580s, including a text on the curation of ulcers based on contemporary and ancient authority and his own experiential practice (1575) and *An Anidotarie Chyrurgicall*

[80] William Brockbank, 'Old Anatomical Theatres and What Took Place Therein', *Medical History* 12, no. 4 (1968): 317–84 (p. 376).

[81] Kate Cregan, *The Theatre of the Body: Staging Death and Embodying Life in Early Modern London* (Turnhout: Brepols, 2009), 19–20.

[82] B/1/2, fo. 3r.

Plate 1 Watercolour by Thomas and Paul Sandby showing a view of Beaufort Buildings, looking towards the Strand, 1760s. (BM: 1880,1113.2854). © The Trustees of the British Museum. (*See also*, p. 34)

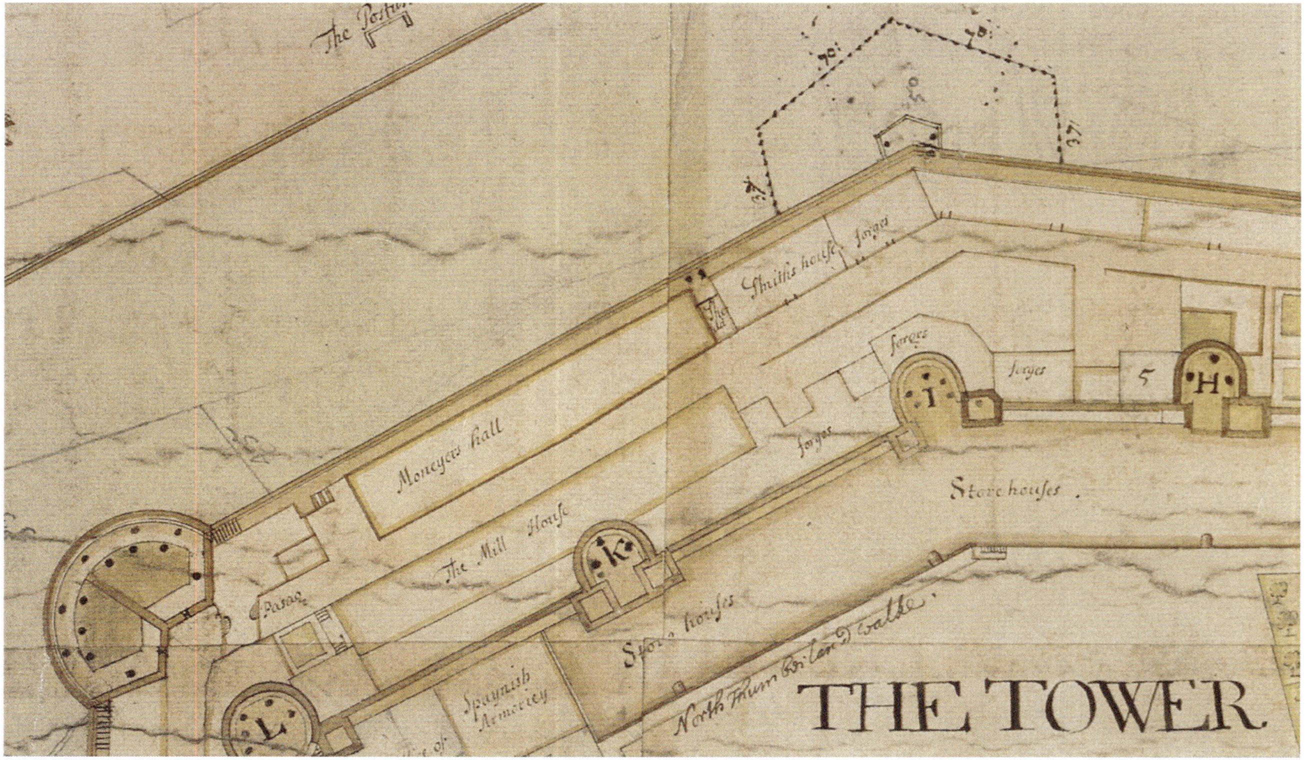

Plate 2 Detail from a 1682 plan of the Tower (RA: TOL MP/TOWER/73/6-8). Courtesy of The Board of Trustees of The Royal Armouries. (*See also*, p. 51)

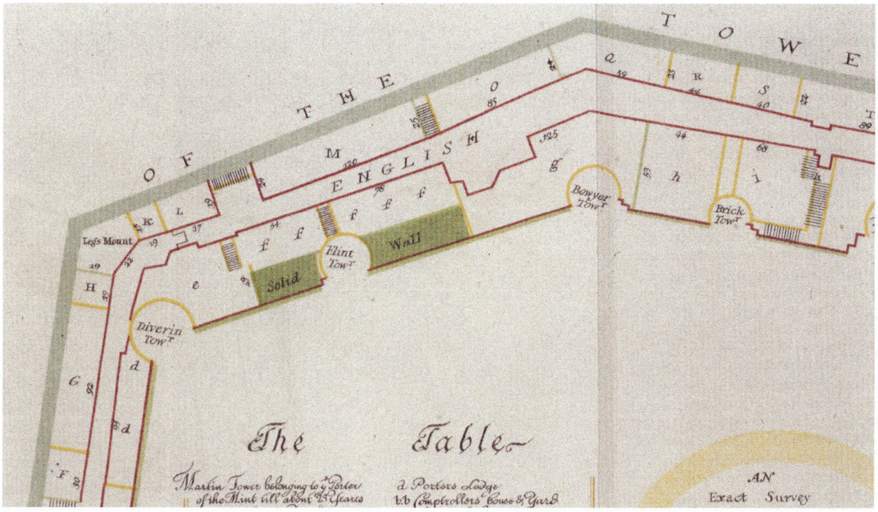

Plate 3 Detail from William Allingham's *Exact Survey of the Ground-Plot or Plan of His Majest[ie]s Office of Mint* (1701). © The Royal Mint Museum. (*See also*, p. 62)

Plate 4 Coat of arms of the Ordnance at the Tower of London. Photograph by Richard Nevell, Wikimedia Commons. (*See also*, p. 74)

Plate 5 Military and civilian workers at the Royal Laboratory fuses, attributed to Gamaliel Massiot, *c.* 1750 (NMM: PAI0745). © National Maritime Museum, Greenwich, London. (*See also*, p. 91)

Plate 6 Page showing drawing of the utensils for driving portfires, 'Process Manual' *c.* 1800. The National Archives, SUPP 5/24. (*See also*, p. 93)

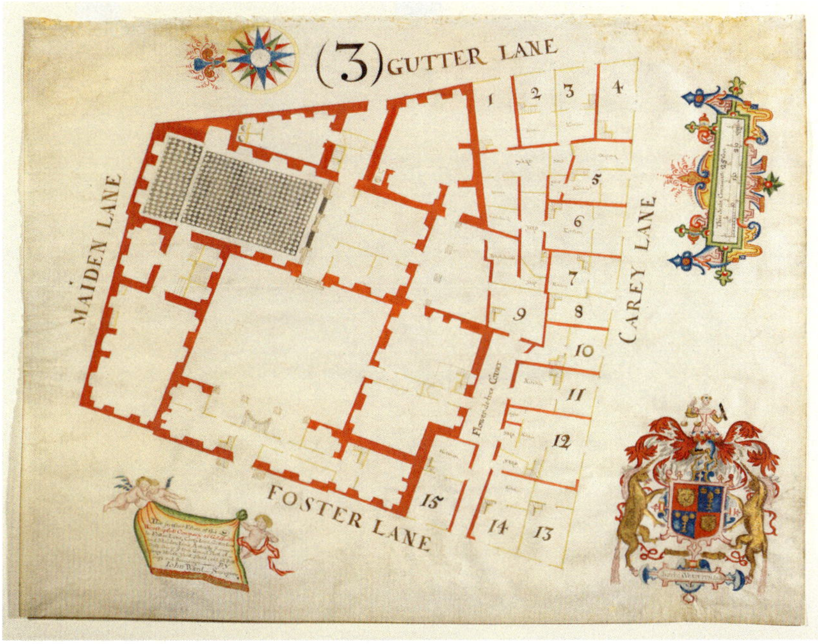

Plate 7 John Ward, Goldsmiths' Hall: ground-storey plan, 1692. Courtesy of the Goldsmiths' Company, London. (*See also*, p. 109)

Plate 8 John Banister delivering a visceral lecture at Barber Surgeons' Hall, *c.* 1580 (MS Hunter364) [University of Glasgow Library]. (*See also*, p. 118)

Plate 10 View of the gateway to Old Trinity House, watercolour, *c.* 1840 (LMA: q804075x). © London Metropolitan Archives (City of London). (*See also*, p. 132)

Plate 9 Cross-section of the Barber Surgeons' Anatomy Theatre (1636). Wellcome Collection: 25372i. (*See also*, p. 123)

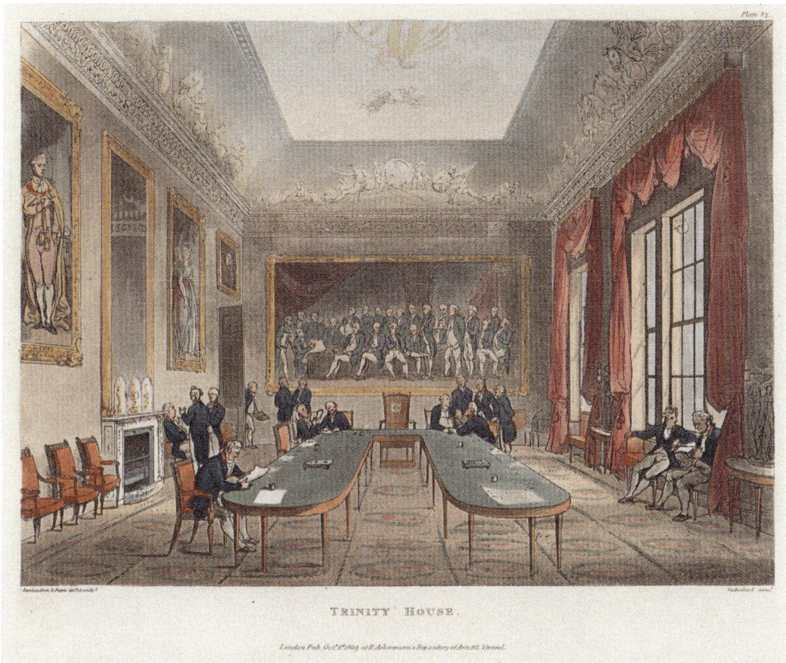

Plate 11 The Court Room of Trinity House, Augustus Pugin and Thomas Rowlandson for Rudolph Ackermann's *Repository of Arts* (London, 1809). The Elisha Whittelsey Collection, The Elisha Whittelsey Fund, 1959, Metropolitan Museum of Art. (*See also*, p. 160)

Plate 12 The East India Company's headquarters, *c.* 1800, by Thomas Malton the Younger. Yale Centre for British Art, Paul Mellon Collection. (*See also*, p. 167)

Plate 13 Portrait of Alexander Dalrymple, attributed to John Thomas Seaton 1765 (NMS: T.2008.168). © National Museums Scotland. (*See also*, p. 191)

(1589). Banister aimed 'to furnish and profit a Chirurgian with knowledge of the body, and with manifold and plaine practise'.[83] A compendium of 'the works of that famous chyrurgian Mr John Banester' was published posthumously in 1633.[84]

Barber Surgeons' Hall had an exceptionally rich visual culture and civic portraits of eminent barber surgeons decorated the hall chamber and parlour.[85] Dominating the internal hall was the imposing group portrait, painted by artists of the studio of Hans Holbein, of the Barber Surgeons receiving their charter from Henry VIII. Composed around 1541, the year the amalgamated Company was formed, the oil on panel painting shows eighteen members of the new guild, with inscriptions identifying the sitters (including William Butts, court physician to Henry VIII, and John Chambre, who served Henry VII and his successor).[86] A guild group portrait was rare in sixteenth-century London, and perhaps it 'impressed upon the Barber-Surgeons themselves the potential of company portraiture in their own hall'.[87] Unusually for London livery companies, the Barber Surgeons also had a 'table of pictures in the Hall' (named in court minutes from 1601) composed of the faces of dozens of living (and deceased) barber surgeons. By 1611 there is record of another set of portraits, 'the Table of Anatomy', located in the more prestigious and less accessible room of the parlour, including Masters of Anatomy and Examiners of Surgeons. Forty-one likenesses featured by the mid-1630s (none of which are known to have survived).[88] These tables might have been sheets of paper, 'specially prepared wooden boards' or 'framed paintings on panel'.[89] Robert Tittler has suggested that the descriptions of these pictures in the archives as 'hanging' and 'standing' imply tables 'which were either hung on the walls or "stood" as hinged panels resting, like bistro menu boards, on the floor'.[90] For the individual citizen, visual representation at the company hall was a means of reflecting and reinforcing civic and professional standing; for the livery company, portraiture on display 'served as constant reminders of the virtues of civic benefaction, civil comportment and sage governance'.[91] Depending on the location, pictures would have been visible to guild governors, the rank and file and select visitors. Professional skill and identity were also likely significant in these representations, especially considering the visual predominance of those involved in surgical demonstrations and teaching at Monkwell Street.[92]

83 *The Workes of That Famous Chyrurgian, Mr Iohn Banester by Him Digested into Five Bookes* (London, 1633), sig. A4v.

84 Ibid.

85 Robert Tittler, *The Face of the City: Civic Portraiture and Civic Identity in Early Modern England* (Manchester: Manchester University Press, 2013), 155; 'this particular company went to great lengths to display images of its masters and benefactors, living as well as deceased'.

86 Tarnya Cooper, *Citizen Portrait: Portrait Painting and the Urban Elite of Tudor and Jacobean England and Wales* (New Haven, CT and London: Yale University Press, 2012), 155–6. Unusually for physicians, William Butts and John Chambre also sat for individual portraits by Hans Holbein (both works dated *c.* 1543).

87 Robert Tittler, 'Portraiture, Precedence and Politics amongst the London Liveries *c.* 1540–1640', *Urban History* 35, no. 3 (2008): 349–62 (p. 354).

88 B/1/4, fo. 139; *The Annals of the Barber-Surgeons*, 509.

89 Tittler, *The Face of the City*, 155.

90 Robert Tittler, 'Portrait Collection and Display in the English Civic Body, *c.*1540–1640', *Journal of the History of Collections* 20, no. 2 (2008): 161–72 (p. 168).

91 Ibid., 166.

92 On the representation of artisans and tools of their craft within company halls: Cooper, *Citizen Portrait*, 196–7; Kilburn-Toppin, *Crafting Identities*, 133–7.

The deeply status-driven and competitive nature of this representation is brought to light by a court of assistants' order of 1601, concerning 'defacing of pictures'. It was commanded that no master or freeman 'shall blot, deface or put out or cause to be blotted defaced or put out any of the pictures of any of this Company ... lawfully placed in any of the tables of pictures hanging in the Common hall of this mystery' (with a very hefty £10 fine for the misdemeanour).[93] There was also regulation as to the costume featured in the Table of Anatomy which stood at the parlour door. It was ordered by the guild governors in August 1611 that 'pictures shall be put out' of the table unless guildsmen were 'pictured in a plain citizen's cap as others therein are placed and in a livery gown'.[94] Appropriate civic dress, including the wearing of gowns, was also to be a sticking point during public dissections.

'Public' dissections at Barber Surgeons' Hall followed a ceremonial script. The masters and stewards of the anatomy were responsible for preparing the body, serving the anatomist during these teaching exercises, ensuring his comfort, and the material decorum of the dissection process. They also organized the dinners which took place in the hall following each lecture day.[95] The stewards provided 'a mat about the hearth in the hall that Master Doctor need not to take cold upon his feet'. As to his apparel, the master and stewards of the anatomy supplied 'two aprons to be from the shoulder downward and two part of sleeves for his whole arms'. Spare aprons and sleeves, 'white and clean', were necessary should the anatomist's appearance become 'unseemly' during the dissection. They were also responsible for keeping the dissection instruments clean. The Reader would use two white rods 'to touch the body where it may please him' and a wax candle 'to look into the body'.[96] From the late 1560s, scaffolding and seats were erected in the hall chamber on dissection days (the 'old standing', used on city streets for civic ceremonial – viewing 'coronations or any noble peer his coming through the city' – was recycled for this purpose).[97] The scaffolding ensured all members of the audience 'have a good prospect' (there had been complaints about visibility at dissections). A curtain attached to the frame with pillars and rods of iron also managed the sight of the spectacle, for the process of dismemberment was not in itself intended to be visible; no member of the crowd 'shall behold the dissection or incising of the body, but that all may be made clean and covered with fair clothes until the Doctor shall come and take his place to read and declare upon the parts dissected'.[98] Perhaps this curtain or screen additionally served to create suspense and make the spectator experience theatrical.[99] The fixed seating arrangement also worked to reflect and reinforce company and civic hierarchies, for audience members were

93 B/1/2, fo. 76.

94 B/1/4, fo. 139.

95 *The Annals of the Barber-Surgeons*, 362.

96 B/1/1, fo. 25v.

97 B/1/2, fo. 24v.

98 B/1/2, fo. 24v. See also Brockbank, 'Old Anatomical Theatres', 372. Until the 1590s anatomy theatres across Europe were temporary, often with seats organized according to rank.

99 There is a substantial literature on early modern anatomy theatres as theatrical spaces. Most useful here have been: Brockbank, 'Old Anatomical Theatres', 371, 'Its purpose was to offer a performance'; Kate Cregan, 'Teaching the Anatomical Body in Seventeenth-Century London', *Medicine Studies* 2 (2010): 21–36 (pp. 22–3).

intended to be seated here 'in their degrees and ranks'. This was the civic ideal, but the lived reality was undoubtably more complicated. The court of assistants expressed deep concern in March 1634/5 at 'remissness' in attendance and 'the disorderliness of those surgeons that do appear for wanting their outward ornament [and] commixing themselves confusedly among the common people'. This behaviour was said to have 'eclipsed' 'the honour and worthiness of this company'. Orders were re-issued for the wearing of gowns and seating on the scaffolding (with fines for disobedience).[100]

In October 1631 the Barber Surgeons' court of assistants decided that a 'private dissection roome' was to be built; 'a farie convenient roome built over the great staire case next the backyard'. Their discussions reveal that, prior to this, dismemberment for private anatomies had often taken place in the kitchens, 'a great annoyance to the tables dresser boards and utensils ... by reason of the blood filth and entrailes of those Anatomyes'.[101] By February 1635/6 the court was also speaking of 'the want of a publique theatre for Anatomycall exercise'.[102] The construction of the anatomy theatre at Barber Surgeons' Hall from the mid- to late-1630s was explicitly intended to establish the guild as *the* preeminent metropolitan experts in anatomical knowledge and instruction. Notably, this was London's first purpose-built anatomical theatre. This is significant, since the London physicians considered themselves to have 'a better rational knowledge and greater social status', which frequently fuelled their (failed) 'attempts to regulate the other two branches of medicine: surgery and pharmacy'.[103] The College of Physicians, 'a cluster of self-conscious humanist intellectuals', constructed their octagonal anatomy theatre, designed by Robert Hooke, on Warwick Lane four decades later, between 1671 and 1679.[104] The built fabric of these institutions reflects jurisdictional rivalries and is a 'story of architectural one-upmanship that took place in the London medical world over the seventeenth century'.[105] Moreover, the design and building of the Barber Surgeons' anatomical theatre, overseen by Inigo Jones, Surveyor of the King's Works, and inspired by anatomy theatres at the Universities of Padua (1594) and Leiden (1597), put the building in dialogue with great European centres of anatomical knowledge and learning.[106]

From the mid-1630s, with Jones as architect, Barber Surgeons' Hall was architecturally enhanced with a new 'great parlour' and an elliptical anatomical theatre. A tablet of stone at the entrance to the new parlour chamber was inscribed

[100] B/1/5, fo. 185. For a broader European context to seating spaces 'assigned in accordance with the dignity of the persons attending' see Rafael Mandressi, 'Affected Doctors', *Osiris* 31 (2016): 119–36 (p. 131). See also Giovanna Ferrari, 'Public Anatomy Lessons and the Carnival: The Anatomy Theatre of Bologna', *Past and Present* 117 (1987): 50–106 (at p. 61).

[101] B/1/5, fo. 129.

[102] B/1/5, fo. 204.

[103] Wear, *Knowledge and Practice*, 217.

[104] Pelling, *Medical Conflicts*, 11. It is probable that prior to the substantial 1670s build the College of Physicians had a space for dissections 'on a small, domestic scale ... It is most likely to have been a smaller space with theater-like seating, built within an existing structure'. Matthew Walker, 'Architecture, Anatomy and the New Science in Early Modern London: Robert Hooke's College of Physicians', *Journal of the Society of Architectural Historians* 72, no. 4 (2013): 475–502 (p. 481).

[105] Ibid., 477.

[106] Brockbank, 'Old Anatomical Theatres', 376.

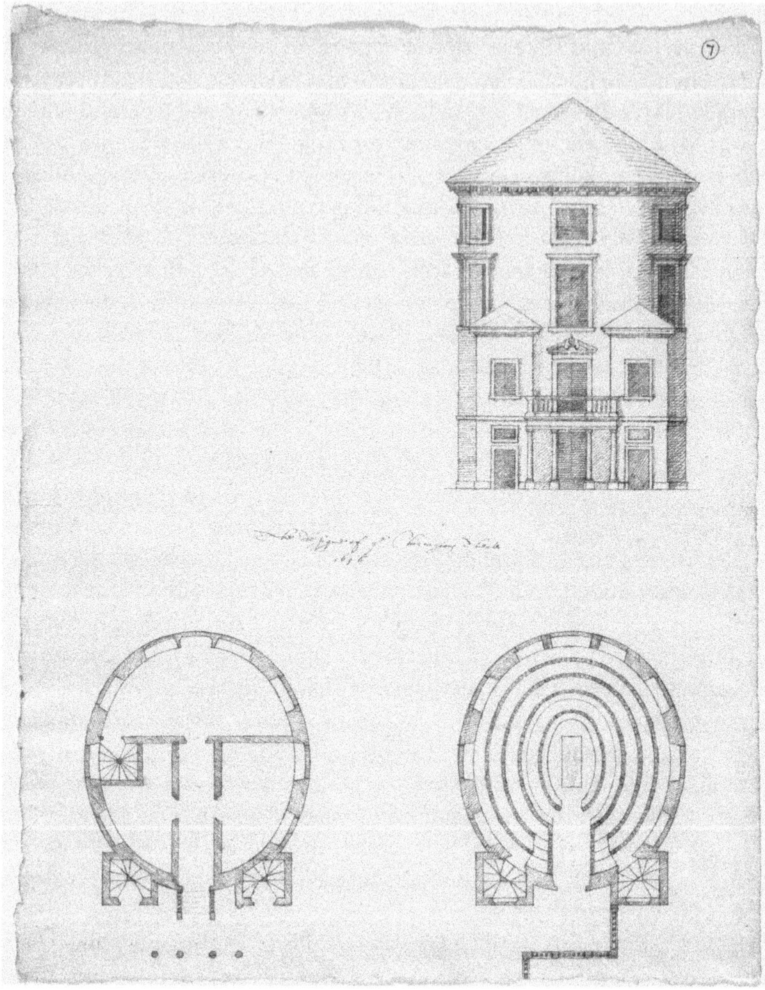

Figure 5.4 Plans and elevation of Inigo Jones's design for the Barber Surgeons' Anatomy Theatre (1636). Worcester College, Oxford.

with the date of construction and the names of the master and wardens.[107] A gallery linked the parlour directly to the theatre.[108] The 1636 interior plans of the anatomy theatre show how the body was placed centre stage, surrounded by four tiers of cedar wood raked seating (Fig. 5.4).[109] An open section opposite the main entrance allowed for a ceremonial arrival by Company governors. Two square turrets external to the theatre, either side of the entrance, housed the staircases.[110] The interior design of the

[107] Kilburn-Toppin, *Crafting Identities*, 125.
[108] B/1/5, fo. 217.
[109] Dobson and Walker, *Barbers and Barber-Surgeons of London*, 81.
[110] Brockbank, 'Old Anatomical Theatres', 376.

theatre was highly ornate and, like its continental counterparts, was decorated with flayed bodies and the skeletons of notorious criminals. The concave ceiling of the theatre was painted with 'the constellacons [*sic*] of the heavens, and the 7 planetts over the 12 signes [of the zodiac], and skeltons to be wrought and set up on every one of the signs of corbells'.[111] The surgeons Edward Arris and Henry Boone organized for the body of 'Canbury besse to be placed on the corbel stone of the Signe Libra', a 'skeleton by them wrought on when they were Masters Anatomysts'. Their names or mottos were additionally set up under Libra.[112] Elizabeth Evans, known as Canonbury Besse, was executed at Tyburn in April 1635 as an accessory to murder. It was said she lured men from London's public houses, playhouses and streets to her chamber, where violent ambush and robbery awaited – conducted by her partner in crime,

Figure 5.5 Cross-section of the Barber Surgeons' Anatomy Theatre (1636). Wellcome Collection: 25372i.

See also colour version, reproduced in plate section 9.

[111] B/1/5, fo. 225.
[112] B/1/5, fo. 237.

Thomas Sherwood (Country Tom).[113] In a sensational contemporary account of their lives, crimes and deaths, Henry Goodcole Ordinary of Newgate recorded how 'the Coy-duck, or divellish allurer to sinne and confusion, was dissected, and her dryed Carkase or Sceleton of Bones and Gristles is reserv'd, in proportion to be seene in Barber Surgeons Hall'.[114] In an act of macabre post-mortem unity, the Company also arranged for the skeleton of Thomas Sherwood to be set up in the theatre.[115] A Hall inventory of 1728 additionally records a 'new skeleton pendent from the Centre of the Roof', 'two muscular figures finely painted in gilt frames' and 'a skeleton in a frame the gift of M[aste]r Knowles'[116] (Fig. 5.5/Plate 9 gives an indication of some of the skins and skeletons on show). As Jonathan Sawday has argued, 'the "culture of dissection" was devoted to the gathering of information and the dissemination of knowledge of the "mystery" of the human body … But the "culture of dissection" also promoted the beginnings of … the "surveillance" of the body within regimes of judgement and punishment.'[117] The display of infamous corpses looks to be a salacious means of extending punishment post-mortem, and post-dissection. Indeed, Barber Surgeons' Hall featured prominently in the contemporary morbid imagination. In Thomas Nashe's *The Unfortunate Traveller* (1594) the fictional narrator imagines how 'if any knocked at the door, I supposed it was the beadle of Surgeon's Hall come for me'. Crispin, in Edward Ravenscroft's play *The Anatomist* (1697), declares: 'I had rather be a Sot than an Anatomy, I will not have the flesh scrap'd from my Bones. I will not have my Skeleton in Barber-Surgeons-Hall.'[118]

The inaugural dissections in April 1638 were grand events attended by the Lords of the Privy Council 'and other lords and persons of state at the dedication of the theatre'.[119] The dinner hosted 'in our great parlour at the publique anatomy' was a lavish occasion and the Company even hired 'Venice glasses' and tapestries from the King's Wardrobe.[120] Who were the 'public' that attended dissections at the new anatomy theatre? The European experience suggests a substantial and socially mixed crowd at public anatomies, including scholars, artists, merchants and 'ordinary townspeople', and thus not simply 'students of medicine and members of the medical profession'.[121] Motivations of the anatomy crowd perhaps ranged from education and fascination about divine creation, to entertainment and voyeurism.[122] We lack definitive evidence for attendees at the anatomy theatre at Barber Surgeons' Hall. Surgeons (and surgeons in training) doubtless predominated, considering that lectures and dissections at Monkwell Street were explicitly intended to be teaching tools for guildsmen. The court

113 Henry Goodcole, *Heavens Speedie Hue and Cry Sent after Lust and Murder* (London, 1635), B1r.
114 Goodcole, *Heavens Speedie Hue and Cry*, C4r.
115 Cregan, *The Theatre of the Body*, 103–4; Jonathan Sawday, *The Body Emblazoned: Dissection and the Human Body in Renaissance Culture* (London: Routledge, 1995), 76.
116 *The Annals of the Barber-Surgeons*, 488.
117 Sawday, *The Body Emblazoned*, 4.
118 Quoted in ibid., 47–9.
119 B/1/5, fo. 238.
120 *The Annals of the Barber-Surgeons*, 453–4.
121 Ferrari, 'Public Anatomy Lessons', 64; Sawday, *The Body Emblazoned*, 42.
122 Ferrari, 'Public Anatomy Lessons', 55, 94–5, 98–9, 102.

of assistant's mid-1630s concern about surgeons 'commixing themselves confusedly among the common people' at public anatomies does suggest a broader crowd (though 'common' might refer to the rank and file of the guild).[123] Since the theatre, though technically a separate building, was located within the regulated complex of a livery hall, it is reasonable to speculate that attendance was monitored and that those of social status or with professional connections to the guild were preferred.

It was a grisly reality that the body of the condemned might itself disrupt the seemliness of proceedings at a public or private dissection. There was a strong enough possibility of resuscitation of bodies brought from Tyburn after hanging that in 1587 the court of assistants established measures for financial accountability for private anatomies: 'It is agreed that if any body … brought to our hall for the intent to be wrought upon by the anatomisers of our company shall revive or come to life again, as of late hath been seen, the charges about the same body so reviving shall be born, levied and sustained by such person or persons who shall so happen to bring home the body.'[124] In November 1740 this fate awaited William Duell, a sixteen-year-old who had been hanged at Tyburn following a conviction for rape.[125] The Company Clerk Joseph Wheller was sufficiently stirred to write a full account of the revival on the last page of the rough Minute Book 1738–42. He noted how Duell 'having hung some time was cutt down and brought to this Company's Hall in order to be dissected where he had not been five minutes before Life appeared in him and being let blood and other means used for his recovery in less than two hours he sat upright drank some warm wine and look'd often round him'. Duell was returned to Newgate and later transported.[126] The Ordinary of Newgate also reported the 'accident' in considerable detail in his popular printed account of the condemned at Tyburn. In support of the Barber Surgeons' Company, presenting them as saviours of the tale, the Ordinary suggested that 'If he [Duell] had fallen into [t]he Hands of the Mob [at Tyburn], instead of the Surgos, in all Probability he could never have came to Life'. His macabre account concludes with a brief description of two infamous stories of revivals on the anatomist's table, including Anne Greene at Oxford in 1650, convicted and hanged for infanticide, whose story was recorded by 'near 40 of the Prime Wits at the University' and printed 'bearing the Title of News from the Dead, the Revival of Ann Green'.[127] The Ordinary's positive spin on a dissection (that Duell 'came to Life' in the hands of the Surgeons) must be read against the backdrop of popular 'brawls, disorders and tumults' which resisted the removal of corpses by surgeons or their agents. It was one of the Company Beadle's responsibilities to collect the bodies of executed felons from Tyburn, and there could be a concerted

[123] B/1/5, fo. 185.

[124] B/1/2, fo. 42r.

[125] *Old Bailey Proceedings Online* (www.oldbaileyonline.org, version 8.0, 28 February 2023), October 1740, trial of WILLIAM DUELL (t17401015-53).

[126] *The Annals of the Barber-Surgeons*, 359.

[127] *Old Bailey Proceedings Online* (www.oldbaileyonline.org, version 8.0, 28 February 2023), *Ordinary of Newgate's Account*, November 1740 (OA17401124). On the case of Anne Greene see also Laura Gowing, *Common Bodies: Women, Touch and Power in Seventeenth-Century England* (New Haven: Yale University Press, 2003), 48–9.

challenge from the crowd.[128] In the eighteenth century, special constables even 'had to be paid to protect the surgeons and their Hall during the four annual lectures'. Following the execution of two participants in the coal-heavers' strike of 1768, and the removal of their bodies to Surgeon's Hall for dissection, 'a crowd of Irish women formed outside the Hall praying their countrymen "might live again"'.[129]

In February 1662/3 Samuel Pepys accepted an invitation to watch a lecture and dissection, and dine at the Barber Surgeons' anatomy theatre. Pepys commented upon the ceremonial aspect of proceedings: the audience were first 'led into the Theatre; and by and by came the Reader, Dr Tearne, with the Maister [Thomas Lisle] and Company, in a very handsome manner'. After a lecture on the 'Kidnys, Ureters, and yard [penis]' Pepys walked into the hall chamber for 'a fine dinner' with 'good learned company, many Doctors of Physique'. He enjoyed his view of 'a very excellent piece of the King [Henry VIII] done by Holben … with the officers of the Company kneeling to him to receive their charter'. After dinner Pepys accompanied Dr Scarborough and some of his friends to see the body; 'he was a lusty fellow, a seaman that was hanged for a robbery. I did touch the dead body with my bare hand; it felt cold, but methought it was an unpleasant sight'. The naval connection perhaps explains Pepys's invitation to this anatomy lecture. Pepys's outing to the anatomy theatre also reveals something of the organization of the space, for at the end of the visit 'we went into a private room, where I perceive they prepare the bodies, and there was the Kidnys, Ureters, yard, stones and semenary vessels upon which he read today'.[130] This separate chamber, out of sight of the audience, tallies with the inventory of 1728 which records a 'Lower Theatre' containing 'a table to dissect on & a hatch for ye Theatre door'.[131]

The lectures and anatomies performed at Barber Surgeons' Hall could have a more substantial audience than the assembled crowd. Lectures might reach a wider public in the form of print, and some Company Readers contributed to the burgeoning market in medical books.[132] Manuals on surgery ranged from broad guides to the art, such as Peter Lowe's *A Discourse of the Whole Art of Surgery* (1654), to specialist works, like those focusing on military field medicine and techniques for treating gunshot wounds.[133] Through print, anatomical and surgical knowledge moved between the Hall

[128] *The Annals of the Barber-Surgeons*, 299. The disruption of proceedings at Tyburn 'by brawls, disorders and tumults' was especially acute in the eighteenth century. 'The surgeons and physicians, their beadles and porters, were the most common targets in these disorders.' The biggest demand for corpses came from private teaching schools and hospitals of London though, not the established corporations. Also, the teaching of anatomy shifted; by the 1740s hands-on 'experience in the dissection of corpses had become conventional practice in surgical training'. See Peter Linebaugh, 'The Tyburn Riot against the Surgeons', in Douglas Hay et al., *Albion's Fatal Tree: Crime and Society in Eighteenth Century England* (New York: Pantheon, 1975), 69–70.

[129] Ibid., 75, 102.

[130] *The Diary of Samuel Pepys*, vol. 4, 1663 (1971), 59–60.

[131] *The Annals of the Barber-Surgeons*, 491.

[132] On the market for vernacular medical print, see Mary E. Fissell, 'The Marketplace of Print', in *Medicine and the Market in England and Its Colonies*, ed. Mark Jenner and Patrick Wallis (New York: Palgrave Macmillan, 2007), 108–32.

[133] Elaine Leong, 'Learning Medicine by the Book: Reading and Writing Surgical Manuals in Early Modern London', *BJHS Themes* 5 (2020): 93–110 (p. 94).

and broader metropolitan sites of training and practice, such as surgical households, workshops and hospitals. Print 'played a key role in surgical instruction, whether in structured father-son and shop-based training practices or in self-directed learning'.[134] One such Reader was Scottish surgeon and physician Alexander Reid (*c.* 1570–1641) who published his early 1630s Barber Surgeons' Hall lectures.[135] The results were *The Chirurgicall Lectures of Tumors and Ulcers* (1635) and *A Treatise of the First Part of Chirurgerie* (1638). From *c.* 1620 Reid was a 'foreign brother' of the Barber Surgeons' Company, and from 1631 a 'free brother'. In December 1632 he was appointed to give the weekly Tuesday anatomy lecture at Monkwell Street; he was given £20 *per annum* and he kept the post for several years. By the time of his appointment, Reid was a medical authority who had studied and practised in France and the Holy Roman Empire, as well as Britain.[136]

In his printed lectures Reid stressed the practical elements of surgery and his own experiential understanding, alongside theoretical knowledge, both ancient and contemporary. He wrote of how 'I have laboured to instruct thy minde, and to direct thy hand'.[137] Surgery was presented as a combination of both intellectual learning and manual skill. As Elaine Leong writes, these surgical texts 'were often penned by practitioners-turned-authors. As such, many drew on a mix of lived experience gained at the bedside and bookish knowledge gathered in the library'.[138] Reid was explicit about his long-standing experience: 'I have not concealed from thee my owne observations [...] and I have practised physicke and Chirurgerie now 42 years'.[139] The writings of ancient authorities, including those by Hippocrates and Galen, were used to bolster Reid's claims for the 'antiquitie and dignitie of Chirurgerie' – a practice which Reid claimed was 'no whit inferiour to Physick it selfe'. He also referenced 'the theorems of the Chimists', most prominently Paracelsus, of whom Reid was critical.[140] On writing for example of the material causes of tumours, Reid stressed how the safest course was to follow 'the foot-steps of the ancient, for their way is plain and easie'. By contrast, the 'Paracelsian doctrine is full of difficulties, which hee maketh more obscure by coyning strange words'.[141] Plain talking was Reid's *modus operandi* and across his corpus of printed writings in the vernacular, which he envisaged as training manuals, he adopted an accessible, practical tone.[142]

[134] Ibid., 96.

[135] Also known as Alexander Read.

[136] Shona MacLean Vance, 'Reid, Alexander (*c.* 1570–1641), Anatomist and Surgeon,' *ODNB*.

[137] Read, *The Chirurgicall Lectures*, A5r.

[138] Leong, 'Learning Medicine by the Book', 95. See also Domenico Meli and Cynthia Klestinec, 'Renaissance Surgery Between Learning and Craft', *Journal of the History of Medicine and Allied Sciences* 72, no. 1 (2016): 1–5.

[139] Alexander Read, *The Chirurgicall Lectures of Tumors and Ulcers* (London, 1635), A5r-v.

[140] Read, *The Chirurgicall Lectures*, pp. 3, 7.

[141] Read, *The Chirurgicall Lectures*, p. 25.

[142] Emily Cock, '"Lead[ing] 'em by the Nose into Publick Shame and Derision": Gaspare Tagiacozzi, Alexander Read and the Lost History of Plastic Surgery, 1600–1800', *Social History of Medicine* 28, no. 1 (2014): 1–21 (pp. 10–11).

This blend of practical knowledge, gained on the job, and propositional knowledge, acquired through reading, was becoming a recognizable feature of contemporary surgical manuals. By creating a 'learned surgery', practitioners like Reid 'imported the theory of physic into surgery ... and emphasised the practical side of surgery'.[143] The objective was 'a surgeon who would be manually skilled but could think like a physician as well as a surgeon'.[144] As with *The Gouldesmythes' Storehowse*, which showcased the learning and skilled nature of the assayer's craft at Goldsmiths' Hall, Reid's printed lectures underscored the scholarly learning and associated status of the surgeon's art, as demonstrated at Barber Surgeons' Hall.[145] Surgeons were keen to establish how their craft was a practice of the 'mind' and the 'hand'. Manuscripts and printed books composed and translated by author-practitioners expanded the audiences for the productive, skilled, knowledgeable activities that took place in livery halls, and whose in-person audiences were restricted to guildsmen, or even select company governors and officials.

Conclusion

London's guild buildings are not usually associated with productive or industrious activities, and yet a focus on the Assay Office at Goldsmiths' Hall and the Anatomy Theatre at Barber Surgeons' Hall demonstrates how livery company halls were prominent sites of knowledgeable and skilled technical practice, well into the eighteenth century. We are mistaken if we diminish the role of livery companies and their institutional headquarters to merely that of ostentatious social hubs with 'derivative processions, ceremonies, fur-and-velvet pomp and circumstance'.[146] The customary view of livery halls essentially as social sites by the seventeenth century misses the significance of these spaces to the development of metropolitan science.

The activities of testing, training and demonstration were essential to corporate identity and external reputation, in that they maintained standards of precious metals and public confidence, and educational benchmarks for surgeons. At both institutional sites knowledgeable practices, whether trials composed of juries of experienced goldsmiths, or guild-sponsored 'reading[s] on the skeleton', were collaborative corporate endeavours.[147] Learning, examination, regulation, knowledge-making were explicitly collective. Exclusion was also built into this culture. In the case of the Barber Surgeons attendance at lectures and anatomies worked to exclude those who were unlicensed from formal training and continuing instruction in medical

143 Wear, *Knowledge and Practice*, 233.

144 Ibid., 220.

145 Leong, 'Learning Medicine by the Book', 101. Similarly, renowned naval surgeon and guildsman William Clowes used vernacular print as a way of 'enhancing the social and intellectual prestige of his occupation'. See Chamberland, 'Between the Hall and the Market', 71.

146 Ian Anders Gadd and Patrick Wallis, 'Introduction', in Gadd and Wallis, *Guilds, Society & Economy*, 1–14 (p. 5).

147 B/1/5, fo. 185.

theory and practice. Livery company halls were regulated institutional spaces and access to workshops of metallurgical testing, or 'private' dissections, might be closely controlled. The Anatomy Theatre at Monkwell Street, specifically designed for public demonstrations, was a highly visible metropolitan venue, as compared to the closely controlled Assay Office at Goldsmiths' Hall where trials were intended to be secret. Nevertheless, the skilled exercises which took place at company halls impacted upon the livelihoods and reputations of many Londoners. Moreover, in public consciousness Goldsmiths' Hall was the most important metallurgical testing site within the walls, even the nation, and Barber Surgeons' Hall a notorious location for human dissection.

6

Experience and expertise at the Trinity House in Water Lane

Introduction

The Masters, Wardens and Assistants of the Guild Fraternity or Brotherhood of the most glorious and undivided Trinity and of St Clement in the Parish of Deptford Strond is, and was, better known simply as Trinity House. It is, and was, a charitable guild for mariners but over its long history has been many other things besides. It has taken on roles more often associated with government activity, overseen a profitable business and been accepted as a place of expertise on maritime matters, especially in and around the river Thames. While long responsible for sea marks, such as lights and buoys, it was only in the nineteenth century that it became the lighthouse authority it is known as today. One of a number of Trinity Houses associated with port cities on Britain's east coast, its location close to and then within the nation's capital gave it unique opportunities and responsibilities.[1] It gave the corporation access to docks, government, royal and law courts and the headquarters of long-distance trading companies, allowed it to benefit from proximity to skilled individuals and their output, and gave its membership access to opportunities for employment and charity. With about three-quarters of England's overseas trade passing through London by the end of the seventeenth century, it was also able to benefit from duties charged to the large numbers of ships obliged to use its services.[2]

Trinity House was founded and operated within Deptford Strond in the sixteenth century but, from the seventeenth, this became its ceremonial home and business was done elsewhere. Annually, on Trinity Monday, eight weeks after Easter Monday, the Corporation's governing Elder Brethren and as many of the ordinary members (Younger Brethren) as could attend would gather at their hall in Deptford to elect or re-elect the Master, Deputy Master and Wardens. This was next to their first almshouses

1 There are Trinity Houses in Hull, Newcastle and Leith, each with medieval or early modern origins. The Court of Lodesmanage in Dover was sometimes also known as Trinity House. The recent anniversary of the Corporation's first charter led to two institutional histories: Andrew Adams and Richard Woodman, *Light upon the Waters: The History of Trinity House 1514–2014* (London: The Corporation of Trinity House, 2013) and Neil Jones and Paul Ridgway, *Light through a Lens: An Illustrated Celebration of 500 Years of Trinity House* (London: Bloomsbury, 2014).

2 Sacks and Lynch, 'Ports 1540–1700', 390.

and close to the parish church, St Nicholas, where a sermon would be read after the meeting. The attending Younger Brethren were given half a crown and, traditionally, a dinner. From the later seventeenth century, however, they had to make do with wine and biscuits, after which the Elder Brethren would leave their maritime community behind and return in barges to London. There they would dine, at their own or the Master's expense, at a location such as the Pontack's Head, the fashionable eating house run by François-Auguste Pontac in Abchurch Lane.

From Abchurch Lane, in Candlewick Ward, it was a ten-minute walk, along Cannon Street and Eastcheap, in Billingsgate Ward, to Tower Street Ward. Running south from Tower Street was Water Lane, a narrow street with a right-angled bend nevertheless described by John Strype as 'a very great Thorough-fare'. This was because it was 'the ready Passage' to Custom House, described by Will Ashworth as 'the most international and cosmopolitan zone in England', full of British and overseas merchants, shipowners and ships' masters as well as the Board of Customs' officers, clerks, examiners of goods and porters.[3] Water Lane was, according to Strype, 'for the generality taken up by Publick Houses', including the Ship Tavern, and warehouses. At the bend was a wooden gate, behind which was a courtyard and the 'good handsome large Building' that was

Figure 6.1 View of the gateway to Old Trinity House, on the west side of Water Lane, watercolour, *c.* 1840 (LMA: q804075x). © London Metropolitan Archives (City of London).
See also colour version, reproduced in plate section 10.

[3] Strype, *Survey*, I.iii.53; William J. Ashworth, *Customs and Excise: Trade, Production, and Consumption in England, 1640–1845* (Oxford: Oxford University Press, 2003), 133.

Figure 6.2 View of the courtyard of Old Trinity House, as rebuilt after fire in 1718, watercolour by Thomas Hosmer Shepherd, 1857 (BM: 1880,1113.3437). © The Trustees of the British Museum.

the home of Trinity House from 1661 (Figs. 6.1/Plate 10. and 6.2). The lease records a site of more than 6,000 square feet that was abutted by tenements and warehouses owned by, among others, 'Widow Smithers Coffee woman', a net-maker, a broker and a wharfinger.[4] It was, as Strype noted, more 'conveniently seated … for the Dispatch of their frequent Business within the City' than their original home in Deptford Strond.[5]

Strype's accounts of Trinity House in his *Survey* are thorough and largely accurate, if perhaps overly generous in their assessment of its role. His description of its governance, privileges and responsibilities was based on their 1685 Royal Charter, while that of its House and location was, presumably, based on first-hand knowledge. The index to a now-lost series of reports and documents kept by Trinity House shows that its secretary had made and kept a copy of Strype's account and, around the same time, the Brethren had produced their own similar description for inclusion in William Maitland's *History of London*.[6] In accordance with tradition, the origin of Trinity House was given as

4 Copy of 1720 lease between William Russell, elder and younger, and Trinity House, 25 November 1720, CLC/526/MS30276 (Papers of the Corporation of Trinity House, LMA).

5 Strype, *Survey*, II.v.286.

6 Index to Select Entries, 1725–9, p. 91, CLC/526/MS30046/003. Maitland, *History of London*, 633.

Henry VIII's 1514 Charter, and its first Master was said to be Thomas Spert, master of the *Mary Rose* and the *Henri Grace a Dieu* and Clerk Comptoller of the King's Navy. Strype noted that Trinity House had set up Spert's memorial at St Dunstan's Stepney in 1622.[7] It was just four years earlier that the Corporation had first moved from its Deptford home, to Stepney, up river and onto the north bank of the Thames.

The mariners who made up the Younger Brethren – pilots and 'every Master, or Mate, expert in Navigation' – largely continued to live around Deptford, Wapping, Limehouse, Shadwell, Ratcliffe and Stepney.[8] So, too, did many of the Elder Brothers, but their business interests increasingly took them into London, and their increasing wealth allowed some to set up homes in the City, or even estates beyond it. Many of the Elder Brothers were self-made men, who had been mariners, were successful as masters and commanders, perhaps branching out as shipowners, merchants and investors, and gaining position as members of parliament and on business and charitable boards. The majority were associated with the merchant marine, especially the East India Company, but Trinity House was a meeting point for the royal and various merchant navies. Among the thirty-one Elder Brethren, there were always some with senior positions in the Navy Office or Admiralty. In the eighteenth century there were, from time-to-time, others appointed who were neither seafarers nor officials but who were brought in as Younger and then Elder Brothers in order to exercise influence in court or parliament. While, for most of the seventeenth century, the Master was chosen annually from among the Elder Brethren, in the eighteenth the terms were extended and Masters were Admirals, aristocrats or otherwise influential individuals, and had less day-to-day dealings with the Corporation. The Deputy Masters remained mariners and became increasingly important to the running of affairs, being in post over a period of several years. While there were twenty Deputy Masters 1660–1700, there were just sixteen for the whole of the following century.[9]

As Strype reported, the meeting days of the Corporation's Court or its Board were at Water Lane, commonly on Wednesdays and Saturdays, or 'as Occasion and Business requires'. He then set out that business:

> Their Service and Use is, that they appoint all Pilots: They set and place the Buoys and Sea-Marks for the safe Direction of Ships in their Sailing. For which they have certain Duties payable by Merchant-Men. They can Licence poor Seamen, ancient and past going to Sea, to exercise the Calling of a Waterman upon the Thames, and take in Fares, tho' they have not been bound to any one Free of the Watermans Company. They do maintain in Pensions at this time two thousand poor Seamen, or their Widows; every one of which have at least Half a Crown paid them every first Monday in the Month, and some more, besides accidental distressed Seamen.[10]

In addition they had, at this date, three almshouses (two in Deptford and one in Mile End) and large revenues derived from gifts of land and money, from the duties paid

7 Strype, *Survey*, I.ii.41.

8 Quotation from Maitland, *History of London*, 633.

9 There are lists of Masters and Deputy Masters in Adams and Woodman, *Light upon the Waters*, 297–8.

10 Strype, *Survey*, I,ii,41.

for their lighthouses (then at Scilly and Lowestoft and soon to be at Casquets), and, particularly, from the profitable business of selling ballast to all shipping going down the Thames.[11] The duty of dredging the Thames, to ensure it remained clear and of sufficient depth for navigation, and the right to a monopoly in selling it as ballast had been set out in an Act of Parliament in 1594 and reconfirmed in 1732. The Ballast Office, also located at the House in Water Lane, was supervised by the Elder Brethren in turn, had its own clerks and hired a considerable number of ballastmen to work some sixty ballast lighters.

Thus, proximity to the Thames and the Pool of London was essential as it was to the neighbouring Custom House, where Trinity House placed a clerk to collect lighthouse dues. At various times Trinity House also had the responsibility of surveying ships, particularly foreign ones, to record their burden and the duties they owed, to government as well as to Trinity House, and to regulate acquisition of ordnance from the nearby Office of Ordnance and Gun Wharf at the Tower. Water Lane was a short walk to other locations of significant interest to the merchant Elder Brethren: East India House and Africa House near Leadenhall Street, and Muscovy House off Seething Lane. Seething Lane was also the site of the Navy Office, with which Trinity House was in constant communication, especially regarding the supply of pilots for the king's ships. The convenience of their location (see Fig. 6.3) is made plain by Samuel Pepys, who lived and worked at the Navy Office and could frequently dine at Water Lane. As he recorded on 4 September 1662, 'at noon we all [from the Navy Office] to the Trinity house, where we treated, very dearly I believe, the 'Officers of the Ordinance'.[12] Pepys himself was first a guest on 18 June 1660, when he travelled to the Stepney house with his patron, the Earl of Sandwich, before being told of the position at the Navy Office that Sandwich had in mind for him.[13] He was made a Younger Brother, at the Water Lane house, on 15 February 1661/2, became an Elder Brother on 24 January 1672 and was elected Master in 1676 and 1685.

Although a convenient site, it was a risky one. It burned down shortly after they first occupied it in the Great Fire of 1666 and again in a fire around Thames Street in 1715. They rebuilt each time, but the increasing significance of location is underscored by the fact that when homeless between 1715 and 1719, they did not return to Deptford, or Stepney as they had in 1666, but remained in Tower Street Ward.[14] Meeting on Friday 14 January 1714/15 immediately after the fire, the Board agreed to advertise that they would meet in the Mitre Tavern in Fenchurch Street, 'where not only the Ballast Office will be daily kept at the usual hours morning and afternoon, but all the other affairs of the Corporation will be transacted till they can accommodate themselves with another

[11] There are useful histories of all the lighthouses and lightvessels associated with Trinity House at https://www.trinityhouse.co.uk/lighthouses-and-lightvessels; on ballast, see Cori Convertito-Farrar and Kenneth Cozens, 'Trinity House Ballast Office', Thames Symposium, 2009; https://www.academia.edu/10865539/Trinity_House_Ballast_Office_Thames_Symposium_Paper.

[12] *The Diary of Samuel Pepys*, vol. 3, 1662 (1970), 187.

[13] *The Diary of Samuel Pepys*, vol. 1, 1660 (1970), 177.

[14] Pepys records on 24 November 1666 going by coach to Stepney 'to the Trinity-house – where it is kept again now, since the burning of their other house in London'. *The Diary of Samuel Pepys*, vol. 7, 1666 (1972), 381. They were still there in May 1668.

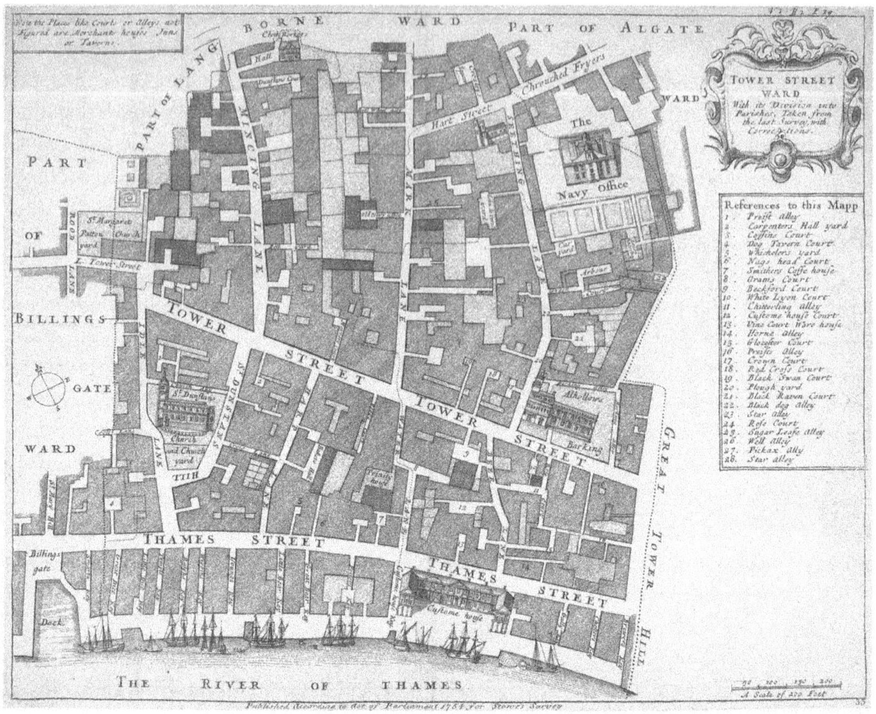

Figure 6.3 Map of Tower Street Ward by Richard Blome, from the 1754–5 edition of John Strype's *Survey of the Cities of London and Westminster*. Trinity House is in the centre, between Tower Street and Thames Street (LMA: k123570x). © London Metropolitan Archives (City of London).

House proper for their business'.[15] Just four days later they had taken a lease on a house in Mark Lane, north of Water Lane and running parallel to Seething Lane.[16] This was another significant building, of three floors, plus attics and vaults, with 'two greate parlours', next to All Hallows Staining.[17] They immediately transferred their business, the Ballast Office and goods and materials recovered from Water Lane, returning only after their home had been rebuilt. There they remained until the final years of the eighteenth century, when a new, classical building was erected on Tower Hill.

Like the Tower and livery company halls considered in previous chapters, Trinity House was a multipurpose site. As well as housing the busy Ballast Office, it hosted meetings of the General Court and those of the smaller By-Boards, and dinners for Elder Brothers and their guests. While still a Younger Brother, Pepys recorded attending twenty-nine Trinity House dinners in his diary.[18] John Evelyn, attending 'a great dinner'

15 14 January 1714/15, Board Minutes, CLC/526/MS30010/003.

16 19 January 1714/15, Board Minutes, CLC/526/MS30010/003.

17 Mark Lane: lease for a year (to be renewed at will) by Wight Woolley, merchant, to the Corporation of a house in Mark Lane in the parishes of St Mary Staining and St Dunstan in the East, CLC/526/MS30275.

18 J. R. Tanner, 'Samuel Pepys and the Trinity House', *The English Historical Review* 44, no. 176 (1929): 573–87 (p. 574).

at Water Lane after their annual election at Deptford, noted that there were 'above eighty at one table'.[19] It was also a domestic space: the chief clerk had chambers within the house, as did other servants, such as 'Betty the Cookmaid', who lived there over twenty years before illness signalled her retirement with an allowance.[20] There was also space for guests: Pepys recorded that Major Robert Holmes had 'good lodgings at the Trinity-house' during a visit in 1662.[21] While much business was carried out elsewhere – Elder Brothers were sent to offer evidence in court, to survey ships and problems on the Thames and beyond, or to press their cause for acts and against petitions in parliament – the corporate authority of their house was important. The General Courts were where decisions were made, and the House was where examinations, certifications and apprenticeship bindings took place. The Corporation had the authority at least since an act of 1565 to certify and regulate Thames pilots, and from 1621 they were also required to examine ships' masters to determine whether they were qualified to take charge of the king's ships and if so, what rate. Added to this, from the 1670s was the task of examining and certifying boys from the Royal Mathematical School as competent in mathematics and navigation and ready to be apprenticed to captains and, from 1702, also those applying to become schoolmasters on naval ships.

To emphasize their corporate significance and authority, Trinity House was a site of display, with portraits and silverware memorializing patrons and the generosity of deceased members. While charitable giving was recorded at their almshouses with statues, plaques and decorations on the exterior and stained-glass panes in their chapels, Trinity House's walls also recorded gifts: 'Divers Tables' were hung 'round this Hall, containing the Names of their Benefactors with the mention of their respective Gifts'. The display also reflected the nature of the corporation. According to Strype, 'In their Hall there is an old Flag hanging up, which they say was taken from the Spaniards by Sir Francis Drake, whose Picture hangeth up there.' He noted too that they had an 'exact Model of a Ship of a great size rigged, enclosed in a Frame glazed round' and 'two large Globes', also enclosed in a glazed case, the gifts of two of Charles II's admirals, Sir Jeremy Smith and Sir Thomas Allen. Portraits, including several of royalty, hung in the two main rooms: the Court Room, or Hall, and the room in which By-Board meetings and examinations took place, identified in a 1680 inventory as the Green Room.[22] Strype also stated that 'their Parlour', which may be the latter room, contained 'five large Plats drawn curiously and exactly by a Pen, describing certain Sea Fights in King Charles II's Time'.[23] Charts, too, were on display, probably including maps of the

[19] Entry for 20 July 1685, William Bray (ed.), *The Diary of John Evelyn* (New York and London: M. Walter Dunne, 1901), vol. 2, 230. The election was delayed from the usual Trinity Monday in 1685 because they were awaiting the production of their new charter.

[20] 17 October 1712, Court Minutes, CLC/526/MS30004/008, p. 141.

[21] 3 February 1661/2, *The Diary of Samuel Pepys*, vol. 3, 1662 (1970), 22.

[22] Strype, *Survey* I.iii.42. The 3 August 1680 inventory is cited by William Chaplin in a typescript account of the portraits at Trinity House (CLC/526/MS30068, p. 6), but it has not been located. On gifting and display of gifts in guilds, see Jasmine Kilburn-Toppin, 'Gifting Cultures and Artisanal Guilds in Sixteenth- and Early Seventeenth-Century London', *The Historical Journal* 60, no. 4 (2017): 865–87.

[23] *Survey* I.iii.42; Chaplin (p. 29) tentatively identified these as grisaille drawings by William Van de Velde the Younger.

western and eastern hemispheres, by Antonio Sanches (1637) and Augustine Fitzhugh (1686), respectively.[24]

All guests and all those who had, or wished to have, business with Trinity House – those hoping to become Younger Brethren or presenting themselves for examination, those summoned for misdemeanours, those acting as agents and officials, or keepers of buoys and lights, those petitioning for positions or support, and more – were obliged to enter the courtyard and house, navigating the Beadle at the gate, the clerks inside and a doorkeeper outside the Hall. While some got no further than offices and waiting room, others might enter one or other of the two main rooms to present their case and be quizzed by the Elder Brethren.

Authority, experience and expertise

While the Elder Brethren were, increasingly, individually wealthy and well connected, the Corporation's authority and rights rested on the charters and acts that outlined their responsibilities and privileges. These, in turn, were based on the belief that Trinity House incorporated individuals who could carry out such tasks effectively. Thus, the 1566 Seamarks Act, which gave Trinity House the right to establish marks to help mariners avoid danger, referred to it as 'a Company of the chiefest and most expert masters and Governors of ships'.[25] While this was the period in which the word 'expert' was beginning to shift to include skill, book learning and particular ability, here it retained the original meaning of personal experience.[26] The Elder Brethren might be expected, by age, to be the people who had the most experience and, therefore, knowledge of maritime matters. This, the earliest surviving archives suggest, included knowledge and experience of how to measure, equip and man ships; decide rates of pay; and apportion blame or excuse conduct. Before the professionalization of the royal navy, the Privy Council, officers of the navy and lords of the Admiralty frequently called on them for advice. They were also, collectively, understood to have knowledge and experience of places overseas and their potential or challenges. For example, in 1610, the Privy Council asked for their opinion on a Newfoundland settlement, to which they could offer a reply based on first-hand experience. They were also asked to opine on the extent of the Levant Sea, based on knowledge of people, trade and terms used in the area. In 1611 they were required to offer their view on the reliability of the accounts of the survivors of Henry Hudson's North West passage expedition in order to determine what pay they might be owed.[27] While the expedition travelled beyond

[24] These survived until another fire, in 1940: Chaplin, typescript, 72–3.

[25] Quoted in C. R. B. Barrett, *The Trinity House of Deptford Strond* (London: Lawrence & Bullen, 1893), 20.

[26] Eric H. Ash, *Power, Knowledge, and Expertise in Elizabethan England* (Baltimore and London: Johns Hopkins University Press, 2004), 10–11.

[27] G. G. Harris (ed.), *Trinity House of Deptford Transactions, 1609–1635* (London: London Records Society, 1983), vol. 1, entries 2, 21, 137, 165, 23, available at British History Online http://www.british-history.ac.uk/london-record-soc/vol19, taken from CLC/526/MS30045/001-2.

seas that they had experienced, the Elder Brethren were assumed to be equipped to make a judgement.

In particular, the Elder Brethren were understood to be experienced navigators, especially in relation to managing vessels around the English coasts and along the Thames, with its shifting sands and changing navigable channels. Because of this, they were expected to know the places in which seamarks, lights and buoys would be beneficial and how best to establish them. When their legal right and duty to do this was challenged in 1617, they responded by insisting on their knowledge and experience, asserting that their challengers lacked their skill in judging and acting in specific circumstances. In an address to the Privy Council, they asked, 'Are we not Seamen, bred even from our childhood? In the knowledge of marine affairs, of Navigable Channels, Sands, ebbing, flowing and setting of Tides, in the knowledge of setting out seamarks is not this knowledge within the compass of our element?'[28] It was Francis Bacon, then Lord Keeper of the Seal, who was asked to decide the case. He supported Trinity House in their claim that the Elizabethan act confirmed 'an authority mixed with a trust settled in that Corporation' to set up lighthouses and other marks, and that this, in turn, was 'grownded upon the skill and experience which they have in marine service'. The trust and authority held by Trinity House were, said Bacon, 'a matter of a high and precious nature in respect of the salvation of shippes and lives, and a kinde of starrelighte for that element'.[29]

In their early transactions, the knowledge that the Elder Brethren had was described as having been gained 'by tradition', 'by experience' or 'ever been so known'. Experience included learning from mistakes, and in one report they quoted the motto, 'no man knows that sand so well as he that hath lost his ship upon it'.[30] While knowledge of the organization of maritime labour, or of ships' loads, was gained over time, that required by the pilots certified by the Corporation was sometimes acknowledged as relatively easy to pick up on the back of other experience, for it was suggested that foreign navigators would know the required channels of the Thames once they had travelled them. This was highlighted as a potential defensive weakness, as was the publication of charts, sailing directions or large numbers of seamarks, and a reason to maintain control of pilotage. However, it was also suggested that without practice the requisite knowledge would be lost within a few years, especially in this ever-shifting environment. Thus, recent and continuing experience of the river and coasts was most valued in those who were to navigate or regulate navigation.[31] The Elder Brethren, traditionally senior by age and, as time went on, by position of wealth and

[28] Quoted in Jones and Ridgway, *Light through a Lens*, 25.

[29] J. V. Lyle (ed.), *Acts of the Privy Council of England Volume 35, 1616–1617* (London, 1927), 201–25 (p. 204). Available at British History Online https://www.british-history.ac.uk/acts-privy-council/vol35/pp201-225.

[30] Harris, *Transactions*, vol. 1, entries 159, 165, 142.

[31] On experience, practice and pilots' knowledge in tension with approaches of 'theory proponents' in the Spanish context, see Alison Sandman, 'Mirroring the World: Sea Charts, Navigation, and Territorial Claims in Sixteenth-Century Spain', in Smith and Findlen, *Merchants and Marvels*, 83–108. On pilots' knowledge and secrecy see *idem*, 'Controlling Knowledge: Navigation, Cartography, and Secrecy in the Early Modern Spanish Atlantic', in *Science and Empire in the Atlantic World*, ed. James Delbourgo and Nicholas Dew (London: Routledge, 2008), 31–52.

influence, often had their seagoing days behind them. Nevertheless, there were always some amongst them who were recently and even actively experienced. Their names recur as the men chosen to sound the river's channels, assess reported problems or work out the positioning of buoys and lights. These were also the most likely to carry out examinations of pilots and masters, deciding whether they should be certified by Trinity House as experienced in and capable of navigating particular routes or qualified to take charge of different kinds of ship.

Despite the focus on particular experience, and knowledge of habit and tradition, in the earlier seventeenth century, Trinity House was also asked to judge novelties. These included new designs of ships, alternative types of, or arrangements for, ordnance, new methods of measuring ships and their burdens, new processes or inventions. The fact that an engine produced by Mr Needham and Mr Browne in 1634 for dredging the Thames was declared a 'mere fancy or toy fallen from an idle head' suggests that they did not look favourably on change.[32] However, while there was a tendency to refer to tradition and 'ancient' precedent, novelties were adopted. In 1626 they gave their opinion of different methods for measuring the tonnage of ships, demonstrating their knowledge of two methods and the experience that led them to reject both the 'old way' and the 'new way' proposed by William Gunter and S. Wells, and to suggest their own, which they said was based on reason, experience, antiquity and art.[33] They also, in 1630, supported the use on merchant ships of drakes (light-weight iron guns recently developed by the king's gun-founder, John Browne) against Charles I's desire to restrict their use. Browne feared that the art of making them would be lost if they were restricted to the king's ships and, asked by the Master of the Ordnance, Lord Vere, to comment, Trinity House was happy to recommend their flexibility of positioning and relative ease of use.[34] Nevertheless, the idea of 'improvement', and the associated judgements of novelties as 'beneficial' and 'useful' to mariners or the wider public, became more noticeable from the end of the seventeenth century and into the eighteenth. Trinity House was considered well placed to judge novelty because of the Brothers' experience of former or existing methods, and understanding of the practicalities and potential benefits of implementing change.

Experience was, of course, an exceptionally important asset, and mariners brought experience of places and activities that it was hard for others to gain. As Philippa Hellawell has recently argued, significant credit was given to mariners by individuals associated with the early Royal Society as a source for trustworthy and useful knowledge: 'their direct experience of novel and remote phenomena and the long-term accumulation of that experience' conferred authority and 'amounted to a form of "expertness"'.[35] As she notes, credit was sometimes given because lives and livelihoods depended on correct judgement. Trinity House likewise suggested that shipowners and masters might be

32 Harris, *Transactions*, vol. 2, entry 461.

33 Harris, *Transactions*, vol. 2, entry 275. This method was dismissed by Gunter and Wells in a State Paper included in M. Oppenheim, *A History of the Administration of the Royal Navy and of Merchant Shipping in Relation to the Navy* (London and New York: John Lane, 1896), vol. 1, 266–7.

34 Harris, *Transactions*, vol. 2, entry 385. See Stephen Bull, '*The Furie of the Ordnance*': *Artillery in the English Civil Wars* (Woodbridge: The Boydell Press, 2008), 44–6.

35 Hellawell, 'The Best and Most Practical Philosophers', 8.

depended on because of their lively interest, rather than gentlemanly disinterest.[36] Hellawell notes how the Royal Society's 'Directions' to 'sea-men bound for far Voyages' entrusted them with performing a range of observations at sea in different parts of the world. Under this plan Trinity House was to act as repository for observations recorded in 'an exact *Diary*'. Two fair copies were to be produced, one for the Lord High Admiral (James, Duke of York) and the other to be given 'to *Trinity-house* to be perused by the *R. Society*'.[37] While there is no evidence that mariners responded, it is clear that the Society regarded Trinity House as the primary storehouse of maritime knowledge. The second set of 'Directions' was addressed specifically to 'Masters of Ships, Pilots, and other fit Persons', in other words those mariners with navigational responsibilities and most likely to be certified by (or Younger Brethren of) Trinity House. This time, as well as the Corporation acting as receiving house, it was hoped that they would distribute the 'Directions': the Society intended to send 'a good number' of copies 'to be Lodged with the Master of *Trinity-house*'.[38] If, as Hellawell suggests, seamen were most likely to be considered credible when observations were repeated and confirmed by different individuals, Trinity House was well positioned to present itself as offering the results of collective wisdom, memory and experience.

The Royal Society's minutes include occasional references to Trinity House as qualified to judge the practicality of maritime inventions. In 1662, for example, Sir Robert Moray read to the Society an account of a novel design for a sounding instrument by Lawrence Rooke (the professor of Geometry at Gresham College, who had drafted the first set of 'Directions') and the Society ordered 'a copy of it to be sent to Trinity-house'. It was for such reasons that Master of Trinity House was one of the *ex officio* positions listed as Commissioners of Longitude in the 1714 Longitude Act. Its ranking after the political and naval positions – Lord High Admiral or First Lord of the Admiralty, Speaker of the House of Commons, Commissioner of Navy, Commissioner of Trade, and the Admirals of the Red, White and Blue squadrons – and before the 'scientific' ones – President of the Royal Society, Astronomer Royal and the Savilian, Plumian and Lucasian professors at Oxford and Cambridge – seems apt. The presumed role of Trinity House was perfectly articulated by Isaac Newton when he was asked for his opinion of a mechanical sea log. He gave his view but added, 'I have no experience in sea affairs nor ever was at Sea and therefore my opinion is not much to be relied on without the opinion of Trinity House.'[39] Elsewhere, he suggested that the practical

[36] A Trinity House report claimed that because shipowners and masters were the most concerned to judge the quality of materials and stores correctly, since their credit, goods and lives depended on it, they were better judges than any officer employed to certify standards: Harris, *Transactions*, vol. 1, entry 142.

[37] Lawrence Rooke, 'Directions for Seamen, Bound for Far Voyages', *Philosophical Transactions* 1, no. 8 (1665): 140–3 (p. 141).

[38] 'Directions for Observations and Experiments to be Made by Masters of Ships, Pilots, and other Fit Persons in their Sea-Voyages', *Philosophical Transactions* 2, no. 24 [1667]: 433–48 (p. 434). The minutes of a meeting of the Council of the Royal Society, 4 April 1667, record the order that 'one hundred copies should be taken off from the printers, at the Society's charge, and lodged with the Master of Trinity-house to be disposed of by him to fit seamen', CMO/1/127 (Royal Society).

[39] Draft letter by Isaac Newton responding to the Lords Commissioners of the Admiralty, n.d., Papers on Finding the Longitude at Sea, Newton Papers, MS.Add.3972, f.41r (Cambridge University Library) https://cudl.lib.cam.ac.uk/view/MS-ADD-03972/81.

usefulness of astronomical methods might, if approved in theory by astronomers, be tested by the 'sea-Captains of Trinity House' to ensure they 'are satisfied that the Observations can be made at sea with sufficient exactness', before they were referred for consideration for a reward by the Commissioners of Longitude.[40]

By the end of the seventeenth century, Trinity House was more frequently having to engage with and adjudicate new approaches to navigation, and to deal with other and contrasting forms of expertise. Trinity House met several of Eric Ash's criteria of expertise: it possessed, and was acknowledged by others to possess, 'a body of specialized practical or productive knowledge, not readily available to everyone' that was 'based at least in part upon experience, acquired both in the learning and in the subsequent wielding of the knowledge in question'.[41] It is harder to see it meeting the definitions that Ash has asserted as were integral to the redefining of 'expertise' in the sixteenth century to include, or be replaced by, learned skill or particular ability.[42] It is unlikely that a guild corporation would claim to have the kind of expertise that was 'distinguishable from common practitioners or artisans' in their field.[43] If the kinds of experts that Ash discussed for the sixteenth century, or the projectors who drew on experimental philosophy in the late seventeenth and early eighteenth centuries discussed by Stewart, staked their claims for government or private patronage, Trinity House, collectively, was more likely to be assessing such claims than making them.[44] Indeed, in a complication of Ash's concept of 'expert mediators', Trinity House's Elder Brethren were both the patrons (merchants) and the practitioners (mariners) that he suggests were mediated between.[45] Assessments might be made with a view to maximize opportunities and profits but they were based on their own knowledge, experience, skill and trials.

Experience, knowledge and utility

There are many examples of the Trinity House Courts, By-Boards and committees dealing with projectors. This became more frequent in the later seventeenth century, after their move into the city, presumably because of their greater accessibility and proximity to other sites where such individuals might make their case. There was also an increasingly friendly environment for investment in schemes and inventions offering utility and improvement, both for private profit and for public and national benefit (often blurred together).[46] Trinity House was not, corporately, investor or

40 Draft letter by Newton, MS.Add.3972, f. 40r https://cudl.lib.cam.ac.uk/view/MS-ADD-03972/79. The Commissioners of Longitude did not meet as such until the need to consider John Harrison's timekeeper prompted a meeting in 1737: Richard Dunn and Rebekah Higgitt, *Finding Longitude: How Ships, Clocks and Stars Helped Solve the Longitude Problem* (Glasgow: Collins, 2014), 73, 88.

41 Eric H. Ash, 'Introduction: Expertise and the Early Modern State', *Osiris* 25 (2010): 1–24 (pp. 5–6, 9).

42 Ash, *Power, Knowledge, and Expertise*, 11.

43 Ash, 'Introduction', 7–8.

44 Stewart, *Rise of Public Science*.

45 Ash, *Power, Knowledge and Expertise*, 138–41.

46 Stewart, *Rise of Public Science*; Slack, *The Invention of Improvement*.

patron but adjudicator of the usefulness of inventions. In 1715, for example, Caleb Bassingwhite, a shipwright, 'attended with a new invented Engine for the defence of ships in close quarters'. It was agreed that a certificate would be provided at the next Court, confirming 'the usefulness of his Machine'.[47]

While Bassingwhite took the initiative (presenting himself at the Mitre Tavern, since this was the brief period when Trinity House was homeless), Trinity House often undertook their assessment of inventions or schemes at the request of the Navy Office or the Admiralty. For example, Josiah Burchett, who was Secretary to the Admiralty and an Elder Brother of Trinity House, wrote in 1717 asking them to 'consider the Instrument and proposal of Mons[ieu]r D Berlain with which he undertakes to solve so may usefull Problems in plain and Mercators sailing'. They were to inspect the instrument, which they did 'on private board days' and 'report their Opinion of the usefullness thereof for the Improvement of Navigation'.[48] Burchett also introduced a proposal by John French 'for discovering some Properties in the Magnetick Needle, by the Observation whereof & its application to Spherical Trigonometry & Astronomy several usefull Problems may be demonstrated relating to the Latitude & Longitude at Sea'. French had evidently been trying various avenues, for John Noyes, the Trinity House clerk, told them that the Town Clerk attended 'to desire some of the Brethren would meet the L[or]d Mayor & Ald[e]rmen Tuesday next at Guildhall on the same Account'.[49]

Other novelties in and proposed improvements to navigation for which reports were returned to the Admiralty included Fatio de Duillier's *Navigation Improved* (published 1728), on a new method for finding latitude, Lieutenant Austin's quadrant and log, and Henry De Saumarez's 'Machine for measuring a Ship's Way at Sea, as better than Log'.[50] On 30 January 1741/2 Burchett forwarded details of 'an Instrument for Shewing at one view without Calculation the Latitude a Ship is in at Sea, with the true times of the Day' submitted by Richard Graham, a fellow of the Royal Society and Comptroller of Westminster Bridge. They heard that Graham had taken the instrument 'with him to Sea in his Last Cruize, where he made Constant trials of it' and had found it as exact as a Hadley's or Davis's Quadrant. After calling in Graham to discuss the instrument in person, it was decided that a trial was necessary, the arrangements for which were referred to the By-Board. On 6 February, the Deputy Master, Captain Richard Stratton, reported that the trials had been completed and concluded that the instrument might be useful if the Sun could not be observed at its culmination. They agreed to write a report that the invention could be of great use.[51] Again, Graham's instrument was

[47] 2 April 1715 and 18 May 1715, Board Minutes, Trinity House, CLC/526/MS30010/003. This individual was presumably the same Caleb Bassingwhite as the plaintiff in a 1724 Bill and answer, a shipwright of Rotherhithe, Surrey (C 11/2008/3, Court of Chancery Pleadings, National Archives), who also worked on the rebuilding of the *Assistance* 1711–13.

[48] 13 July 1717, Board Minutes, CLC/526/MS30010/003; 31 July 1717, Court Minutes, CLC/526/MS30004/008, p. 190. Burchett, unusually, resigned from Trinity House long before his death, citing 'more weightier affairs', 22 February 1722/3, Court Minutes, CLC/526/MS30004/009, p. 20. He asked to be replaced by his son-in-law, Captain Charles Hardy, which was allowed, but with the proviso that this was 'not to be drawn into a precedent'.

[49] 28 September 1717, Board Minutes, CLC/526/MS30010/003.

[50] Volumes containing extracts from court and by minutes, etc., CLC/526/MS30046/003, p. 91.

[51] 30 January and 6 February 1741/2, Court Minutes, CLC/526/MS30004/010, pp. 116, 118. See also Adams and Woodman, *Light upon the Waters*, 118.

seen by several institutions, having already been published in the Royal Society's *Philosophical Transactions* and considered by the Board of Longitude.[52]

Trinity House's role in this economy of assessment is revealed in a publication by Saumarez, recounting his attempts to gain recognition for the invention mentioned above. He first wrote to the Royal Society but was told that the machine was neither relevant to their business nor new, and that it would hinder a ship's course. He then tried the King, who referred his petition to the Admiralty. They, in turn, referred it to Newton, who thought it might save some labour but be less exact than existing logs. Newton declined to give a final opinion and indicated that they should refer to Trinity House.[53] Saumarez then attended Trinity House 'several times' with 'Draughts upon Paper, and Models in Wood and Copper', presenting his petition on 17 March 1715/6. After inspecting the models, the Elder Brethren 'were pleased to propose some Difficulties', based on experience of how ships might behave in certain conditions, which could 'kick up the said Machine out of the Ship'. Saumarez responded at length but Trinity House reported briefly to the Admiralty on 8 May that the machine was 'wholly Impracticable'. Saumarez did not initially see the report and so several times waited at the Admiralty fruitlessly before presenting another petition on 5 July. He claimed this delay put his priority at risk, and that the pronouncement of impracticability was solely due to a desire to avoid the costs of a trial. He complained of his treatment by all, including that Trinity House 'were pleased to make a long Examination of the Matter' but 'a very short Report of it'.[54] However, while Saumarez was disappointed, the Admiralty was sufficiently content that the judgement was sound.

Such business was also carried out beyond Trinity House, for Elder Brethren might be asked to comment on inventions or proposals at parliament, just as they were asked to supply information in courts on maritime disputes relating to pay, property, liability and conduct. In 1755 they were quizzed about an invention by Richard Lyddell 'relating to the Machines or Vessels invented' by him 'for the Removal of Earth, Ballast, Sand, or any other Kind of Matter, from any Port, River, Harbour, Creek, or Inlet'.[55] The Parliamentary Committee questioned two Elder Brethren – Captains John Major and Thomas Bennett – who both approved the utility of the machine and offered information about how ballast should be dealt with, alongside a shipbuilder, a captain who had sailed in the vessel, a Custom House surveyor and others. Two years later, Trinity House was asked to present to the House of Commons 'Copies of such Surveys, Reports, and other Papers' as they possessed relating to Rye and Dover harbours

[52] Graham's description of his instrument was printed in *Philosophical Transactions* 38, no. 435 (1733): 450; 16 January 1741, Minutes of the Board of Longitude, RGO 14/5, p. 9; https://cudl.lib.cam.ac.uk/view/MS-RGO-00014-00005/13 (Royal Greenwich Observatory Archives, Cambridge University Library). The previous Master of Trinity House, Charles Wager, was also present, as First Lord of the Admiralty. The result of their deliberations is not recorded: the next meeting was not until 4 June 1746.

[53] Henry De Saumarez, *An Account of the Proceedings of Henry De Saumarez of the Island of Guernsey, Gent* (London, 1717), 7–8. Newton's draft is at MS.Add.3972, p. 41r. https://cudl.lib.cam.ac.uk/view/MS-ADD-03972/81. This was the occasion on which he stated that his opinion was not to be relied on without that of Trinity House.

[54] Saumarez, *An Account*, 9–18.

[55] 17 March 1755, *Journals of the House of Commons*, vol. 27 (London: 1803), 235.

to support plans for their repair. On 21 March 1757, their clerk, John Shuttlewood, presented six sets of relevant documents of reports, petitions, surveys and minutes, dating between 1661 and 1743.[56] The importance of such materials to the Corporation and its activities was underlined by the clerk Richard Noyes who, in the 1715 fire, 'sacrificed his own goods … to the preservation of the Corporation's Books'. His successor, John Wharton, however, recalled with sadness having seen 'a large press of papers in the old lobby, with labels pointing to sundry materials transactions and disputes, which that fire consumed entirely'.[57]

While these examples were largely offering opinions on which others might act, in the second half of the eighteenth century, Trinity House also offered some encouragement of their own. In 1761 they received 'A Memorial from Mr Jeremiah Sisson, Mathematical Instrument Maker' on his newly invented quadrant, which he said 'in several Experiments, particularly on board His Majesty's Ship Winchester, had been found very useful'. He was there in hope of 'the Consideration & Encouragement of this Corporation', which they provided by giving him five shillings.[58] They agreed to a rather more generous financial reward of 5 guineas for Robert Erskine, a Scottish inventor and engineer, 'as a Mark of this Corporation's Encouragement to his Ingenuity'. He had attended the House in person, with a 'Model of a Machine for drawing Water out of Ships with greater Facility & in larger Quantities than hitherto invented'. He and the model were examined, evidently with satisfaction, and he was also asked to let them know 'when he shall have a practicable Machine ready for actual Service'.[59] Another form of encouragement was subscription. While many Brethren subscribed to publications individually, it was also done Corporately, for example in 1769 for two copies of William Falconer's *A Universal Dictionary of the Marine*, and in 1778 they subscribed 5 guineas to Robert Waddington's *Epitome of Theoretical and Mathematical Navigation*. This had been angled for by the author, who dedicated the book to them, suggesting that his 'Theoretical and Practical Experiments' met 'the immediate Concerns of your respectable House'.[60]

Most of the books, schemes and instruments put before Trinity House were referred to the By-Board and, occasionally, to one or more individuals for trials or further scrutiny. The By-Boards were made up of a smaller, active and experienced group of Brethren, though this did not always serve to make an informed decision. One invention referred to Newton by the Admiralty, and thereafter to Trinity House, required them to outsource the investigation. This was an observing instrument

[56] 11 March 1757, *Journals of the House of Commons*, vol. 27, 772, 789–90.

[57] John Whormby, *An Account of the Corporation of Trinity House, Deptford Strond, and of Sea Marks in General* (London: 1746), xvii. Noyes's widow was eventually recompensed for her husband's £242 2s 8d worth of goods lost when 'Saving their Cash and Goods from the Fire': 17 May 1721, Court Minutes, CLC/526/MS30004/009, p. 7.

[58] 7 November 1761, Board Minutes, CLC/526/MS30010/12, p. 84.

[59] 5 November 1763, Board Minutes, CLC/526/MS30010/12, p. 162.

[60] 4 February 1769, Board Minutes, MS 30010/12, p. 357; 4 April 1778, Court Minutes, MS 30004/013, p. 11. Dedication, Robert Waddington, *An Epitome of Theoretical and Practical Navigation* (London: 1777). Jim Bennett's account of Waddington's search for mathematical employment and patronage is instructive: '"The Rev. Mr. Nevil Maskelyne, F.R.S. and Myself": The Story of Robert Waddington', in *Maskelyne: Astronomer Royal*, ed. Rebekah Higgitt (London: Robert Hale, 2014), 59–88.

designed by Captain Jacob Rowe, an experienced mariner, salvager and designer of instruments. Newton considered Rowe's 'instruments for taking altitudes ~~of the Sun~~ <at sea>', which he thought 'well contrived' and 'deserve to be tryed at sea'.[61] Presumably as a result, Burchett asked Trinity House 'to Inspect into an Instrument invented by Cap. Row for taking an Observation by the Sun at any time of the day'. At a By-Board meeting on 27 November 1725, they were 'inform'd that Mr. Row attended w[i]th the said Instrum[en]t' and was 'Call'd in'. They found that they could not, by themselves, come to a conclusion: 'after viewing [Rowe's] Instrum[en]t & hearing what he had to Say relating thereunto It was Resolv'd that Mr. Weston do View the said Instrum[en]t & report his Opinion thereof'.[62] Weston was Thomas Weston, former assistant to the astronomer royal, John Flamsteed, at the Royal Observatory, who ran an academy in Greenwich, which specialized in mathematical and navigational teaching.[63] He was also, from 1719, mathematical examiner to Trinity House.

Trinity House and mathematical navigation

The increasingly complex mathematics associated with navigation tested the expertise claimed by Trinity House. While many techniques based on astronomy and geometry could be learned through practical experience, with or without a clear understanding of the theory underlying them, moves towards formalizing mathematical training for mariners and the increasing availability of mathematical texts and instruments were a challenge to established routes of training and certification.[64] The encroachment of mathematical practitioners into the working world of navigation was a significant challenge that the Elder Brethren had to find ways to equip themselves to deal with. Ash has pointed to the mathematization of navigation as creating a gap between patrons and practitioners and an opportunity for mathematically adept expert mediators. Trinity House was both patron and practitioner and, by employing mathematicians to fill its internal gap, ensured that the Corporation could maintain its established roles.[65]

This mathematical gap was first exposed as a result of Trinity House's relationship to the Royal Mathematical School, founded in 1673. This consisted of forty promising scholars of Christ's Hospital, who were selected to be 'further educated in a Mathematical Schoole' at the King's expense. They were to be

[61] Draft letter from Isaac Newton to the Lords Commissioners of the Admiralty, 26 August 1725, MS.Add.3972; https://cudl.lib.cam.ac.uk/view/MS-ADD-03972/59; Jacob Rowe, *Navigation Improved: In Two Books* (London: 1725). On Rowe's exploits, see John Ratcliffe, 'Bells, Barrels and Bullion: Diving and Salvage in the Atlantic World, 1500 to 1800', *Nautical Research Journal* 56 (2001): 35–56 (pp. 49–50), and on his 'fluid quadrant', see Katy Barrett, 'The Wanton Line: Hogarth and the Public Life of Longitude' (University of Cambridge, unpublished PhD thesis, 2014), 56–60.

[62] 27 November 1725, Board Minutes, CLC/526/MS30010/4.

[63] On Weston see Kim Sloan, 'Thomas Weston and the Academy at Greenwich', *Transactions of the Greenwich and Lewisham Antiquarian Society* 9, no. 6 (1984): 313–33 and Higgitt, 'Mathematical Examiners at Trinity House'.

[64] See Margaret E. Schotte, *Sailing School: Navigating Science and Skill, 1550–1800* (Baltimore: Johns Hopkins University Press, 2019) and Ash, *Power, Knowledge, and Expertise*, Chapter 3.

[65] Though see Higgitt, 'Mathematical Examiners' on competition for the role and its changing significance over time.

taught and instructed in the Art of Navigacon and the whole Science of Arithmatique until their age and competent proficiency in these parts of the Mathematiques shall have fitted and qualified them in the judgment of the Master of the Trinity House for the tyme being to bee initiated into the practices of Navigation and to bee bound out as Apprentices for seaven yeares to some Captaines or Comanders of Shipps.[66]

Differences arose between the school, Trinity House, their leaders and advisors over the curriculum, selection of mathematical masters and examination, with differing emphasis on theory or practice.[67] These issues arose as early as 1675, when Pepys, already an MP and Secretary to the Admiralty, became both a Governor of Christ's Hospital and Master of Trinity House. He was well aware of the need for the Corporation to play its part in ensuring that this scheme to improve the skills and educational standards of the King's navy, adopted with enthusiasm by Charles II, was a success.

Pepys was neither a mathematician nor a mariner but, as Master, he sought to represent the views of the experienced Elder Brethren. He asked them 'to propose what parts of Arithmetique and Mathematticks ought to be taught', a request he had to repeat the following month.[68] Elsewhere, there was increasing concern about what the boys should be taught and how they should be examined, and efforts were made by both Trinity House and the master of the Mathematical School, John Leake, to clarify matters.[69] However, in October 1677 it was reported at Trinity House that 'they have mett with difficulty in examining' the boys, and Pepys reported a whole list of shortcomings to the governing board of Christ's Hospital.[70] It was, however, difficult to work out if the problem was the curriculum, the master or the examination. It was also proving to be difficult to find apprenticeships for the boys, and Pepys advised reducing the number of scholars.[71] Although he criticized the curriculum and teaching, he was also concerned that mathematical navigation could not properly be tested by the Elder Brethren who, 'through want of use and overlooking their Books', were 'not themselves best fitted for Examining the Children'.[72] Nevertheless, he agreed that input from mariners was required, and he urged that masters at the School should be teachers

66 E. H. Pearce, *Annals of Christ's Hospital* (London: Methuen & Co., 1901), 101.

67 These debates have been explored by Rob Iliffe, 'Mathematical Characters: Flamsteed and Christ's Hospital Mathematical School', in *Flamsteed's Stars: New Perspectives on the Life and Work of the First Astronomer Royal, 1646–1719*, ed. Frances Willmoth (Woodbridge: Boydell Press, 1997), 115–44. See also Schotte, *Sailing School*, 95–113 and Jason Grier, 'Navigation, Commercial Exchange and the Problem of Long-Distance Control in England and the English East India Company, 1673–1755' (unpublished PhD thesis, York University, 2018), 29–85. The episode is examined from the perspective of Trinity House in Higgitt, 'Mathematical Examiners'.

68 15 July 1676, 12 August 1676, Court Minutes, CLC/526/MS3004/005, pp. 7, 10.

69 On 27 September 1676, the Court agreed that they should not examine the boys until Leake had certified what parts of mathematics the boys were skilled in: Court Minutes, CLC/526/MS30004/005, p. 15.

70 15 October 1677, Court Minutes, CLC/526/MS30004/005, p. 65; Iliffe, 'Mathematical Characters', 122.

71 Difficulties were reported by the Master, John Prowd, on 1 August 1677 and Pepys said he would investigate concerns: CLC/526/MS30004/005, p. 54; Iliffe, 'Mathematical Characters', 122.

72 Pepys Manuscripts on Christ's Hospital, quoted in Grier, 'Navigation', 63.

of practical navigation who had been to sea, as opposed to the university-trained mathematicians that Newton and others assumed to be appropriate.

When Leake's successor as mathematical master, Peter Perkins, died, a host of possible replacements were named. On 6 January 1680/1, Sir Matthew Andrews told his fellow Elder Brethren at a General Court meeting that they should 'consider of a fitt p[er]son to be presented for the Instructing the Children of the new Royall Foundacon at Christs Hospitall' as the King had 'declared he would have one Skilfull not onely in the Theory but also in the practice of Navigacon'. The Earl of Craven agreed, saying 'his Ma[jes]tie had declared he would take the approbacon of Tr: house before any p[er]son should be settled in that Employm[en]t'. Andrews, who had only recently become an Elder Brother, was a merchant and shipowner, formerly based in India, now an MP and evermore active committeeman. At the same meeting he had declared his intention 'not onely to profess himself a Member but really to endeavour the Interest & advantage of the Corporacon' within Parliament.[73] However, they seem to have failed to pass their message to the king effectively, for Andrews reported at the next meeting that two other contenders, Robert Wood and Thomas Streete, were under consideration, and their own nominations, for Matthew Norwood and Mr Yates, came too late. He reported 'that Street had obtained the Kings Letter for it', although it was, in fact, Wood who was appointed.[74]

It is interesting to note that, shortly before this, the Elder Brethren had created an opportunity to display themselves to the King as possessing the more theoretical and mathematical skills of the seventeenth-century navigational expert. This involved a precision positional observation carried out when two Elder Brothers, Captains Hugh Till and Simon Bayly, were sent to establish a lighthouse at St Agnes in the Isles of Scilly. They had taken 'a large Quadrant' and 'made an exact Observacon' of the latitude. While not a particularly difficult observation, it was useful knowledge for at least two reasons. Firstly, it could be used to correct existing charts, which were known to show the islands at too high a latitude, and 'might be of great benefit to Shipping'. Secondly, its precision could be boasted of, and they agreed that the King 'should have knowledge of it'.[75] A report was drawn up and, on 9 November 1680, the Master (the Earl of Berkely, chosen for having the King's favour, which he had demonstrated by successfully lobbying for the Scilly light patent, rather than his 'Skill and Knowledge in Maritime Affairs')[76] and Deputy Master, Simon Nichols, waited on the King. They judged his interests well, for he 'graciously rec[eiv]ed' the report and 'declared himself to be very well pleased therewith and freely Discoursed about it Enquiring whether they had made a certaine Observacon thereof & <how>'. He spoke, too, of the inaccuracies of the Dutch charts that were commonly used, and said that he 'designed a more exact

[73] 6 January 1680/1, Court Minutes, CLC/526/MS30004/05, pp. 194–5.
[74] 11 January 1680/1, Court Minutes, CLC/526/MS30004/005, p. 196.
[75] 19 October 1680, Court Minutes, CLC/526/MS30004/005, p. 172.
[76] *The Earl of Berkeley's Speech to the Corporation of Trinity-House at Deptford, Upon Trinity-Munday, 1680. When His Lordship Was Elected Master by a General and Unanimous Vote* (London: 1681), 3.

Observacon of the head lands'.[77] Till and Bayly, who had previously received £200 for their work on the Lowestoft lighthouse, were rewarded another £70 for the 'great care and pains' they had taken over the Scillies project.[78] The Brethren were evidently well pleased that they were able to offer this useful and carefully produced knowledge to the King.

Despite being able to call on skills like those of Till and Bayly, when difficulties with mathematical examinations arose again in 1681, Trinity House found it necessary to obtain the advice of a teacher of mathematics in order to ensure the continuance of the King's pet project. After repeated refusals by the Elder Brethren to certify candidates taught by Robert Wood, and as Christ's Hospital carried out their own enquiries, Trinity House asked Richard Norris, a practical mathematician whom Pepys referred to in his notes as 'Old Norris, our Mathematician', to investigate.[79] Despite his blaming 'a Tide' against him at Trinity House Wood seems to have been unsatisfactory to all and was removed.[80] Trinity House sought to smooth the waters, on the one hand identifying and approving ships' masters who would take apprentices and, on the other, finding a way to ensure confidence in the abilities of the boys that passed their examinations. In March 1682, Pepys warned the Elder Brethren of 'the necessity th[a]t lay upon this Board to be more circomspect than formerly' in their examinations and criticisms. He noted that

> there were some Comand[e]rs of Ships that did endeavour to asperse Trinity House to his Ma[jes]tie for giving Certificates upon a Slight Examinacon of these Boyes; And That the King Sett his heart more upon this foundacon than upon any other of that kind having contributed very bountifully towards their maintenancies.[81]

He suggested that, if necessary, he would examine, or observe the examinations, himself.

As part of his investigation, Norris had undertaken examinations of the boys, and on 19 December 1682 it was suggested by Nichols that they pay him for his pains. Further, he said the Corporation might 'admitt him into that Service' on a regular basis. There were never more than ten boys a year to examine and so it was agreed to pay him £10 per annum. They squared the circle of expertise and authority by resolving

[77] 9 November 1680, Court Minutes, CLC/526/MS30004/005, pp. 177–8. This last comment presumably relates to Greenvile Collins's coastal surveys, which received royal support, beginning in 1681. Pepys said that he had urged the Elder Brethren to survey coasts themselves – feeling it should properly be in their hands and not the observations of a single individual – but, since they were not undertaking this, they should make sure they had some ownership of Collins's survey, which they supported with greater than usual generosity on 10 April 1682, Court Minutes, CLC/526/MS30004/006, pp. 42–4.

[78] 11 January 1680/1, Court Minutes, CLC/526/MS30004/005, p. 197.

[79] Samuel Pepys, *Samuel Pepys's Naval Minutes*, ed. J. R. Tanner ([London]: The Naval Records Society, 1926), 124; 2 February 1681/2, Court Minutes, CLC/526/MS30004/006, p. 34. Flamsteed was also carrying out examinations and investigations, on behalf of Christ's Hospital: copy of a letter to Sir John Frederick, President of Christ's Hospital, 17 September 1681, in *Correspondence*, vol 1, 812–13.

[80] Grier, 'Navigation', 69.

[81] 21 February 1681/2 and 27 March 1682, Court Minutes, CLC/526/MS30004/06, pp. 37–8.

> That the Brethren of this Corporacon Tho: they are the most proper Judges of the Qualifications of those Children for Sea Affaires, yet they may not be so well acquainted with the School Ma[ste]rs Rules leading thereunto, w[hi]ch ought to be Supplyed by one whose dayly practice it is.[82]

In effect, however, they delegated the responsibility. In the 1690s Pepys noted that 'our present Tr[inity] H[ouse] are so conscious of their own disability to examine the Christ's Hospital boys committed to their examination by the King's Institution, that they professedly allow 10*l.* per annum out of the poor's money' to pay a mathematical teacher 'to do it for them in their presence'.[83] Norris was, however, removed within months by unanimous vote because of ongoing criticism, which risked 'ill Consequence both to the Hospitall & this House'.[84] He was replaced by Lancelot Colson, a physician, astrologer and alchemist based at Great Tower Hill, and, subsequently, by his kinsman John Colson, who had a considerable reputation as a mathematical teacher and had long acted as an examiner for the Visitors of Christ's Hospital, determining whether or not there were boys ready to proceed to their Trinity House examination.[85] He kept his old role as he took on the new, which, as Pepys wrote in 1694, 'seems to make the examination in either place the greater mockery'.[86]

A final significant clash occurred in 1708, with questions again raised about whether the boys were being adequately prepared for their future roles. Captain John Merry, a warden of Trinity House and a Governor of Christ's Hospital, reported that he had sat through the examinations of five boys carried out by the mathematical examiner at Trinity House, which he said took two to three hours per day over nine days, and had found them 'more ignorant in their business than any others that have of late years come before the Brothers of Trinity House'. This was despite the fact that the Trinity House examiner, John Colson, had previously, as examiner for the Mathematical School, found the boys 'competently knowing'.[87] This led to yet another change of master at Christ's Hospital, with James Hodgson, another former assistant at the Royal Observatory, replacing Samuel Newton. Thereafter, the Elder Brethren seem to have been content, perhaps because they began to stay away from the mathematical examinations, ceding full control to the examiners. Certificates were, however, still

82 19 December 1682, Court Minutes, CLC/526/MS30004/006, p. 71.

83 Pepys, *Naval Minutes*, 401.

84 21 March 1682/3, Court Minutes, CLC/526/MS30004/006, p. 86. Lancelot Colson was proposed and voted for at the same meeting.

85 On the Colsons see John Gascoigne, 'Sensible Newtonians: Nicholas Saunderson and John Colson', in *From Newton to Hawking: A History of Cambridge University's Lucasian Professors of Mathematics*, ed. Kevin C. Knox and Richard Noakes (Cambridge: Cambridge University Press, 2003), 171–204 (p. 191).

86 Pepys, *Naval Minutes*, 401–2.

87 On Merry, see his Hudson's Bay Biographical Sheet, Archives of Manitoba https://www.gov.mb.ca/chc/archives/_docs/hbca/biographical/m/merry_john.pdf. Excerpts from the Minutes of the Committee of Almoners, 8 December 1708 and 16 September 1708, in Nerida F. Ellerton and M. A. Clements, *Samuel Pepys, Isaac Newton, James Hodgson, and the Beginnings of Secondary School Mathematics: A History of the Royal Mathematical School within Christ's Hospital, London 1673–1868* (Cham: Springer, 2017), 76.

issued in the name of the Master, Wardens, Assistants and Brethren, and examinations were required to take place at Trinity House, the site of their corporate control and authority.

An exception that proved the rule was given when a later mathematical examiner, John Weston, received permission to examine prospective naval schoolmasters 'at his own house at Greenwich' in 1728.[88] The schoolmasters, who would be certified by the Admiralty and employed on Royal Navy ships, were examined by Trinity House's mathematical examiners from 1702, for which the latter were occasionally allowed an additional £10, when times of war increased the numbers.[89] Weston was not allowed to examine the mathematical scholars at home, and the other examiners carried out all their examinations in Water Lane. The unusual example of the candidate Nathan Prince – who had been a scholar and tutor of mathematics at Harvard for twenty years – shows the importance of the site. Prince's mathematical knowledge was not in doubt but he took the advice of the Trinity House doorkeeper to visit the mathematical examiner, William Mountaine, at his home to discuss what topics and books he should revise. The examination, however, necessarily took place at Trinity House. On two occasions they met at Mountaine's home, just over the river in Gainsford Street, Shad Thames, but then proceeded together to Water Lane for the examination, despite it being a private and discursive process. Undertaken in the building's second chamber, the candidate was surrounded by portraits of Charles II, James II as Duke of York, 'and other great men'.[90]

Reasserting experience

Despite the tribulations over certifying mathematical competence, the brethren of Trinity House continued to insist on their own collective expertise in relation to practical navigation. In the late seventeenth and early eighteenth centuries, this included mathematical instruments and techniques but not necessarily the kind of precision and complexity linked to astronomical methods of finding longitude that were being taught at the Royal Mathematical School. Rather, they projected confidence in their skills and did not seek additional methods. It had been mathematicians rather than mariners who pressed for the passing of the Longitude Act in 1714: Newton, in a deleted comment in a draft letter, commented that the Chair of the Parliamentary Committee which approved the Longitude Bill 'being a seaman represented that they did not want the longitude, & so far as I can observe the sea-men are generally of that

[88] 21 August 1728, Board Minutes, CLC/526/MS30010/4.

[89] The additional sum was granted after appeal by Colson, 5 February 1707/8, Court Minutes, CLC/526/MS30004/008, pp. 70–1 and by William Mountaine, 16 March 1744/45, Court Minutes, CLC/526/MS30004/010, p. 304. On the examination and appointment of naval schoolmasters, see H. W. Dickinson, *Educating the Royal Navy: Eighteenth- and Nineteenth-Century Education for Officers* (London and New York: Routledge 2007), 10, 13.

[90] William L. Sachse, 'The Journal of Nathan Prince, 1747', *The American Neptune* 16, no. 2 (1956): 81–97 (quote at p. 94); Laura Morris, 'Biographical Note on Prince', Harvard Library (2011), https://hollisarchives.lib.harvard.edu/repositories/4/resources/4037.

opinion'.[91] This chairman was, according to William Whiston, close to throwing out the idea altogether.[92] It is likely that he was Sir John Leake, a commander in the Royal Navy, Elder Brother of Trinity House, its Master in 1709–11 and in 1713 a Lord of the Admiralty.[93] Leake was, after all, said (by his kinsman and heir) to be 'one of the best seamen this island has produced, being perfect master both in theory and practice'.[94]

For Trinity House, navigating successfully was about handling ships, recognizing the dangers of sandbanks and knowing routes, the processes of sounding depths and the compass directions required around particular coasts and seamarks. This was the core knowledge on which the Elder Brethren examined pilots, who were quizzed on and found qualified to take charge of ships for particular routes. Likewise, in the examinations of those seeking certification as masters of Royal Naval ships, the Elder Brethren established whether they were sufficiently experienced to navigate ships of a particular rating as far as around Britain, the Mediterranean, both sides of the Atlantic or the Baltic. These candidates had to have at least three years' experience at sea and demonstrate their knowledge, by experience, of the role of master and the specific hazards and details of their chosen routes. They also, however, had to be able to answer some more general navigational questions, including on what had been novel mathematical knowledge and instrumental practice in the sixteenth century: finding latitude and magnetic variation, predicting tides and making use of charts. Individuals could fail these examinations, whether for lack of time at sea, experience of the specific class of ship or routes or the more general knowledge required for navigating open sea.[95]

The Brethren of Trinity House took care to maintain and project their mastery of such knowledge. As discussed above, charts, globes and models were part of the display within Trinity House at Water Lane. They also represented their informed practical skill at their almshouses, for example with the statues of Captain Richard Maples and Captain Robert Sandes, two Elder Brothers who had left substantial legacies to the Corporation for its charitable activities (Figs. 6.4 and 6.5). They were presented as active practitioners of navigation, rather than simply as men wealthy enough to disperse Indian diamonds or Lincolnshire estates.[96] The former, whose inscription

91 Draft letter from Newton to the Admiralty, MS Add. 3972; https://cudl.lib.cam.ac.uk/view/MS-ADD-03972/79; see Dunn and Higgitt, *Finding Longitude*, 34–65.

92 Whiston's account is discussed in Richard S. Westfall, *Never at Rest: A Biography of Isaac Newton* (Cambridge: Cambridge University Press, 1980), 835.

93 Samuel Clark noted the committee members in his historical account of the discovery of longitude in *An Easy Introduction to the Theory and Practice of Mechanics* (London: J. Nourse, 1764), vol. 1, 33–4. Sir Thomas Hardy (1666–1732), at the time a Warden of Trinity House, was also on the parliamentary committee.

94 'Leake, Sir John (1656–1720), of Mile End, Stepney, Mdx. and Beddington, Surr.', in *The History of Parliament: the House of Commons 1690–1715*, ed. D. Hayton, E. Cruickshanks and S. Handley (2002); https://www.historyofparliamentonline.org/volume/1690-1715/member/leake-sir-john-1656-1720.

95 Details of a 1670 masters' examination are discussed in Schotte, *Sailing School*, 94. These techniques as mathematical novelties are key to Ash's discussion of Elizabethan expertise in mathematical navigation: *Power, Knowledge and Expertise*, Chapters 3–4.

96 Margaret Whinney, *Sculpture in Britain, 1530–1830* (New Haven: Yale University Press, 1992), 138, 223; C. R. Ashbee, *Survey of London Monograph 1, Trinity Hospital, Mile End* (London: Guild & School of Handicraft, 1896), 18–21, available at British History Online, http://www.british-history.ac.uk/survey-london/bk1/, pp. 18-21. The statue of Maples was moved to the Mile End when the almshouses at Deptford were demolished in 1875. Both are now displayed at Trinity House.

Figures 6.4–6.5 Statues of Capt Richard Maples, by Jasper Latham, 1683, and Capt Robert Sandes by Peter Scheemakers, 1746, in Trinity House. Photograph by Rebekah Higgitt.

reveals that he was a Commander of an East Indiaman at the time of his death, rests a hand on a back-staff and, at his feet, are a globe, dividers, mariner's compass and lead and line. Sandes, who was Deputy Master 1719–21, holds a telescope and likewise has a globe at his feet, alongside an anchor. His left foot is raised, suggesting a readiness for action. A plaque at Deptford, perhaps mounted when the almshouses were rebuilt in 1788, presents a whole range of instruments and objects relevant to Trinity House: as well as anchor, lantern, the Red Ensign flag of British merchant ships and canon, there is a compass, telescopes, an octant, dividers, parallel rule, dredging net and lead weight for sounding.[97] While such monuments would be seen by the almspeople, who, as retired mariners and their widows, would have had little need to be reminded of the maritime experience they represented, they were also seen by visitors and guests invited to the Trinity Monday ceremonies.

The Elder Brothers' skills and knowledge were performed when they undertook surveys of the Thames Channels. This activity was depicted on a chart of the Thames Estuary and surrounding coast that Greenvile Collins dedicated to Trinity House in 1698 (Fig. 6.6). A vessel, with the Trinity House arms on its sail, is shown surrounded by buoys, lighthouses and other marks. The figures shown on board are, we may imagine, busy with compass and sounding lead. While the Elder Brethren were perhaps less assiduous in carrying out regular surveys than they sometimes claimed, they were undertaken from time to time.[98] The resulting reports and drawings were stored within Trinity House, collected in a now lost series titled Survey Book and Select Entries. This guild knowledge, shared amongst Elder and Younger Brethren and the communities they served, could travel up or down the Corporation's hierarchy. Surveying activity could introduce or reinforce knowledge among Elder Brethren or pilots. In 1677, for example, Captain Browne was asked to survey and buoy the channels, making use of 'six of the ablest London Pylotts', while in 1729 the Elder Brethren decided to send some of their certified pilots to take soundings down the North Channel, up the Coast towards Yarmouth and then up the South Channel 'for their better Information'.[99] In the 1740s they made use of their Buoy Sloop, proposing to make more frequent surveys 'for the Improvement of our Pilots & other Publick Services', taking as many pilots as they could 'for their better Instruction'.[100] A library of maritime and navigational texts held out the possibility of self-education but had less institutional focus than the exchange of information by experience, observation and practice.[101]

Pilots and masters were, more broadly, an important source of knowledge and information for the Brethren of Trinity House, as well as being individuals they

[97] This was also moved to Mile End and then to the new Trinity House, where it is still on display.

[98] Pepys noted in 1683 the Elder Brethren 'speaking as if they did examine the sands in the River afresh every year or two', which he may have doubted: *Naval Minutes*, 209.

[99] 7 June 1677, Court Minutes, CLC/526/MS30004/005, p. 41; 20 December 1729, Court Minutes, CLC/526/MS30004/009, p. 85.

[100] 26 January and 2 March 1744/5, Court Minutes, CLC/526/MS30004/010, pp. 195, 201.

[101] Admiral Thomas Allen in 1681 left £100 'to be laid out in books for the use of this Corporation', quoted in Adams and Woodman, *Light Upon the Water*, 32; an inventory of the books was taken 27 September 1721, Board Minutes, CLC/526/MS30010/4, n.p.; on 6 June 1761 it was resolved that no books should be taken out of the house 'without Leave of the General Court', Board Minutes, CLC/526/MS30010/12, p. 67; books gifted or subscribed to would be added to the library.

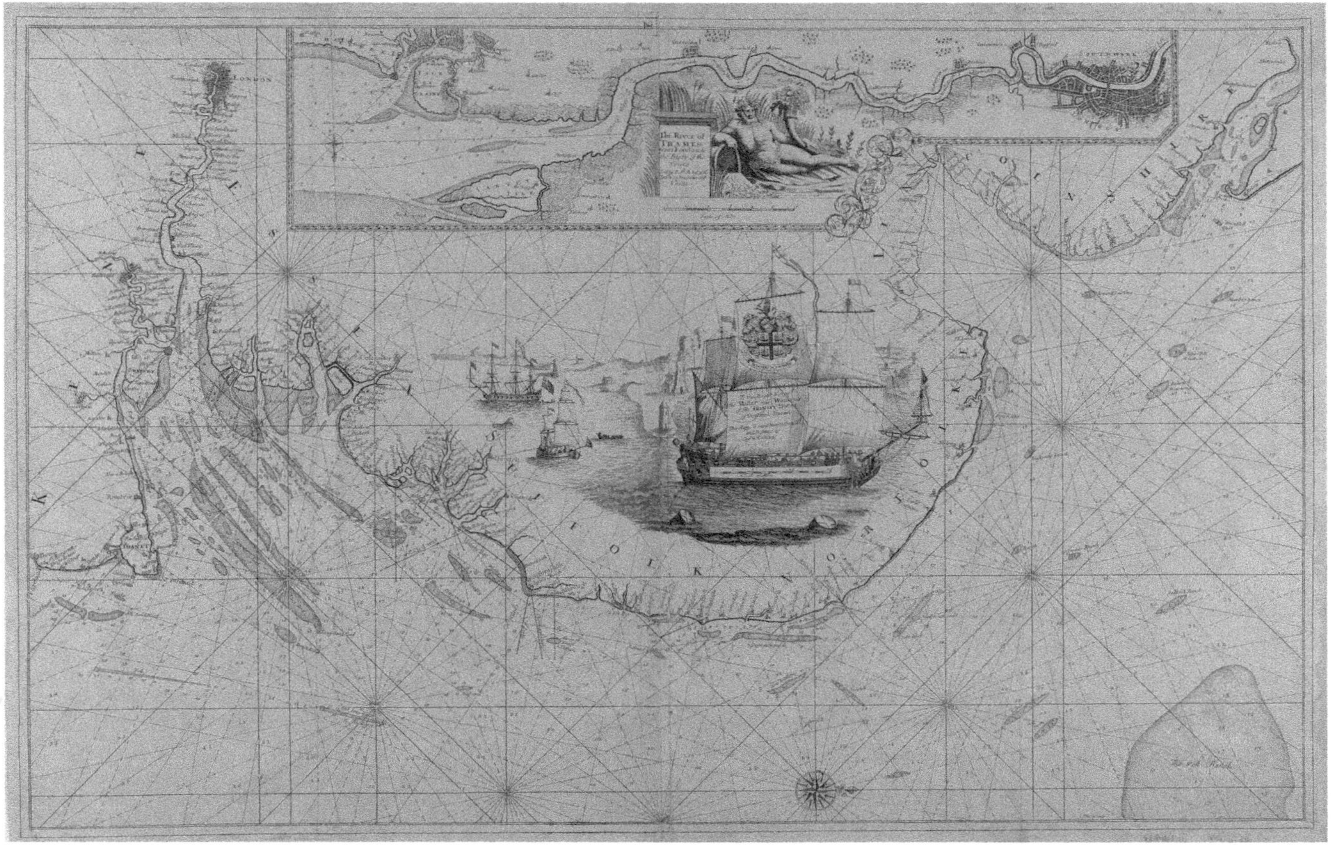

Figure 6.6 Greenvile Collins, Chart of the Thames Estuary to the Wash, 1698, showing the Trinity House yacht and a range of seamarks (NMM: G218:11/1(1)). © National Maritime Museum, Greenwich, London.

might have to deal with in relation to complaints or accidents. Active mariners were their eyes around the river and coasts, especially in reporting the loss of marks or movement of sandbanks, and their informants or witnesses in relation to proposed or actual changes. Their network also consisted of their buoy and lightkeepers, as well as agents and clerks dealing with light duties or ballast. Nicholas Boult, the buoy keeper at Yarmouth appointed in November 1715, was a particularly assiduous informant, an approach that proved to be mutually beneficial.[102] He wrote letters and attended Trinity House in person to report on necessary repairs or replacement for buoys, informed the Brethren of the death of the lightkeeper at Caster and passed on details of the local inhabitants' preferred replacement, and ultimately gained authority to oversee a wider area and other employees. By 1738 he was an agent as well as buoy keeper and was inspecting the repairs to Caster lighthouse and overseeing the lightkeeper's decisions.[103] It was through such individuals, based on or travelling round the British coasts and beyond, that Trinity House was able to assert its knowledge and authority well beyond its city base.

However, it was clearly important for the Elder Brethren to be able to claim and exercise expertise amongst themselves. The minute books suggest that particular Brethren tended to be more regularly engaged in such activities, their names appearing repeatedly as both examiners and those undertaking surveys or investigations. They usually had a significant amount of experience as masters and commanders and were often also the more active attendees of the By-Boards, to which these matters, like the consideration of proposals for schemes and instruments, were often referred. These individuals – internally recognized as the most capable of (and willing to do) this work – were formalized as a group of four in 1789, when a Committee of Examiners was formed. As the name suggests, the committee members were to undertake pilots' and masters' examinations and, in time of war, they were clearly expected to be busy and were to receive £20 per annum. To them were also 'referred for consideration all communications on improvements in nautical matters, and they report an opinion on such references, as well as on the inspection of charts, sailing directions, discoveries of sands, shoals, and on maritime suggestions'.[104] As with other posts within the Corporation – Supervisor of the Ballast Office, Wardens and Assistants – the positions were appointed annually and by order of precedence (in this case the two eldest and the two youngest), but Brethren were entitled to refuse and in practice it was done by those most capable of the work.

The first four members of the Committee were Captains John Bartholomew Lanty, Timothy Mangles, Joseph Cotton and James Strachan. The first two had been elected Elder Brothers in 1776 and long been active as examiners and surveyors. In that year Lanty had retired, aged forty-eight, from fifteen years' of service as master of the royal yacht *William and Mary*.[105] Mangles had been master of a transport ship and later went into partnership with his brothers as shipowners in Wapping, establishing the

[102] 16 November 1715, Court Minutes, CLC/526/MS30004/008, p. 175.

[103] 2 September 1721, 31 March 1722, 10 July 1725, Board Minutes, CLC/526/MS30010/4; 22 July 1738, Board Minutes, CLC/526/MS30010/8.

[104] Joseph Cotton, *Memoir on the Origin and Incorporation of the Trinity House of Deptford Strond* (London: J. Darling, 1818), 174–6 (details from By Board minutes, 12 November 1789).

[105] Navy Board Out Letters, ADM 354/193/88 (TNA).

Southern Whale Fishery.[106] Cotton, who had been a lieutenant in the Royal Navy before moving to the merchant marine, was the recently retired and wealthy commander of an East Indiaman, who later became a director of the East India and East India Docks Companies, and governor of the London Assurance Company. He was to become Deputy Master of Trinity House 1803–25 and a Fellow of the Royal Society in 1810.[107] Strachan, in the 1750s a ship's master sailing between Scotland and the American colonies, was based in London by the 1760s and a merchant by the 1770s. Based in Mincing Lane, in Tower Street Ward, he became an Elder Brother in 1788. He showed a personal interest in advancing navigation: as well as subscribing to Andrew Mackay's 1793 *The Theory and Practice of Finding the Longitude*, he left £100 to Trinity House in his will, 'in Case a School of Mathematical Science & particularly of navigation and nautical Astronomy … shall be established by the Corporation', and a gold ring for each Elder Brother.[108]

Trinity House collectively, led by such members particularly, adopted a self-consciously experimental approach to lighthouses in the later eighteenth century. Trial had always had a role to play in making decisions about whether the provision of lights would be beneficial and how to make them effective. Elder Brethren were tasked to travel to view the situation of proposed lights, or the effects of those in use. They also drew on the opinions of their local officers and mariners more widely. Positioning had to be decided, and the shapes of grates, flues and lanterns and types of fuel used were sometimes adapted, whether through local or central initiative. Even before the more intense developments of the late eighteenth and nineteenth centuries, lighthouses saw frequent change. Structures at Lowestoft, for example, were built in 1609 to display two candle lights, higher being rebuilt in 1675 to display a coal-fired light, and the lower converted in 1730 to a whale oil lamp.[109] Local changes were not necessarily approved. In 1677, Captains Browne and Collier reported that the lights at North and South Foreland had become dimmer because of 'laying aside the antient way of fire Cradles to save expence of Coales'. The keepers had used 'a new invention of Furnaces which holds <not> a quarter part of Coales that the Cradles did' and would 'burne cleere no longer than the Coales in them are continually stirred'.[110] However, some suggestions were taken up: Mr Harding, the lightkeeper at Castor (Caistor), wrote in

[106] Convertito-Farrar and Cozens, 'Trinity House', pp. 5, 8–9. Timothy Mangles was named as Master of the *Charming Jenny* Transport in 1759: ADM 354/162/124 (TNA). Identified as a gentleman of Leytonstone, he left £100 in stock to Trinity House, with the interest to go to the poor: Will of Timothy Mangles, 20 December 1800, PROB 11/1351/239 (TNA).

[107] Election Certificate for Joseph Cotton, EC/1809/14 (Royal Society). J. S. Cotton and H. V. Bowen. 'Cotton, Joseph (1745–1825), Merchant', *OBNB*. While eighteenth-century Deputy Masters were often self-made mariners-turned-merchants – Cotton, less typically, was the son of a physician and poet, Nathaniel Cotton. In 1814, he was awarded a medal by the Society for the Encouragement of Arts and Manufactures for the introduction to Britain rhea, or China grass, a particularly strong and fine fibre.

[108] Will of James Strachan of Tooting, 12 May 1808, PROB 11/1480/58 (TNA); Andrew Mackay, *The Theory and Practice of Finding the Longitude at Sea or Land* (London, 1793) – Mackay became Trinity House's mathematical examiner in 1805: see Higgitt, 'Mathematical Examiners'.

[109] Trinity House, 'Lowestoft Lighthouse' (2016); https://www.trinityhouse.co.uk/lighthouses-and-lightvessels/lowestoft-lighthouse.

[110] 7 June 1677, Court Minutes, CLC/526/MS30004/005, p. 41. The Court ordered that this furnace be removed and be replaced with 'such fire Cradles' as formerly used.

1706 'signifying that it would be of more use to Navigation if the small Light there was further removed to the WSW'. He was asked to supply an estimate, which was agreed, despite the move actually being made east-northeast. The Brethren either trusted, or considered from experience, that the move would be 'for the Publick good'.[111]

From the 1770s, experiments, or trials, were carried out more centrally, as Elder Brothers responded to external developments and encouraged the Corporation to take them up collectively. In 1777, Michael James, one of the more active surveyors and examiners, 'acquainted the Board that he had applied to Mr Edward Nairne, Mathematical Instrument Maker, relative to the Construction and Use of Reflectors', who recommended Mr Everard of Lynn, the patentee of a reflecting light in use there. James and another Elder Brother went to view it, reported to Trinity House, and had one made to 'carry the Experiment into Execution' at Lowestoft. This experimental exhibition of the light was advertised in the press, requesting 'the Opinion of Navigators' of its usefulness.[112] Thus, Lowestoft's coal fire was replaced by a new system consisting of a circle of oil lamps and cylindrical reflector system of 4,000 small mirrors, known as the 'spangle light'.[113] In the same year, three copper lanthorns 'and proper Reflectors to be used therein' were ordered from James Lukin for the Casket lighthouse while, the following year, the Board recommended that the Court 'direct Reflectors with proper Lamps to be made for the Light Houses at Portland'.[114]

In the 1780s they turned to the exploration of parabolic reflectors, this time with the Deputy Master, Sir William James, telling the Court that he had 'sometime seen' an improved reflector by Lukin. This he wanted to have tried at Scilly 'with all the Dispatch possible'.[115] Lukin was brought in to set up the new lights and in 1785 was rewarded for his services 'in making Experiments of reflecting Lights, and in proceeding to examine the Light-houses at Scilly the Caskets, Portland, the Needles, &c. by Order of this Corporation at various Times'.[116] In 1788, 'several Experiments of Lights' were 'exhibited at Shooter's Hill', the highest point over Greenwich. Their success led the Elder Brothers 'to try the Experiment of a Coal Light in Comparison with the Lamps & Reflectors' at Portland.[117] The full report of the Portland experiments – as directed by the Deputy Master, Captain Hector Rose, and two Elder Brethren, 'attended by Mr Spence Maritime Surveyor, and Mr Lukin Manger of the Works' – was read, and Spence's sketches shown, at the General Court of 28 August 1788.[118] Installations of

[111] 21 August and 9 November 1706, Court Minutes, MS 30004/008 pp. 38, 42.

[112] 13 August 1777, 14 January and 17 June 1778, Board Minutes, CLC/526/MS30010/16, pp. 68, 119, 195.

[113] D. Alan Stevenson, *The World's Lighthouses before 1820* (Oxford: Oxford University Press, 1959), 1776–8, 1898–9.

[114] 12 June 1779 and 22 April 1780, Board Minutes, CLC/526/MS30010/013, pp. 348, 467.

[115] 1 November 1783, Court Minutes, CLC/526/MS30004/013, p. 195. William James, who had been a royal naval officer with the Bombay Marine, made a fortune from prize money and private enterprise before becoming a Director of the East India Company in 1768. He died just a month an half after this. T. H. Bowyer, 'James, Sir William, first baronet (1722–1783), naval officer and director of the East India Company', *ODNB*. James Lukin may have been related to Lionel Lukin (1743–1834), a lifeboat designer whose first boat was named the *Experiment*, and James Lukin (1828–1917), author of many popular mechanical books.

[116] 1 September 1785, Court Minutes, CLC/526/MS30004/013, p. 255.

[117] 14 March and 23 April 1788, Court Minutes, CLC/526/MS30004/013, pp. 349, 355.

[118] 28 August 1788, Court Minutes, CLC/526/MS30004.013, p. 362.

reflectors continued at other sites, although by 1796 Lowestoft's system had changed again, this time using a parabolic reflector alongside an Argand oil lamp.

There were also trials in relation to fuels. John Hill, their agent at Wells in Norfolk, was sent a new oil to try in 1778, which he reported to be no better than the former. However, he took the initiative of mixing it with rape oil, 'which made it burn and give a very good Light'. The Court then ordered that rape oil 'be sent when wanted to Wells, Caster, & Lowestoft, to be tried in the Light Houses there'.[119] In the same year, the By-Board considered trials of cannel coal, sent from Liverpool to Scilly, which was found to 'maintain a much clearer Light than any other Coals with fewer Cinders and less Waste'.[120] In 1779 tests were carried out in relation to the spangle light to see if fuel use could be minimized by reducing the numbers of burners lit without compromising safety. The lightkeeper at Lowestoft, James Mayes, had reported on oil consumption but the By-Board members were dissatisfied and asked him, again, 'to take the Opinion of some well judging Persons' about whether leaving off all the land-side burners would still leave the light brighter than the coal light and 'to desire that a circumstantial and intelligible Opinion may be obtain'd' about how much brighter it might be. This time Mayes returned a report on 'the several Persons who observed the Reflect for three Nights successively' about the brightness and reported a use of five quarts of oil in twelve hours with the reduced number of burners. He was then asked to try again with a further reduction in the number of lamps.[121]

While Mayes seems not to have been fully trusted to carry out these trials effectively, other agents clearly were. The Brethren were increasingly likely to pay for and delegate responsibility to individuals with particular skills. Commonly, like Spence and Lukin, they were contracted for particular tasks, differentiated from the local workmen used previously by being brought in from elsewhere. A notable example is that of John Smeaton who built lighthouses at Eddystone and Spurn Point in the 1750s–60s. His name had been suggested when the president of the Royal Society had sought a recommendation for an engineer. The hydrographic surveyor Lieutenant Murdoch Mackenzie carried out work commissioned by the Corporation in the 1770s and 1780s in a much longer-term arrangement than had existed with Greenvile Collins a century earlier.[122] Mackenzie was a naval officer and not paid as a contractor but in 1785 the Court instructed the By-Board to consider compensation for 'his Employment by this Corporation in Surveys of the Queens & other Channels from 1774 to 1777; and also in Surveys of the Needles Channels from 1781 to 1784 also in drawing Draughts of the said Channels for the Use of the Corporation at various Times &c.'[123] Changing expectations about the use of expertise and Trinity House's increasing wealth both encouraged a regularizing of such arrangements. In 1792 they appointed the architect and engineer Samuel Wyatt as the Corporation's surveyor, the first in a series of such positions – including, from

119 11 April 1778, Court Minutes, CLC/526/MS30004/013, p. 13.

120 9 December 1778, Board Minutes, CLC/526/MS30010/13, p. 285.

121 13 and 27 March 1779, Board Minutes, CLC/526/MS30010/13, pp. 318, 323.

122 A. W. Skempton, 'Smeaton, John (1724–1792)', *ODNB*; Adams and Woodman, *Light Upon the Waters*, 92–5; Andrew David, 'Lieutenant Murdoch Mackenzie and His Survey of the Bristol Channel and the South Coast of England', *The Cartographic Journal* 40, no. 1 (2003): 69–78.

123 1 September 1785, Court Minutes, CLC/526/MS30004/013, p. 255. Mackenzie discovered the Queen's Channel in 1774.

Figure 6.7 The Court Room of Trinity House in the early nineteenth century, showing the group portrait on the far wall and a glimpse of the decorated ceiling. Augustus Pugin and Thomas Rowlandson for Rudolph Ackermann's *Repository of Arts* (London, 1809). The Elisha Whittelsey Collection, The Elisha Whittelsey Fund, 1959, Metropolitan Museum of Art.

See also colour version, reproduced in plate section 11.

the 1830s, scientific advisors. In this role until his death in 1807, Wyatt erected several lighthouses as well as the new Trinity House on Tower Hill.[124]

Conclusion

Wyatt was included in the large group portrait by Gainsborough Dupont of the Merchant Elder Brethren, (visible in the background of Fig 6.7/Plate 11) commissioned to decorate the splendid new classical Trinity House completed in 1798.[125] Standing

[124] Trinity House surveyors, scientific advisors and engineers are listed in Adams and Woodman, *Light upon the Waters*, 299; G. H. L. Le May and John Martin Robinson, 'Wyatt, Samuel (1737–1807), Architect and Engineer', *ODNB*.

[125] John Hayes, 'The Trinity House Group Portrait', *Burlington Magazine* 106, no. 735 (1964): 309–16. By the 1770s the Merchant Brethren were a distinct group of 20 Elder Brethren, holding their own meetings and putting forward nominations for Master and Elder Brothers.

next to the only other figures not in dress uniform (the Corporation's secretary, David Court, and solicitor, Ambrose Weston), he offers his plans for the new building to the Deputy Master, Captain Hector Rose. Seated on the far right, conversing over a paper, are Captains Joseph Cotton and Joseph Huddart. The latter, already a Fellow of the Royal Society, had as an East India Company commander established a reputation as an excellent surveyor, making use of the latest techniques to achieve accuracy. Trinity House thus retained its own claim to expertise, with Huddart surveying for the Corporation and overseeing lighthouse projects.[126] As in the seventeenth century, so at the dawn of the nineteenth, Trinity House presented its members as expert in maritime matters. Over that period, however, it had become increasingly obvious that it could most effectively do so by drawing on the particular skills of just some of their number – to undertake examinations of pilots and masters, surveys and trials – or by co-opting the skills of others.

The collective nature of Trinity House, and its focus of activities within Water Lane, had ensured that these skills could be incorporated into the projected authority of the whole. Its new building, on Tower Hill, was a statement of current wealth and significance but also drew on institutional history, bringing the objects and images displayed in the old House with them. Dupont was initially brought in to clean 'the antient and valuable Pictures now in the Court Room, many of them former Kings and Queens of this Realm', ready to hang in 'the most conspicuous Part of the New Building'. He was also commissioned to paint four individual portraits (the current monarch and consort, Earl Howe and William Pitt, Master of Trinity House and Chancellor of the Exchequer), and the canvass, described in the *Morning Herald* as 'a large CONVERZATIONI PIECE of the whole *effective Trinity Board*'.[127] Apart from the uniforms, there are few clues as to their particular skills, despite the claim of Cotton that the Merchant Brethren were, as a result of 'extensive experience and knowledge in every branch of nautical practice and theory', considered 'the first maritime authority in the empire'.[128] However, just discernible in the background are a lightship, what appears to be the dome of St Paul's (symbolizing London), a terrestrial and a celestial globe. On the ceiling above the portrait and the Brethren was painted a vast armillary sphere, supported by putti, others of which wield navigational instruments: telescope, compass, cross staff and octant (Fig. 6.7).[129]

The plans of the new building reveal much about how it conducted its business. The key officer, the secretary, had both an office and room assigned in the basement, while the principal storey had the Deputy Master's room. Accessible only via this was a small room labelled 'Archive', containing the Corporation's books and records. The principal storey retained the old pattern of two larger rooms, the Court Room and Board Room. Leading to the former are a suite of waiting room, anteroom and withdrawing room, suggestive of the number of people hoping or expecting to be called into meetings

[126] Susanna Fisher, 'Huddart, Joseph (1741–1816), hydrographer and engineer', *ODNB*.
[127] Quotes in Hayes, 'The Trinity House Group Portrait', 310.
[128] Cotton, *Memoir*, 129.
[129] The ceiling decoration was originally by J. F. Rigaud and restored by Holman Hunt in 1850: Adams and Woodman, *Light upon the Waters*, 129. The coves contain allegorical representations of rivers, the sea, Britain and its royal and merchant naval might.

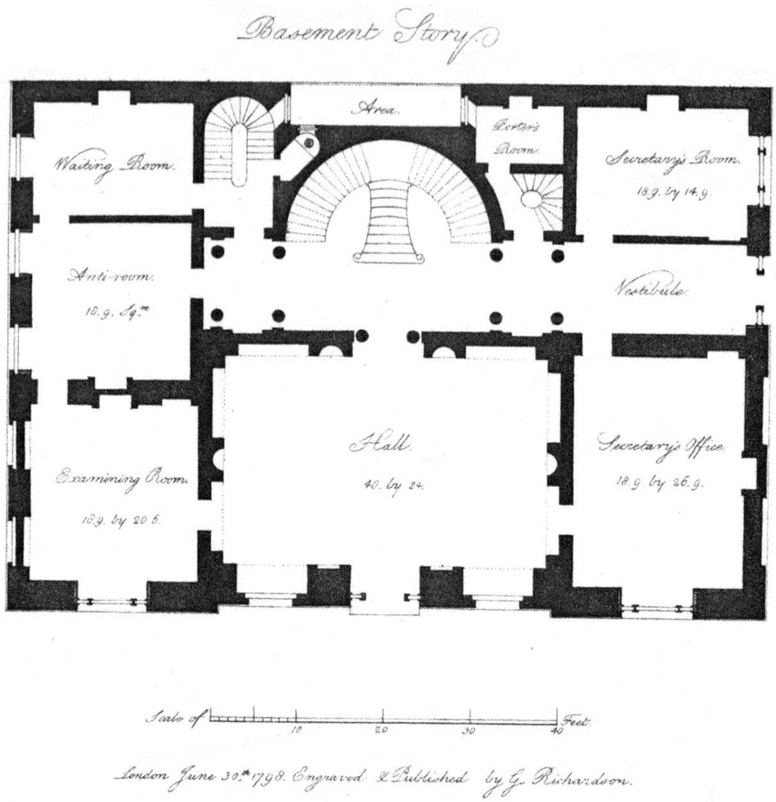

Figures 6.8 Plan of the Basement Storey (ground floor) of Trinity House, by George Richardson, 1798 (LMA: SC/PZ/CT/01/0061). © London Metropolitan Archives (City of London).

of the Court. There was a similar set-up of waiting room and anteroom on the floor below, this time leading expectant candidates to the Examining Room (Fig. 6.8).

Trinity House, as a charitable entity, received criticism for profiteering from its monopolies and splashing out on its showy new residence. In defence they became more transparent regarding finances – pointing out the large sums of money spent on charity – and asserted their vital role in the safety and prosperity of the nation. By the early decades of the nineteenth century, it was politic to link their efforts to science and to underline the selfless nature of the Elder Brothers' efforts. In 1818 Cotton ended his *Memoir* by claiming 'the pride, as well as the gratification of the Brethren, has been to give every encouragement to scientific pursuits, and practical maritime improvements, so that the Corporation may be fully entitled to the motto – Prodesse quam conspici': to accomplish rather than to be conspicuous.[130]

[130] Cotton, *Memoir*, 160.

7

Regulating practice and using knowledge at East India House

Introduction

From 1600 to 1813, the monopoly of legitimate trade between subjects of the English crown and all territories east of the Cape of Good Hope was in the hands of the Governor and Company of merchants of London trading into the East Indies. This cumbersomely named body was more commonly called the East India Company (EIC). It consisted of an elected Governor and his deputies, a treasurer and twenty-four committee men, who together met and managed the Company as the Court of Committees, later known as the Court of Directors. These were responsible to the Generality, or General Court of Proprietors, an initial hundred or so venturers who put up the money for the new Company's first trading voyage, later rising to several hundred, as they subscribed to further voyages.[1] Planning and financing had begun at least eighteen months before the Company obtained, at the second attempt, a charter of incorporation from Elizabeth I on 31 December 1600.

The new Company was headquartered at the house of its first governor, Sir Thomas Smythe, in Philpot Lane on Fenchurch Street, from its establishment until Smythe retired in 1621. No other arrangement is known to have been made for the EIC at this period even when Smythe was absent, as he often was. In 1601 he was held in the Tower on suspicion of complicity in the Earl of Essex's abortive rebellion against Elizabeth I, or from June 1604 until summer 1605, when he travelled to Moscow to negotiate new trading privileges for the Muscovy Company.[2] From 1621 to 1638 the EIC leased Crosby House in Bishopsgate from the Earl of Northampton. When he tried to raise the rent the Company moved to Leadenhall Street for the first time, where it would remain until its formal winding-up in 1858. There it occupied a succession of houses; initially the house of its Governor, Christopher Clitherow, and then, from 1648, the adjoining Craven House rented from its owners until 1712 when an arrangement was finally made to purchase it. This house was demolished in 1726 and re-edified over the

[1] On the organization and governing of the EIC, see Rupali Mishra, *A Business of State: Commerce, Politics, and the Birth of the East India Company* (Cambridge, MA: Harvard University Press, 2018), 37–46.

[2] Basil Morgan, 'Smythe [Smith], Sir Thomas (*c.* 1558–1625)', in *ODNB*.

next three years in fashionable Palladian style.[3] Finally, between 1796 and 1800 it was remodelled and extended to contain, besides the grand Sale Room and large rooms for the Courts of Proprietors and Directors to meet, office space for 400 clerks, thousands of volumes of the Company's records, a library and a museum.[4]

We do not claim that the Committees or Directors sitting within East India House had effective control over the wide range of knowledge-using and -making activities that took place across, within and at the boundaries of its corporate world.[5] Nevertheless, within this series of buildings, and over the two centuries spanned by this book, were many types of knowledge-community. The concerns and activities around which these gathered are divisible into three broad categories: those that were basic to the logistics of its primary function, that is, sailing ships to Asia and back with valuable cargoes intact and in good, saleable condition; those that supported the Company in establishing the extent, improving the administration and maximizing the exploitation of its Asian territories, and which led to the creation of either new markets or alternative sources of supply for existing ones; and those linked to its role as a primary producer, or vector, of the raw materials of science, including data and specimens. The first were fostered throughout the Company's history as a trading concern, from 1600 to 1833, while the second were of more pressing concern after its acquisition of vast territories in India, following the battles of Plassey and Buxar (1757 and 1764). Activities relating to the third category were exercised sporadically throughout its history, initially on behalf of other individuals and associations but subsequently also on its own account, and were pursued more systematically and purposefully from the end of the eighteenth century onwards.

These categories are, of course, not wholly separable. They all overlap to some degree, and especially the first with the second and the second with the third. But we suggest that they are meaningfully distinct, partly because they were more or less important at different times, and thus are useful means of making the chronological sweep of the EIC's knowledge-making history manageable. Another kind of distinction we can helpfully make is between those forms of knowledge that the EIC, or a specialized community within it, had to master in order to operate effectively as a commercial organization, and those that it was notably well-placed to produce. Again, this is not a perfect distinction but, in general, the further back we go in the Company's history the more likely we are to find its knowledge-gathering activity narrowly linked to its

3 William Foster, *The East India House: Its History and Associations* (London: John Lane the Bodley Head Ltd, 1924), 2, 5, 18; Malini Roy, 'East India Company headquarters on Leadenhall Street', British Library Asian and African Studies Blog, 25 January 2017, https://blogs.bl.uk/asian-and-african/2017/01/east-india-company-headquarters-on-leadenhall-street.html (accessed 20 July 2023).

4 John Timbs, *Curiosities of London: Rare and Remarkable Objects of Interest in the Metropolis* (London: David Bogue, 1855), 264; Jessica Ratcliff, 'The East India Company, the Company's Museum, and the Political Economy of Natural History in the Early Nineteenth Century', *Isis* 107, no. 3 (2016): 495–517.

5 Anna Winterbottom sees the wide range of knowledge creation and exchange within EIC territories and sphere of influence as essentially ungovernable from the centre: *Hybrid Knowledge in the Early East India Company World* (Basingstoke: Palgrave Macmillan, 2016), 8–15. Jason Grier, exploring navigational knowledge and the EIC, states that 'any strong claim for European long-distance control does not stand up to historical scrutiny': 'Navigation', 22.

commercial objectives. We suggest that these different types of knowledge can be productively analysed using the same set of fundamental questions: How were these types of knowledge acquired and compiled? Who had access to them and how were they distributed? How was expertise assessed and valued within the Company?

Much of the historiography touching on knowledge and the EIC focuses on activities carried out in its overseas territories but, though clearly connected, our focus is on decisions and activities in and around the Company's London headquarters.[6] This chapter is concerned with how the East India Company identified, assessed, secured and transmitted such knowledge and skill, and how its methods for doing so evolved as the Company grew and its functions transformed. Despite concerns to protect commercially sensitive knowledge, the Company also relied on other businesses – booksellers, chart makers, scientific instrument makers and teachers of mathematics and navigation – to ensure that its ships could reach and return safely from overseas. Its requirements worked to shape businesses and premises in the area around its headquarters. Over time, and for reasons both internal and external to the Company, its attitudes towards the acquisition of navigational skill, including its support by astronomy and survey, and of new knowledge about their colonized territories, became increasingly public in nature. We are not primarily concerned with moments at which the Company, or its servants and the data or materials they could supply, were of interest to or aimed directly at scholars and natural philosophers, but with the Company's changing motivations around the deployment of such networks.[7] As we show, it was only in the following century that the Company by policy and at scale invested in people, institutions and projects that could be labelled 'scientific'.

The first site we consider in this chapter is Crosby House, but the first depiction of an East India House dates from its period at Craven House. This building was described by Stow in 1603 as a 'faire house' and by Daniel Defoe over a hundred years later as 'an old, but spacious Building; very convenient, though not beautiful' (Fig. 7.1).[8] This was despite the attempts to 'beautify the front of this house' for the coronation of Charles II, 'to represent the Companies loyall gratitude to His Majesty'.[9] On the morning of 17 April 1661, Pepys saw both the coronation arches and 'the picture of the ships and

6 See e.g. Deepak Kumar, 'The Evolution of Colonial Science in India: Natural History and the East India Company', in John M. MacKenzie, *Imperialism and the Natural World* (Manchester: Manchester University Press, 1990), 51–66; Kapil Raj, *Relocating Modern Science: Circulation and the Construction of Knowledge in South Asia and Europe, 1650–1900* (London: Palgrave Macmillan, 2007); Matthew H. Edney, *Mapping an Empire: The Geographical Construction of British India, 1765–1843* (Chicago: University of Chicago Press, 1997); Winterbottom, *Hybrid Knowledge*.

7 This is in contrast to a study of another trading company, the Hudson's Bay Company, which emphasizes contributions to knowledge in the public realm: Binnema, *Enlightened Zeal*. Our approach has commonalities with Jennifer Howes, *The Art of a Corporation: The East India Company as Patron and Collector, 1600–1800* (London: Routledge India, 2023), which considers how the Company's changing self-image and understanding of its internal and external audiences are revealed in its commissioning and collecting of art.

8 John Stow, 'Limestreete warde', in *A Survey of London. Reprinted from the Text of 1603*, ed. C. L. Kingsford (Oxford: 1908), 150–63. *British History Online,* http://www.british-history.ac.uk/no-series/survey-of-london-stow/1603/pp150-163 (accessed 15 June 2023); Daniel Defoe, *A Tour Thro' the Whole Island of Great Britain, Divided into Circuits or Journies* (London: 1725), vol. 2, 133.

9 Quoted in Foster, *East India House*, 41.

Figure 7.1 The Old East India House in Leadenhall Street, 'From a painting in the possession of Mr. Pulham of the India House', *c.* 1830 (BM: 1978,U.3639). © The Trustees of the British Museum.

other things … set up before the East Indy-house'. They were, he said, 'well done'.[10] The ships were still visible at the end of the century, when Ned Ward's *London Spy* (1698) commented on them, and were still present, if repainted, when represented in a drawing dated to around 1711.[11] Decoration included the Company's and the Royal coat of arms, the figure of a merchant adventurer and, in two earlier views, dolphins. The monarch was represented by his arms, but the main theme was, clearly, maritime. Representation of the Indies, while present among the coronation arch described in Chapter 2, was entirely absent. It is only after the demolition of Craven House that

[10] *The Diary of Samuel Pepys*, vol. 2, 1661 (1970), 77.

[11] Forster, *East India House*, 129; this image, attributed by Foster to George Vertue, *c.* 1711, is British Library, WD1341.

the Company 'expressed its power and authority through architecture and design of its buildings'. This was when the Company also began to collect and display art and artefacts that made its headquarters 'as an institutional focal point for illustrating and conveying ideas about the empire in India'.[12] Externally, the 1720s building presented solidity and dependability, while decoration and artworks inside literally and figuratively represented the wealth that Britain gained through the Company's monopoly.[13]

The frontage of the final East India House was on significantly more imposing scale, with prominent sculptural reliefs and statuary, befitting the large territorial empire acquired by the Company (Fig. 7.2/Plate 12). Now fully entwined with the British State and Empire, the Company displayed on its pediment George III, in Roman armour, surrounded by Liberty, Order, Religion and Justice, defending the commerce of the East. This scene was surmounted by Britannia, flanked by figures

Figure 7.2 The grand new frontage of the East India Company's headquarters in Leadenhall Street, *c.* 1800, by Thomas Malton the Younger. Yale Centre for British Art, Paul Mellon Collection.

See also colour version, reproduced in plate section 12.

[12] John McAleer, 'Exhibiting the "Strangest of All Empires": The East India Company, East India House, and Britain's Asian Empire', in *The MacKenzie Moment and Imperial History*, ed. Stephanie Barczewski and Martin Farr (Cham: Springer International Publishing, 2019), 25–45 (p. 26).

[13] Mildred Archer, 'The East India Company and British Art', *Apollo* 45 (1965): 401–9.

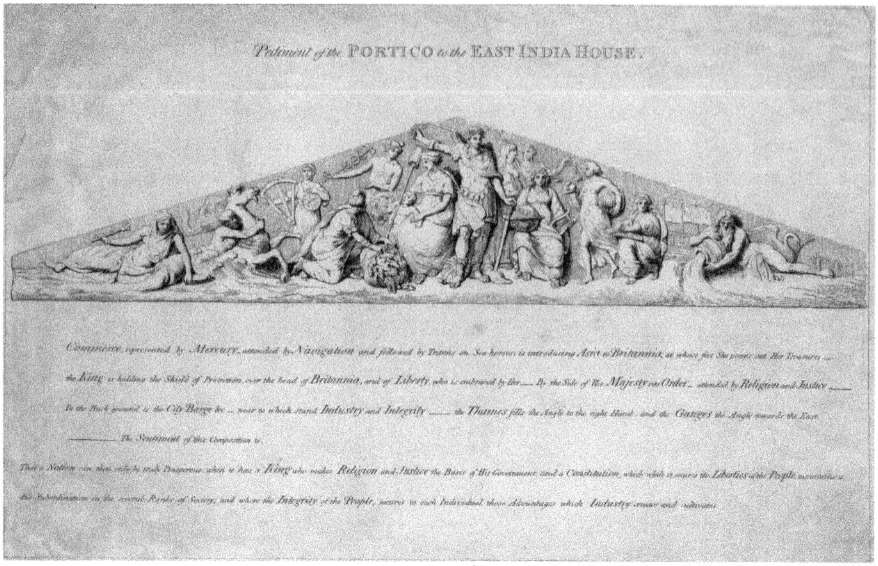

Figure 7.3 Print showing the design of the pediment of the portico to the East India House, after John Bacon (BM: 1852,0328.33). © The Trustees of the British Museum.

of Europe and Asia.[14] The message of the portico's decoration (see Fig. 7.3) was given by its designer as follows:

> That a Nation can then only be truly prosperous, when it has a King who makes Religion and Justice the basis of His Government, and a Constitution, which while it secures the Liberties of the People, maintains a due Subordination in the several ranks of society; and where the Integrity of the People, secures to each Individual those advantages which Industry creates and cultivates.[15]

While this, and the classical architecture, made grand statements about nation, monarchy and empire, the Company's reliance on resources of London and skills of its mariners was not forgotten. To the right of the central group, the figures of Industry and Integrity connect the scene to the river Thames and the City Barge. On the other side, Mercury represents Commerce, introducing Asia and her riches to Britannia. The figure of Navigation bridges Commerce and the river Ganges, taming the waves represented by Triton on a sea horse. She carries a compass and, prominent in the design, a large Hadley's quadrant, or octant.

[14] Joan Coutu, *Persuasion and Propaganda: Monuments and the Eighteenth-Century British Empire* (Montreal: McGill-Queen's University Press, 2006), 287–8.

[15] From the inscription to the print showing the pediment's design (British Museum: 1852,0328.33).

Information and security at East India House

The Company's London establishment in the early seventeenth century bore similarities to other spaces considered in this volume. East India House was multifunctional, being both a working and a residential space. It was also a space that, like the Mint's Assay House at the Tower, had to be secure and private where called for, but quasi-public at other times, for example when members of the generality attended meetings of the General Court, or when individual employees (members of neither the generality nor the Court of Committees) or their families wanted to petition the Company. The curiosity of the public was an acknowledged threat: the Company was governed by a body of men who were acutely conscious, by virtue of considerable experience in the management of long-distance trade, of what might be called the metropolitan hazards of its operations, and of the need to secure every step of the process of building ships, loading them, and getting them into the open ocean. From that point the management of the Company's enterprise was committed into the care of its merchants, factors, pursers, captains and masters.

Besides directors and proprietors, the EIC's *Lawes or Standing Orders*, published in 1621, identified forty-two types of employees and outlined their responsibilities, including mariners, porters, shipwrights, carpenters, pilots, watchmen and clerks of various kinds. Within twenty years of its establishment, therefore, the EIC had mobilized a varied and skilled workforce and set out formal rules for its recruitment and employment.[16] They operated over several different sites and a considerable range; as well as the Fenchurch Street headquarters, the Company owned or operated docks and shipyards at Deptford and Blackwall, warehouses in the City, timber-yards in Reading, the ships themselves and a storehouse in the Downs (the East Kent anchorage that was also the last point on English soil at which outbound company ships were supposed to dock).

The regulations reveal detailed attention to all aspects of the Company's work in England, defined chains of responsibility within the EIC, and reflect how seriously ideas of security and work discipline were taken. The Company's Deptford yard was patrolled at night, and each piece of timber brought into it was numbered and audited. The standing orders provided for a taphouse within the precincts to sustain the workmen and specified what food and drink it should sell and at what prices. (Beer consumption at breakfast and between meals was limited to six pints per man, for example.) The Clerk of the Yard was to take particular care that no 'Strangers, or Loyterers' should lurk in the taphouse. Refreshment taken other than at mealtimes was to be carried out to the workmen, so that they would have no excuse for being indoors during ordinary working hours.[17] The yard was guarded during the day by a porter and at night by watchmen. The porter's responsibilities included ensuring that the workmen kept regular hours, that all shipbuilding materials brought into or taken out of the yard were accounted for, and preventing theft by workmen. He was also instructed under no circumstances to admit 'Gentlemen, Strangers or any others' to

[16] East India Company, *The Lawes or Standing Orders of the East India Company* ([London]: 1621).

[17] *Lawes or Standing Orders*, 32.

the yard for money, a prohibition specific enough to suggest that the yards were an object of genteel curiosity and possibly of espionage.[18]

The EIC also put strong protections in place around its 'house'. The publication of the *Lawes or Standing Orders* in 1621 coincided with the Company's move to Crosby House in Bishopsgte, and the instructions were probably specific to the new quarters. There, too, a porter was responsible for monitoring and opening the street door and back gate, and for checking every night that all the open rooms in the house were empty and that their doors were shut and locked. This was to be done in the company, or under the supervision, of a named officer, John Blount, who may have been a relative of Sarah Blount, wife of the Company's first governor. The keys were then surrendered to the Company's Secretary, while the external door was also to be 'shut and chayned euery Euening' after the officers had left. The street door and back gates were secured 'with Two lockes of seuerall openings', with the porter and Blount each holding a separate key. Several of the Company's employees had lodgings in the house, and they would be shut into the building between ten at night and six in the morning.[19] The EIC's bullion, charters, treaties and official correspondence were kept in a secure room under the supervision of the Treasurer, and subsequently a committee of three, who held its keys.[20] When the Company had occupied part of Sir Thomas Smythe's house that, too, was fitted with a strongroom. Much later, in the 1680s, the Company discussed the idea of refitting part of its house with an assay room, to assess the soundness of the bullion that was the Company's primary seventeenth-century export to Asia.[21]

Many of the early members of the Court of Directors had been involved with the Levant, Muscovy and Virginia Companies, among others, but, for all their knowledge of the practicalities of trade, no-one in the government of the EIC at the time of its establishment had any direct experience of the sea route to Asia. Almost nobody in England at the time did. How to muster the knowledge and skills necessary to sailing large ships and their valuable cargoes safely to India and back was the Company's most pressing problem, after securing the financing necessary to undertake the voyage in the first place.

Since the Company's commercial charter was founded on the principle of exclusivity – it existed precisely in order to monopolize trade between England and the East Indies – its early approach to knowledge-gathering was in many respects equally exclusive. All of the principal seamen and merchants on the Company's early voyages were required to keep detailed accounts of the maritime and commercial conditions they encountered and to provide those accounts to the Company's London leadership. For example, it was a condition of George Waymouth's commission for an exploratory voyage on the Company's behalf to discover the North-West Passage that he submit his report within ten days of his return and make himself available for detailed questioning for the next forty.[22] The *Lawes or Standing Orders* stipulated that the masters and mates of Company ships should deliver 'faire Copies of their Iournaell Bookes, concerning

[18] *Lawes or Standing Orders*, 27.
[19] Ibid., 65–6.
[20] Ibid, 9; Foster, East India House, 35–6.
[21] Morgan, 'Smythe, Sir Thomas'; Foster, *East India House*, 46–7.
[22] George Birdwood and William Foster (eds), *Register of Letters etc., of the Governor and Company of Merchants of London Trading into the East Indies, 1600–1619* (London: Bernard Quaritch, 1893), 22.

their Navigation, and other worthy observations in their course of their whole voyage' to the Court of Committees, likewise within ten days of the ship's return to England.[23]

There was an abiding tension between the wish for exclusivity and the Company's need to mobilize and create navigational expertise. Recruiting mariners with experience of sailing to the East Indies was difficult because England had so few of them at the end of the sixteenth century. At least two senior members of the English Company's first voyage (1601–3) *had* previously sailed to the East Indies – John Davis, the fleet's Pilot-Major, and James Lancaster as General in overall command. If the initial pool of navigational knowledge of the East Indies was somewhat shallow, it was reasonable to hope that the voyage itself would generate more for future use. Davis himself illustrates the difficulty of maintaining an exclusive hold on individual expertise, however, since he subsequently joined Sir Edward Michelborne's interloping voyage to the East Indies in 1604–5. Michelborne's voyage was granted a charter by James I in apparent defiance of the EIC's exclusive licence for the East India trade.[24] This was seen by the Company, correctly, as an enormous danger. If an English rival managed to establish itself in the India trade, there was every chance that the Crown, from whom the Company held its monopoly, would simply recognize the *fait accompli*. Asian trade, as Rupali Mishra points out, was viewed as a public benefit; that was the basis of the Company's licence, and if it failed to deliver that benefit, or if somebody else could do it as well, there was no reason to continue its charter.[25] The Company therefore had reason to be careful about how knowledge of its activity circulated. Multiplying knowledge of East India navigation and trade might mean that future voyages were smoother and more prosperous but it also risked facilitating possible competition.

The specific challenge the EIC faced was how to transmit knowledge selectively without disseminating it. Margaret Schotte has recently emphasized the role of print in improving the spread and systematization even of 'small' (i.e. coastal) navigation, a skill which until the sixteenth century was overwhelmingly acquired by practical experience. As Schotte points out, the spread of printing by the last quarter of the sixteenth century made secrecy in navigational knowledge virtually impossible to maintain.[26] This insight makes the EIC's avoidance of print in its early years all the more striking. The Company made a point of gathering as much information as possible about East India navigation and trade, but it was to be gathered centrally at the Company's headquarters and made available only to a select few. While the Dutch East India Company (VOC) began to produce its own charts and set up an office for the purpose very soon after its establishment, eventually adding local cartographic services at its major Asian entrepôts, there was no hydrographic office at the East India Company, even after it began to make regular payments to a hydrographer from the 1770s.[27] How then did the EIC expect to build up the body of knowledge needed to

23 *Lawes or Standing Orders*, 45.

24 Charlotte Minh Hà Pham, 'European Navigation, Nautical Instructions and Charts of the Cochinchinese Coast (16th–19th Centuries)', *Moussons. Recherche En Sciences Humaines Sur l'Asie Du Sud-Est* 27 (2016): 101–29.

25 Mishra, *A Business of State*, 9–12 and Chapter 5.

26 Schotte, *Sailing School*, 31–8, 10.

27 W. F. J. Mörzer Bruyns, 'Navigation on Dutch East India Company Ships around the 1740s', *The Mariner's Mirror* 78, (1992): 143–54 (p. 143).

facilitate the successful sailing of ships from London to the East Indies and back? Can policy be inferred from practice, or were they – in practice – at odds? And what was the significance of London as a repository of knowledge and centre of calculation?

Statements of policy are few and far between. The Company specified that all the senior mariners and commercial men aboard its ships should keep detailed journals of information relevant to trade, navigation, local politics and the resupplying of ships during the voyage. As we have seen, fair copies were to be turned over to the Company promptly upon their return – the originals their compilers were implicitly allowed to keep. In principle this system allowed individual mariners to retain the benefit of their acquired experience while giving the Company the capacity to draw upon, collate and distribute it for the benefit of sailors without prior experience of the East India voyage. Some of this information evidently found its way into the instructions given to the third voyage (1607–10), particularly concerning places for the ships to resupply (as well as dire warnings against the evil physical and moral effects of overindulgence in fruit on those occasions).[28] Company captains and masters were issued with rutters, that is, sets of written sailing directions for the voyage to the East Indies. Of the surviving documents from the maritime records of the Company's early voyages, there is one compendium of notes on early voyages; but this document is in multiple hands and significant parts of what it contains could have been compiled from printed sources. The section described as 'a Rutter, or breif and plaine direction for ready sayling into the East Indies digested into a plaine method by experience thither and home againe', for instance, is that of John Davis of Limehouse, whose rutter for the East Indies was published as part of Samuel Purchas's *Purchas his Pilgrimage* in 1625.[29] That this book was a source for the manuscript compendium, rather than vice versa, is supported by the inclusion of other materials that likewise appeared in Purchas.

This, however, points to some of the potential complexity and circularity surrounding the collation and transmission of navigational knowledge of the East India voyage in the early seventeenth century. Purchas had enjoyed some access to the Company's archive and acknowledged the Company's assistance in allowing him to borrow journals in order to compile his history.[30] It is recorded in the minutes of a General Court meeting that he was, for example, allowed to view and make extracts from the account of Richard Roe's embassy to the Mughal Court in 1615. However, he was instructed to limit his notes to 'that w[hi]ch is proper to a History and not prejudiciall to the Companie', and they were to be carefully inspected before he was allowed to carry them out of the building.[31]

Within twenty-five years of the Company's establishment, detailed sailing directions to the East Indies were available in printed form, and some of its employees were making use of them. They may have made extracts from Purchas – at the time the largest book ever seen through an English press – in order to avoid having to carry the bulky

[28] The EIC's commission to Henry Middleton, in Birdwood and Foster (eds), *The Register of Letters*, 51–66.

[29] 'Extracts from several journals, April 1613 to July 1617', East India Company Marine Miscellaneous Records: IOR/L/MAR/C/3, No. 11 (India Office Records, British Library).

[30] Samuel Purchas, *Purchas his Pilgrimage* (London: 1626), 484.

[31] East India Company Court Minutes, 27 February 1622, IOR/B/7, p. 349.

and expensive printed version, although both may have been used. It is certain that copies of Richard Hakluyt's *Principal Navigations* and of the 1598 English translation of Jan Huyghen van Linschoten's *Itinerario* accompanied early EIC voyages.[32] Since it is likely that Purchas's text was derived from journals in the Company's possession, it would appear that, to prepare themselves for voyages, Company employees were taking personal copies of a publicly available book that was originally derived from a Company text. Soon afterwards, in 1628, the Company explicitly discontinued the practice of including specific sailing instructions in the commissions it issued to its commanders, while the issuing of rutters apparently declined.[33] Whatever the combination of factors that produced this decision, its implication is clear: within three years of the appearance of Purchas's account, the Company considered that knowledge of the sailing routes to India and beyond could be taken for granted, and that what had previously had to be codified and taught could henceforth be tacitly assumed. It seems likely that the publication of Davis's and other accounts in Purchas had something to do with this – from that point, information that the Company had made significant efforts to acquire and to pass on in a private and limited way had the status of public knowledge.

The Company was also a storehouse of knowledge in another important respect – knowledge of commanders, past, actual and potential. Indeed, its early court minutes shows more attention being paid to this than to navigational practice. In October 1617 the Company changed its policy regarding the command of its ships, resolving to employ no more 'Captaines, but Maisters that are maryners'.[34] Even before the determination to put expert navigators in undisputed command of individual ships, the Company had paid careful attention to its choice of commanders, weighing the experience and reputations (for courage, discretion and obedience, as well as navigational skill) of the men under consideration. Yet there is also evidence that individual commanders seeking employment with the Company did not consider themselves bound to it and endeavoured to negotiate their own terms, for instance over the extent of permissible private trade, or employment of their relatives. Thomas Best, chief commander of the EIC's tenth fleet and later Master of Trinity House, was 'approved for an excellent Navigator and good Comander' but it was 'doubted hee will nott bee rulde, beinge vngratefull, couoetous and prowde' for he would 'nott to bee drawne from his pruote [private] trade'. He nevertheless was in a position to name several men 'willinge to bee enterteund for Maisters' (one of which the mariners refused to sail with) and requested that the Company employ his son as factor.[35] Early commanders were not simple extensions of the Company's will, nor simply following its guidance.

32 Richmond Barbour, *The Third Voyage Journals: Writing and Performance in the London East India Company, 1607–10* (New York: Palgrave Macmillan, 2009), 59–60; Birdwood and Foster, *Register of Letters*, 419.

33 K. N. Chaudhuri, *The English East India Company: The Study of an Early Joint-Stock Company, 1600–1640* (London: Frank Cass and Co., 1965), 108.

34 Court Minutes, 24 October 1617, IOR/B/6, f. 45.

35 Court Minutes 8 and 21 October 1617, IOR/B/6, ff. 30, 43. See also ff.44–6, and R. C. D. Baldwin, 'Best, Thomas (1570–1639), Sea Captain and Master of Trinity House', *ODNB*.

Of the surviving journals of early EIC voyages, it is noticeable that the most technically detailed tended to be those kept by masters and master's mates – that is, by those most directly involved in the navigation of the ship. Since survivals from this period are erratic, we should not read too much into any apparent patterns emerging from this material. Nevertheless, it is noticeable that some of the most detailed reports, revealing some of those most at pains to specify their use of navigational instruments and the lengths they have gone to in order to be sure of their observations, were made by master's mates – that is to say, by junior officers and possible aspirants to future commands. This suggests that while masters could prove their skill and competence by successfully bringing their ships home, master's mate, with more limited responsibilities, had to find other ways of demonstrating their skills. Their journals may have been one of the most effective ways for them to do this. If this was their motive, it would show that they believed that the journals they delivered up to the Company's London headquarters were likely to be scrutinized there.[36]

The evidence suggests that the early East India Company acted to encourage and, in some ways, to facilitate the navigational expertise of its senior mariners by making sailing directions available and requiring them to keep, and supply, detailed journals of their voyages. But the types of knowledge that the Company sought to promote and acquire were as a rule-specific knowledge of the conditions, routes and hazards of the East India voyage, not general skills for navigation in all conditions. There was no systematic training, no examinations to demonstrate competence in particular navigational skills. There was, however, in the access granted to Purchas to examine Company journals, some encouragement of publications that might promote knowledge of the sailing routes or foster navigational skill. What we know of negotiations about access to information held at East India House indicates that the Company was concerned about the publication of commercially sensitive information. It is therefore striking to note that quite precise navigational information about positions, currents, hazards, anchorages and supply was allowed to be published. The precise nature of the Company's concerns has to be guessed at, but the evidence suggests they were more preoccupied with the secrecy of commercial than navigational knowledge and practice.

It also appears that the Company discriminated between types of information; sailing directions were one thing, charts another. We know that ships were equipped with what was known as the 'English Plot' – an English chart that was presumably in manuscript, since there is no evidence that the Company ever sought to have such a chart printed – as well as commercial atlases and proprietary Dutch and Portuguese maps, presumably obtained by stealth.[37] We also know that these foreign charts were sources of anxiety. The possibility that they had been scattered with false information in order to sabotage rival commercial ventures to the East Indies was taken seriously.

[36] See for example IOR/L/MAR/A/20 and IOR/L/MAR/A/20, detailing the *Hector*'s voyage as part of the second joint-stock voyage of 1614–15, and kept by John Monedon, master's mate.

[37] See IOR/L/MAR/A/41, for the voyage of the Eagle in 1624, which carried Mercator's *Atlas* – possibly the updated Hondius-Mercator Atlas of 1622 – and a Portuguese map which were constantly mutually compared by a master's mate in order to establish and correct errors in them.

The journal of Edward Dodsworth, travelling to and from Surat in 1614–15, recorded that as they passed the northern end of the Maldives 'we founde manie shoaldes and ilandes laide in the platts most false, and eronious'. His conjecture was that these were 'lade downe by the Portingalls to make those seas seme more daingerous vnto vs'.[38]

In short, we can see that the East India Company continued to be concerned to observe proper protocols about the collection and management of commercial and navigational knowledge into the 1620s, but that it also developed a more relaxed attitude to the dissemination of at least some part of it. This, notably, included sailing routes and directions to the East Indies, though not charts. We can see steps taken to encourage the acquisition of navigational knowledge and experience by individuals, careful attention paid to the qualifications of responsible mariners on board its ships, and some efforts to recruit and even publish detailed accounts of English India voyages. This appears to indicate that there was more concern with the observance of proper authority than an overriding preoccupation with keeping knowledge of East Indies navigation and trade out of the public sphere. This was, in other words, a concern about central control and with the Company government's *capacity* to maintain secrecy if wanted, rather than with secrecy in any absolute sense. As discussed in the case of the Mint, institutional statements about controlling access to key spaces or information are often a projection of authority and hierarchies of power.

With this in mind, it is also worth enquiring who the Company was concerned to keep out. It seems significantly more likely that their concern related to potential English interlopers rather than with other polities establishing themselves in the East India trade. Trying to throw obstacles in the path of, say, Dutch or Portuguese knowledge of the India trade would be a foolish preoccupation, since they were already established in it and had their own bodies of skill and knowledge. Far from being able to constrain them, indeed, the EIC had liberally borrowed from them when making its own start. Rather, what constantly preoccupied the EIC was a distrust of its own servants. The risk of being outperformed by London or international rivals was in many ways less pressing than that of being undermined by its own employees, as the many precautions and layers of oversight detailed in the 1621 *Lawes and Standing* Orders attest. Concern over proper protocol and the authority of the London headquarters had as much to do with keeping the EIC's own people in line as with the activity of its rivals. In Mishra's analysis, this distrust extended not just to the Company's employees but also to its stockholders; the EIC's executive authority, the Court of Committees, did its best in many respects to keep the General Court at bay, despite being nominally answerable to it. The 'generality' was identified with the 'public', and the Company's leadership tried to limit its authority by maintaining secrecy where possible.[39]

The threat of employees spoiling the Company's trade, and resistance from the Company's administrative and executive authority to the influence of its shareholders, remained structural issues for the Company deep into the eighteenth century. How to maintain its authority from London and ensure the trustworthiness of its servants was a

[38] 12 March 1614/15, Edward Dodsworth's Journal (25 February 1614–2 November 1615), IOR/L/MAR/A/XIX, f. 9v.

[39] Mishra, *A Business of State*, 45–6.

persistent problem. The Company continued to acquire fair copies of its Commanders' logs, and individual captains and officers in Company ships continued to take notes of meteorological phenomena, ocean currents, harbours and coastal profiles, for their own future reference and instruction as much as the Company's wider benefit. Unfortunately, there is very little surviving information about the Company's maritime practice during the later seventeenth century. Copies of the minutes of the Committee of Shipping, the responsible Committee within the Company, have mostly been destroyed for the period up to the nineteenth century, with only a volume of summaries and extracts covering the period 1739–99 and a chance survival of a volume of minutes for 1685/6 to inform us of its activity. The volume for 1685/6 testifies to the careful attention paid to the provisioning of medicine chests specific to the Company's various factories in Bombay, Surat, Tonkeen, Madras, Bengal, Matsulipatam and St Helena, but has little else to say regarding the Company's navigational knowledge and practice in the later seventeenth century.[40]

Circumstantial evidence suggests that the Company relied at this period on commercially produced charts and atlases, partly derived from the information of East India captains and partly from Dutch charts.[41] Commanders were expected to equip themselves with charts and instruments for navigation and the Company made no prescriptions about what, or whom, they should use to do so. In practice, there were comparatively few suppliers of printed charts and atlases in London, with individual publications dominating the market. John Seller's compilations of charts and sailing instructions, *The English Pilot*, first published from 1671 with subsequent revisions by John Thornton, were the only locally printed materials available for many decades. On the other hand, there is strong circumstantial evidence that Company commanders commissioned their own sets of manuscript charts from London chart-makers, including Thornton. A set of Thornton's charts for East India navigation survive together in Paris and it seems likely that these are the seventeen charts taken out of the *Canterbury*, an English East Indiaman captured by the French navy in 1703. Several of these drew on Dutch charts, and ten are source material for his third book of *The English Pilot* (1703), containing charts and sailing directions for 'Oriental Navigation'.[42]

While Thornton styled himself on his chart of the Ganges and Bay of Bengal, dedicated to the Company (Fig. 7.4), as 'Hydrographer to Hono[ra]ble East India Company' this is better understood as a description of the work he undertook, perhaps with permission or under specific commission, than the title of any kind of official

[40] Minutes of Committee of Shipping, 1685–6, 2 December 1685, IOR/L/MAR/C/27A.

[41] Andrew S. Cook, 'Establishing the Sea Routes to India and China: Stages in the Development of Hydrographical Knowledge', in *The Worlds of the East India Company*, ed. H. V. Bowen, Margarette Lincoln and Nigel Rigby (Woodbridge: Boydell Press, 2003), 119–36 (pp. 122–3). See also Monique de La Roncière, 'Manuscript Charts by John Thornton, Hydrographer of the East India Company (1669–1701)', *Imago Mundi* 19, no. 1 (1965): 46–50.

[42] La Roncière, 'Manuscript Charts', 47, 46; John Thornton, *The English Pilot: The Third Book* (London, 1703), Facsimile edition, with introduction by Coolie Verner and R. A. Skelton (Amsterdam: Theatrum Orbis Terrarum, 1970).

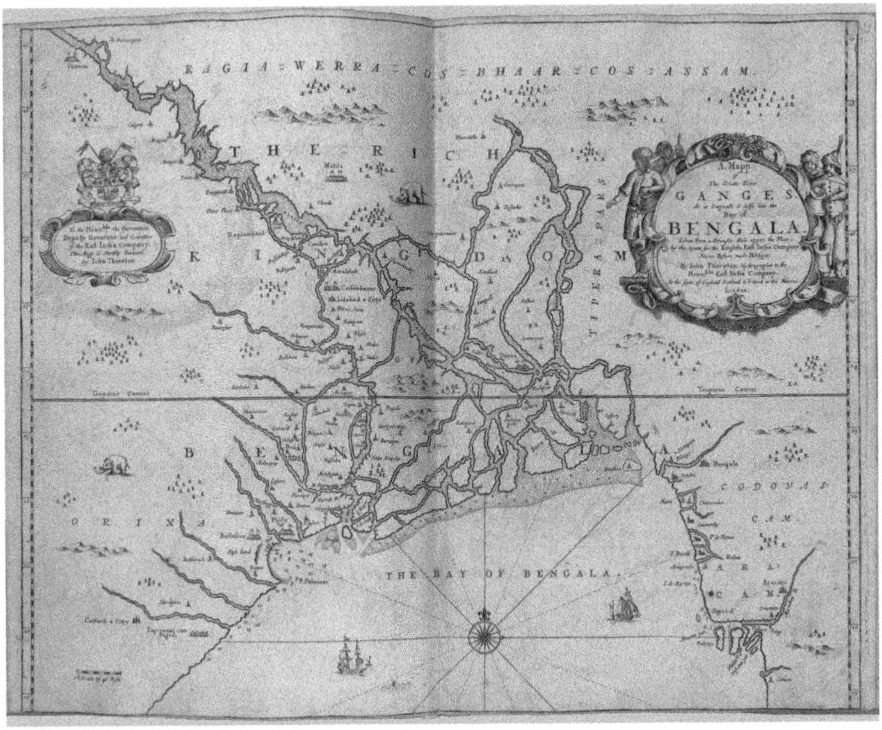

Figure 7.4 John Thornton, 'A mapp of the greate River Ganges ….'. Library of Congress, Geography and Map Division, G1059.T4 1700.

position.[43] On other published charts he described himself simply as 'Hydrographer' at his premises in the Minories: the EIC was one opportunity and potential market among a number of others.[44] As far as it is possible to tell, for the later seventeenth century and for most of the eighteenth, the English East India Company did little to patronize cartographic skill or navigational instrument-makers directly and nothing to specify the nature of the navigational equipment commanders should carry. Nor is it evident that the Company did very much to facilitate the commercial production of the charts that were vital to its trade, though Thornton's dedication suggests that they weren't deliberately obstructive either.

[43] 'A mapp of the greate River Ganges …', National Maritime Museum (G255:1/1). Versions of this chart, said to be 'Taken from a Draight Made upon the Place, by the Agents for the English East India Company', were published in Thorton's *Atlas Maritimus* (1685) and *The English Pilot* (1703). The name of John Thornton's son, Samuel (1665–1715), similarly appears on some of the charts. Their names do not appear in EIC Minutes.

[44] Charts by Thornton in the Lowery Collection place him at 'At the Sign of England, Scotland, and Ireland' and 'At the Signe of the Platt', both in the Minories: Woodberry Lowery, *The Lowery Collection: A Descriptive List of Maps of the Spanish Possessions …*, ed. Philip Lee Phillips (Washington, DC: US Government Printing Office, 1912), 212–13.

Regulation of navigational practice and its impact

Some of the circumstances that created the situation described above changed in the eighteenth century. As the Company became more bureaucratically complex, following the annexation of Bengal, its annual fleets got larger; and it is also from this decade that we begin to find the Company laying down rules about the experience and abilities of its maritime officers and establishing systems of certification and examination to enforce them. In 1751 the Committee of Shipping resolved to 'consider of a method to examine the qualifications of Persons proposed for Commanders and report'.[45] This seems to have been acted on fast, because the extract for the following week has Captains being proposed for particular ships and the Committee of Shipping being ordered to consider and report. Two were approved on the basis of the Committee's recommendations the week after. In 1756 the basis of these qualifications was clarified:

> No Person to be deemed qualified for the command of any Ship in the Company's service but such as have been examined and approved for Chief or second Mate and been one Voyage in such capacity nor to be under 25 years of age, these Orders not to exclude the third Mate of any Ship who by death or otherwise during the Voyage shall succeed to the command[,] and behaved well in the Company's service.[46]

The emphasis, we should note, was on experience at sea and in the India voyage, not on expertise in the abstract, at least for Commanders. The qualifications for mates (junior officers) are not stipulated, but subsequent regulations governing their appointment indicate that these were also to be based on experience, combined with certification of their age. The examinations were to be carried out by the Master Attendant (the superintendent of the Company's dockyards), his chief assistant and the Clerk to the Committee of Shipping wherever possible.[47] What the technical parts of their examination consisted of we do not know. Surviving records of examinations point only to the certification of the candidates' age and seagoing experience. There are plenty of instances of individual candidates being rejected, but in every instance their seagoing experience fell short of the stipulated threshold. Whether they were also found deficient in the technical aspects of navigation and seamanship is left unrecorded.[48]

In general, the technical requirements for sailing East Indiamen safely to Asia and back were gradually more carefully specified as the eighteenth century progressed. The specified skills were to be acquired through experience – officers qualified for promotion through the ranks of the Company's maritime service by having served a

[45] 'Abstract of Proceedings of Court of Directors Regarding Shipping, 1739–1799', 5 June 1751, IOR/L/MAR/C/21 p. 24.

[46] Ibid., 4 February 1756, p. 37.

[47] A resolution to this effect, made at a meeting of the Court of Directors 31 August 1787, is recorded in Regulations and Appointments, IOR/L/MAR/C/644, f. 9.

[48] For a summary of the evolution of qualifications for officers in the Company's maritime service, see IOR/L/MAR/C/644. See IOR L/MAR/C/652 for the results of examinations of prospective commanders and officers by the Committee of Shipping.

given number of voyages to India or China at a stipulated rank. There was some scope for different permutations but in principle this meant that nobody could command a Company ship until they had completed at least three voyages in the Company's service and moved up at least one rank each time.[49]

From 1768 a new stipulation entered the regulations for the Company's maritime officers. This was summarized as 'New Method for finding the Longitude as recommended by the Astronomer Royal to be practised by the Company's Officers and Mates to produce Certificates previous to their Examination'.[50] This was the method of finding longitude by observing lunar distances from fixed stars pioneered by Nicolas-Louis de Lacaille in France, who developed the method and published a specimen of it, and (as the Company's note indicated) the Astronomer Royal, Nevil Maskelyne, in England, whose more comprehensive *Nautical Almanac* was first published in 1766.[51] In fact, the EIC had been indirectly involved in the development of the lunar distance method, as Maskelyne and his assistant Robert Waddington worked on its feasibility aboard an East Indiaman that was carrying them to observe the 1761 transit of Venus. As Jim Bennett has shown, Maskelyne and Waddington worked closely with the officers of the East India ship to refine their method, comparing their results to the mariners' dead reckoning and probably relying on their assistance for carrying out lunar observations.[52] That, when the Company formally recommended the method to its officers, it required candidates to produce their certificates 'previous to their Examination', suggests that this material was not covered by the Company's own examination.

The most frequent certifier of the 1770s and 80s was Robert Bishop, a former master in the Royal Navy who kept a private mathematical academy.[53] There were no official directives of who the accepted certifiers were – it was up to the judgement of the Company's examining committees whether to accept their certificates. However, in 1768 the Board of Longitude had recommended that those applying to be masters on Royal Navy ships should produce certificates from either the headmaster or mathematical usher of the Royal Naval Academy in Portsmouth or from 'Mr Robert Bishop or Mr Samuel Dunn both living in London (who are appointed by this Board

49 Regulations as they stood in October 1783, in wartime, were rather looser, particularly on the age requirement, as experienced seamen were lost to the Navy's impressment: see 'Qualifications of Commanders and Mates of East India Company's Own Ships, 1756–1783', in IOR/L/MAR/C/644.

50 17 October 1768, IOR/L/MAR/C/21. This was less than a month prior to the recommendation by the Board of Longitude to the Admiralty that they order 'Captains & Commanders' of the Royal Navy 'to take care that their present respective Masters do perfect themselves at the Royal Academy there in the Knowledge & Use of the said [Nautical] Almanacs & also of Hadley's Quadrant in making Lunar observations; and produce Certificates that they have done so … '; 12 November 1768, Minutes of the Board of Longitude, RGO 14/5, pp. 173–4; https://cudl.lib.cam.ac.uk/view/MS-RGO-00014-00005/177 (accessed 20 July 2023).

51 Derek Howse, *Nevil Maskelyne: The Seaman's Astronomer* (Cambridge: Cambridge University Press, 1989), 85–91.

52 Jim Bennett, 'Mathematicians on Board: Introducing Lunar Distances to Life at Sea', *British Journal for the History of Science* 52, no. 1 (2019): 65–83.

53 Robert Bishop, *Instructions and Observations Relative to the Navigation of the Windward and Gulph Passages* (London: Printed for the Author, 1761) stated that Bishop was 'Many Years a Master in the Royal Navy'.

to instruct such as shall apply)'. Given that Naval masters were examined by Elder Brothers of Trinity House, very often EIC captains, this synchronicity is not surprising. Bishop had also worked with Maskelyne, then the key figure on the Board of Longitude, to produce forms to facilitate lunar distance calculations.[54] Dunn's 1775 *The Navigator's Guide, to The Oriental or Indian Seas* was dedicated to the EIC's Court of Directors whom he thanked for granting access to log books at East India House.[55] However, in both royal and merchant navies, it seems that such initiatives failed to have the desired results. Company directives in the early 1800s, concerned about the competence of its captains to conduct lunar observations in particular, advised them specifically to seek out the instruction of Lawrence Gwynne of Christ's Hospital Mathematical School. David Miller suggests that this concern was raised by Alexander Dalrymple, acting as hydrographer to the EIC from 1779, who was particularly keen to secure sets of lunar *and* chronometric observations of longitude, for better mutual correction and thus more accurate information to support chart compilation.[56]

By the 1790s, the standing orders to captains and masters in the Company's service required them to take longitudes by lunar distances wherever possible. Special instructions from the committee of shipping – which was otherwise overwhelmingly preoccupied with making sure that ship's stores were handled and housed for maximum economy, and with the prevention of private trade – re-emphasize the order about working lunars. Mates were instructed to 'embrace every opportunity of taking and working the Lunar observations; observing the variation of the compass by azimuths and amplitudes, finding the latitude by double altitudes, and that the same be entered in the Log Book and your Journal'.[57]

Company ships were supposed to use both chronometric and lunar methods of finding longitude, as well as dead reckoning (estimating the ship's position from the hour-by-hour record of its course and speed). From 1791, a new standard form of logbook had spaces for all three, although it is apparent from surviving logs in the years following that chronometer readings had significantly displaced lunar observations as the principal method of keeping track of longitude at sea. In this respect the practice of the EIC was, as has been noted by Jane Wess and others, far ahead of that of the Royal Navy.[58] However, lunar distance observations might still be used as a check on the performance of chronometers, especially on certain routes in places where knowing

[54] 12 November 1768, Minutes of the Board of Longitude, https://cudl.lib.cam.ac.uk/view/MS-RGO-00014-00005/178. Lunar-distance observations form by Robert Bishop, G298/1/3, National Maritime Museum, https://cudl.lib.cam.ac.uk/view/MS-G-00298-00001-00003/1. Bishop's manuscript abstracts of EIC voyages were kept among Maskelyne's papers at the Royal Observatory: RGO 4/155–8, https://cudl.lib.cam.ac.uk/view/MS-RGO-00004-00155/1. He also published charts and sailing instructions for both the West and East Indies.

[55] Samuel Dunn, *The Navigator's Guide, to The Oriental or Indian Seas: or, the Description and Use of a Variation Chart of the Magnetic Needle, Designed for Shewing the Longitude* ... (London: 1775), 12.

[56] David P. Miller, 'Longitude Networks on Land and Sea', in *Navigational Enterprises in Europe and Its Empires, 1730–1850*, ed. Richard Dunn and Rebekah Higgitt (Basingstoke: Palgrave Macmillan, 2016), 223–47 (pp. 234–5).

[57] 'Memoranda of Directions & Admonition Delivered by the Chairman of the Committee of Shipping to the Commanders & Officers in the Company's Service', 18 January 1796, IOR/L/MAR/C/644.

[58] Jane Wess, 'Navigation and Mathematics: A Match Made in the Heavens?', in Dunn and Higgitt, *Navigational Enterprises*, 201–22.

where you were really mattered.[59] They were also, when undertaken in good conditions and repeated several times, considered to produce more reliable position readings. Nevertheless, the reliability and availability of chronometers did change practice in ordinary navigation over time. In the period from 1775 to 1785, between 52 per cent and 72 per cent of ships took at least occasional lunar observations; after 1791 that proportion declined considerably while the number of ships taking at least occasional chronometric readings rose to 80 per cent.[60]

Company officers were, then, expected to acquire relevant knowledge, get it certified, apply it in the field and return the resulting data to East India House. This work acted both as proof of their own performances and honesty and as material for improving the charts which they and their successors would use in the future. None of the material that supported this effort was, as far as it is possible to tell, supplied by the Company. The instruction, the certification, the logbooks, charts and the instruments required to carry it out were acquired by individuals. Rather than East India House acting as a depot or repository, what emerged around it was a string of businesses that set themselves up to supply the Company's wants – or rather, the wants of the Company's officers as generated by the Company's requirements. In particular, we might note the emergence of the 'navigation warehouse' as a form of business in the late eighteenth century, which might be considered a manifestation of the increasing complexity of those requirements and a reflection of the size of the market.

This was not in itself anything new. As Philip Beeley and others have shown, and as was evident in the case of the businesses in the Bulwark of the Tower of London discussed in relation to the Office of Ordnance, there was a significant overlap between mathematical bookselling, instrument making and teaching in London in the second half of the seventeenth century.[61] It is not surprising to find, in the eighteenth century, scientific instrument makers, particularly those specializing in navigational instruments, selling navigational books and charts. An example is Henry Gregory, an instrument maker of Leadenhall Street, who around 1763 traded at the sign of 'the Azimuth Compass, near the India-House'.[62] Gregory was one of the retailers, along with David Steel, of Robert Stevens's *Complete Guide to the East India Trade* (1766). His

[59] For example, if travelling to India a 'flurry of astronomical observations' might be made when a decision needed to be made about turning north to reach the Mozambique Channel or the east Indian coast, after a long period of relying on dead reckoning in open sea: Miller, 'Longitude Networks', 237–8.

[60] Simon C. Davidson, 'Marine Chronometers: The Rapid Adoption of New Technology by East India Captains in the Period 1770–1792 on Over 580 Voyages', *Antiquarian Horology* 40 (2019): 76–91.

[61] Philip Beeley, 'Practical Mathematicians and Mathematical Practice in Later Seventeenth-Century London', *British Journal for the History of Science* 52, no. 2 (2019): 225–48; and e.g. Kevin Tracey, 'Calculating Value: Using and Collecting the Tools of Early Modern Mathematics' (PhD thesis, Swansea University, 2019).

[62] Trade card: Heal Collection 105.48 (British Museum). The SIMON (Scientific Instrument Makers, Observations and Notes) database records many instrument makers and businesses in Leadenhall Street and the surrounding area. As early as 1694 there was a John Redhead advertising as 'In Leadenhall St, over against East India House', although he was a scale maker rather than navigational instrument maker and perhaps served the needs of those dealing with goods rather than shipping. This database is currently being remodelled and augmented as part of the AHRC-funded Tools of Knowledge project and will be made available online.

trade card advertises him as maker of 'all Sorts of Mathematical Instruments for Sea or Land; Viz. Hadley's & Davise's [*sic*] Quadrants, Theodolites, Plain Tables, Gauging & Drawing Instruments. Telescopes at ye most Reasonable Rates' as well as selling 'All Sorts of Navigation Books and Charts'.[63] There is evidence from the last third of the eighteenth century that booksellers and instrument-makers were coming together – full collaboration is probably too strong a term – to supply the developing navigational wants and needs of the East India Company, in particular, as well as the wider area that included the Navy Office and Trinity House.

The term warehouse began to be deployed for retail premises that offered a wide range of goods in the mid-range of price, and the possibility of a leisured and conversational shopping experience from at least the late 1720s.[64] We have, however, yet to find mention of a 'Navigation Warehouse' in London any earlier than 1788.[65] The earliest belonged to David Steel, who already had a reputation as a specialist maritime stationer at Union Row, Little Tower-Hill. He published and sold nautical charts and books of all sorts, for sailors, merchants, shipowners, shipbuilders, even specialist maritime painters. His business was aimed at both royal and merchant navies, the former were offered several specialist naval publications, including the *Navy List* and the *Naval Chronicle*, but *The Ship-Master's Assistant* offered 'Complete Information, as well to MERCHANTS, MASTERS of SHIPS, and Persons employed in the MERCHANT SERVICE, as to OFFICERS and others in the ROYAL NAVY'.[66] The business passed to Steel's wife, Penelope, on his death in 1802, and then to their son John. By 1808 it had moved from Tower Hill to Cornhill, placing it further from the docks and Trinity House and former site of the Navy Office (which completed its move from Seething Lane to the newly completed south wing of Somerset House in 1789) but closer to Leadenhall Street and the East India Company.[67] Penelope Steel's trade card shows the shop windows filled, on one side, with navigational, mathematical and optical instruments and, on the other, with books and charts. The anticipated clientele was represented, either side of the door, by two figurines of midshipmen, and also in the text as 'Captains, Merchants and Shop-keepers' (Fig. 7.5).

The next Navigation Warehouse to open was William Heather's at 157 Leadenhall Street, first listed in a 1794 London directory.[68] These became, in effect, one-stop shops for everything a sea-officer might need – charts, logs and instruments. Latterly, too, Heather's business went into partnership with John Norie, the son of the James Norie who had previously operated a private academy at Burr Street, Wapping.[69] Heather published Norie's *The Description and Use of Hadley's Quadrant* (1796) and, from the

63 British Museum, Heal Collection 105.48.

64 Alun Withey, 'Warehouses and Shopping in Georgian England' (8 February 2017), https://dralun.wordpress.com/2017/02/08/warehouses-and-shopping-in-georgian-england/ (accessed 20 July 2023); Walsh, 'Shop Design and the Display of Goods', 173.

65 A maker called Egerton Smith was located at the Navigation Warehouse, Pool Lane, Liverpool, from 1774 to 1783: Clifton, *Directory*, 255.

66 David Steel, *The Ship-Master's Assistant and Owner's Manual* (London: Printed for D. Steel, at the Navigation-Warehouse, 1788).

67 These moves are traced in the title pages of Steel's publications.

68 *Kent's Directory for the Year 1794* (London: 1794).

69 Susanna Fisher, 'Norie, John William (1772–1843)', in *ODNB*.

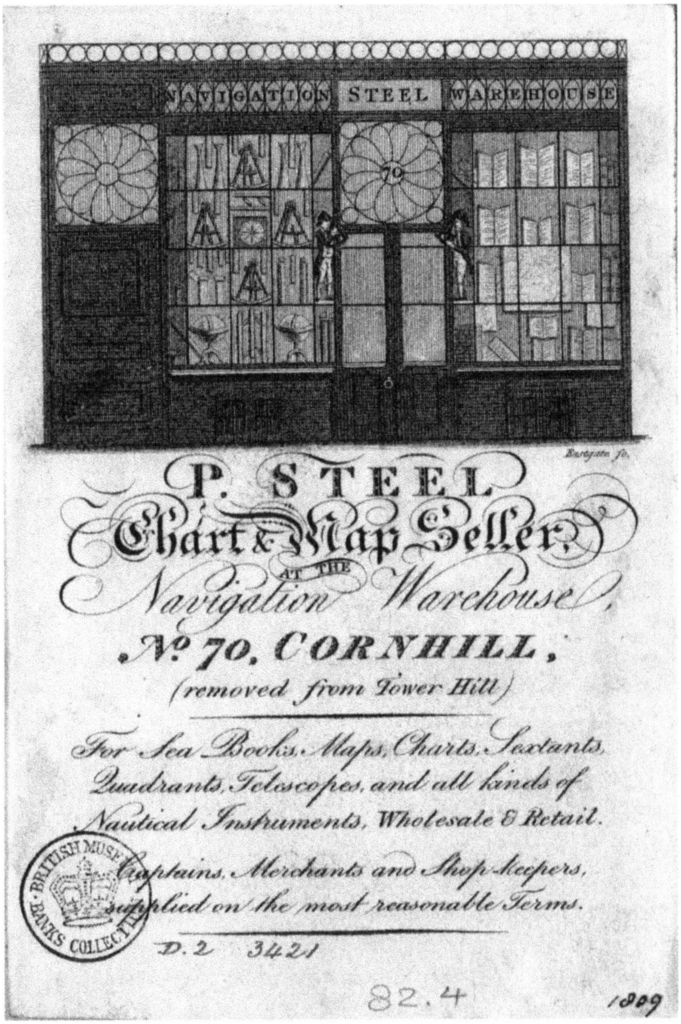

Figure 7.5 Trade card for P. Steel's Navigation Warehouse at 70 Cornhill (BM: D,2.3421). © The Trustees of the British Museum.

following year, Norie was drawing charts for Heather and teaching navigation. In 1800, his 'nautical academy' was listed at 157 Leadenhall Street. Thus, from the end of the eighteenth century, the warehouse offered navigational instruction as well, providing all three of what David Miller calls the hardware, software and wetware of navigation (the instruments needed to navigate a ship; the methods, charts and logs used to plot and record its course; and the skills required to use them).[70] Norie eventually took over the business from Heather and made a significant success of it, publishing an

[70] Miller, 'Longitude Networks', 224. Data extracted from the SIMON database by the Tools of Knowledge project shows that Navigation warehouse as the 'Advertised trade' of fifty-one individual makers or businesses across Britain and Ireland up to 1914.

important update to the *Tables Requisite to be used with the Nautical Ephemeris* for finding the latitude and longitude at sea, a publication that became standard (and is in fact still published under the name of *Norie's Nautical Tables*). The shop on Leadenhall Street also advertised itself with a wooden sign of the Little Midshipman, still present at this address until the business moved in 1878.[71]

As we have noted, the instrument-makers, stationers, booksellers and navigation-warehouse operators – to say nothing of slop-sellers and ships' chandlers – in the surrounding streets did not sell direct to the East India Company; they sold to its mariners and in particular its officers and shipowners. From the last quarter of the eighteenth century until the Company's maritime service was in effect retired in 1833, their place of resort was the Jerusalem Coffee House at Cowper's Court, off Cornhill nearly opposite the Royal Exchange. This was also the official headquarters of the Society of East India Company Commanders, a mutual society and pension fund formed on the Company Captains' own initiative and intended to strengthen their hand in negotiations with their employers and to relieve members and their families in distress. This subscription association made the coffee house a quasi-public centre for news of the Company, and it attracted many allied trades. London directories include a significant number of merchants, insurance brokers and shipbuilders who gave this as their address. An example is Charles Young, a maker of marine chronometers who was to win awards in the Royal Observatory's chronometer trials in 1827 and 1828.[72] Several other chronometer makers – Paul Phillip Barraud, John and Miles Brockbank, later Atkins & Son – had shops in Cowper's Court or Cornhill and are known to have supplied Company Captains with chronometers. There were at least five instrument makers, of various classes, operating on Cornhill and Leadenhall Street during the last third of the eighteenth century, plus a number of specialist stationers.

The EIC did not try to make East India House into a centre of calculation or knowledge-making space in an obvious or public way until the beginning of the nineteenth century but, in the quarter century before that, the surrounding streets accrued a variety of businesses designed to service the wants of its commanders. This was partly an effect of the way that the Company's navigational requirements and instructions to commanders had accumulated gradually over time. It was around these that a new form of business came together – the navigation warehouse – gradually incorporating all the elements that the Company's rules for sailing and qualifications for command in its ships specified. Both the official presence of East India House on Leadenhall Street and the emergence of the formal-informal social space at the Jerusalem Coffee House for the Company's captains played their role in shaping the surrounding streets. These businesses responded to Company stipulations but also to knowledge of where the people who would actually use their products would regularly be. The Company's acceptance of private supply and enterprise as an adequate means of supplying its navigational needs and the skills of its officers thus held until deep into the eighteenth century.

71 See catalogue entry for 'Little midshipman' figure (AAB0173) at the National Maritime Museum.

72 Clifton, *Directory*, 308.

The uses of knowledge

So far, we have considered forms of knowledge and skill that were intrinsic to the Company's ordinary commercial operations, and which were of shared interest with other individuals, institutions and businesses in the wider area. While we have concentrated on charts, sailing directions and navigational instruments and instruction, similar stories can be told about, for example, the Company's relationship to the exchange of goods and information with apothecaries and other medical practitioners.[73] However, it is evident by the last third of the eighteenth century that the Company was also choosing to appoint men of science to what were newly created posts. In many ways mirroring the Company's approach to navigational knowledge, these innovations were typically the outcome of individual and often external interest, effort and enterprise. Much of this effort was undertaken far from London but required – and increasingly could rely on support from – the London-based leadership of the Company. The establishment of institutions such as the Madras Observatory and the Botanic Gardens in Calcutta were the result of individual proposals or personal benefactions which the Company then agreed, in most cases rather tentatively, to support for a time. From this period onwards, we find direct and indirect support also being offered to publishing projects, such as surveys, histories or charts, and ultimately to the opening up of collections within East India House itself, with the creation of both a museum and a library. While none of this knowledge-producing or -disseminating activity was disinterested – and we could hardly expect it to be so for a commercial organization – the concepts of national and public utility and benefit were ones that the EIC could draw on rhetorically when it was useful to them.

Of course, from the very earliest days, the Company had strategically deployed its access to exotic objects and new information, and taken advantage of the curiosity of influential individuals, to gain support and forward helpful diplomatic efforts. The Company made a note, for Henry Middleton of the first voyage, about the King of Bantam's present to King James of two bezoars. In the commission to those undertaking the third voyage in 1606, as well as asking that journals be kept and presented on return, they were asked to 'remember to doe yo[u]r best to bringe for the Lorde of Salisburie [Robert Cecil, Secretary of State], some Parrattes, Mounckeys Marmesaattes or other straundge beastes & fowles that you esteeme rare & delightfull'.[74] As with the sites considered in our other chapters, East India House was of interest to enquiring minds, and some sort of collection was kept at hand to engage visitors. One such, in 1669, was Cosimo III de Medici, who reported that it was 'full of curious things both animal and vegetable' and these were 'kept here to gratifie the curiosity of the public'.[75] This was, evidently, a limited public, probably chiefly consisting of the EIC's own investors, among which was fellow of the Royal Society Robert Boyle. He wrote, in words that echo the earlier comments, of how he had handled rough diamonds, with 'a

[73] Winterbottom, 'An Experimental Community', 330–1.

[74] Birdwood and Foster, *Register of Letters*, 67, 130.

[75] Jessica Ratcliff, 'The East India Company, the Company's Museum, and the Political Economy of Natural History in the Early Nineteenth Century', *Isis*, 107, no. 3 (2016): 495–517.

Parcel of between 100 and 150 (if I misremember not the Number,) put into my Hands at one time in the *East-India* House to gratifie my curiosity'.[76]

Curiosities were also part of the private trade or exchange networks of those who travelled for the EIC, meaning that naturalia and artificialia were added to collections at the same time as, through the Company's larger scale official trade, Indian goods became available as commodities for food, medical treatment, clothing or interior decoration.[77] Winterbottom suggests that the development of social networks in London between 'scholars like Boyle, [Hans] Sloane, [John] Hunter, [Joseph] Banks and others helped grant legitimacy to the objects and animals that the EIC imported'. These connections and the engagement of scholars with such materials helped to demonstrate 'that such things could aid in the improvement of knowledge and, potentially, benefit the nation'.[78] This was far from a simple and straightforward process, and it was one that in our period worked through individual interests and connections more often than as a result of Company initiative or policy. Nevertheless, for the EIC, as other overseas trading companies, the information that they had access to was an important resource in supporting their claims to monopoly or the value of their products, although attitudes towards the openness or secrecy surrounding use of and access to this knowledge varied over time and according to need.[79]

Attitudes towards experimentation also varied. There is evidence that EIC directors did, at times, directly encourage enquiries of an experimental nature for internal use. Those on the third voyage were, for example, asked to bring 'for the Companie some small quantitie of the East India Tynn for the triall & proffe thereof'.[80] Of particular interest was saltpetre, a profitable commodity that the EIC hoped to produce and control itself, rather than relying on Indian producers. In 1683, Company agents at St Helena were asked to test different methods to 'make trial of the earth' there, to see if it contained sufficient nitre to produce saltpetre. While not a success, this is an example of the London-based EIC having 'discoursed with many persons' to gather information, which it used to direct their employees in a new activity. These may have included fellows of the Royal Society, where saltpetre was much discussed, though the detailed instructions mentioned 'Chymists' and 'the people of Pattana [Patana] upon the Riger Ganges'. Perhaps more typical, though, are the other examples highlighted by Winterbottom alongside this one, which show individual initiative, such as assistant surgeon James Kerr producing a report on the cultivation of opium in 1773, or the use of private and social networks to advise, report, try and collect information on tea specimens. Looking at the long history of such interactions, Winterbottom argues for 'a basic continuity within the knowledge culture of the EIC' to the end of the

76 Robert Boyle, *Experimenta & Observationes Physicæ* (London: 1691), 33.

77 For medical and botanical networks, see Winterbottom, *Hybrid Knowledge*; for the impact of EIC wealth and goods on British country houses, see Margot Finn and Kate Smith (eds), *The East India Company at Home, 1757–1857* (London: UCL Press, 2018).

78 Winterbottom, 'An Experimental Community', 336.

79 See Aske Laursen Brock, Guido Meersbergen and Edmond Smith, 'Trading Companies and Travel Knowledge: An Introduction', in *Trading Companies and Travel Knowledge in the Early Modern World*, ed. Meersbergen Brock and Smith (Abingdon and New York: Routledge, 2022), 1–20 (p. 12).

80 Birdwood and Foster, *Register of Letters*, 134.

eighteenth century that laid the ground for a nineteenth-century Company that 'was able to employ, on a formal basis, a range of experts to advise them and to conduct surveys covering everything from mapping to botany to mineralogy'.[81]

Despite this argument for continuity, Winterbottom indicates that Warren Hastings's period as governor and governor-general of Bengal, from 1772 to 1785, was significant in shifting the Company's attitude towards scholarly endeavour, coinciding with its expanding role in territorial governance, and requirement to act as a government and civil service.[82] This was, however, also the period that Binnema associates with changing attitudes towards natural knowledge and practice within the Hudson's Bay Company (HBC): after 'an initial century of secrecy', the HBC's 'contributions to public knowledge' grew to the extent that it developed a reputation for generous scientific patronage in the nineteenth century.[83] Binnema sees this as arising from networks connecting HBC directors and employees to 'metropolitan virtuosi and scientific organizations', but links it most directly to the Company's support of the Royal Society's 1769 transit of Venus expeditions, which brought scientifically trained individuals and precision instruments to Hudson's Bay. As much as any useful of profitable results produced, it was public interest in the explorations of Captain James Cook that followed the transit that account for the increasing value that was perceived in being linked to such activity. Indeed, being seen to contribute to knowledge-producing activities that appeared to lack a profit motive helped maintain 'an invaluable corporate image' of generosity and enlightened disinterest. This was particularly useful when charters and monopolies were subject to criticism, if rivals needed to be outmanoeuvred, and should recruitment and retention of Company servants prove problematic.[84]

Similar influences were at play within the EIC, where there was a crossover between individual, business and social networks. Asserting benefits to the nation and to learning could be a useful counter to criticism or, internally, instil a sense of what we might now call corporate pride. A 1777 Despatch from the London Directors to Madras invoked, encouragingly, if somewhat noncommittally, the notion of 'liberal and useful enquiries and the Company's willingness to take them into consideration'.[85] For Bowen, the Company's increasing interest in and support of a whole range of scientific or scholarly endeavours reflected a sense of their utility to the administration of an expanding empire and in fighting 'propaganda wars' around charter renewals, allowing 'the Company's friends in Britain to proclaim a virtuous and enlightened commitment to the advancement of the nation's knowledge and learning'. [86] Drayton also places the EIC's new patronage of economic botany from the 1780s within a political context: the heavily indebted Company 'needed to be seen as co-operative' by the British crown

81 Winterbottom, 'An Experimental Community', 336–42 (quotations pp. 337, 342).

82 Winterbottom, 'An Experimental Community', 342. See also Joshua Ehrlich, *The East India Company and the Politics of Knowledge* (Cambridge: Cambridge University Press, 2023), Chapter 1.

83 Binnema, *Enlightened Zeal*, 7.

84 Binnema, *Enlightened Zeal*, 8–13 (quotations at pp. 8 and 10). See pp. 79–93 for the impact of the 1769 transit of Venus expedition and pp. 95–112 on how public knowledge could be of little risk or beneficial in the context of trade rivalries.

85 Despatches to Madras, 1776–7, IOR/E/4/867.

86 H. V. Bowen, *The Business of Empire: The East India Company and Imperial Britain, 1756–1833* (Cambridge: Cambridge University Press, 2006), 153.

and government in light of criticism, from the public and Whig Opposition. It was hoped that the cultivation of new commodities and new approaches to the government of India would improve the trade balance, encourage support from Indian rulers and ensure stability. Focusing on agriculture, the Company 'sought vigorously to attach itself to the cause of "Improvement"'.[87] Ehrlich, who has recently explored the political uses of orientalist scholarship by the Company, argues that from the last quarter of the eighteenth century learning took on 'an ideological aspect', as a means of seeking conciliation with its own officials and Indian elites, and 'rehabilitating' individual and collective reputations.[88]

Looking back from the perspective of the nineteenth century, with the foundations of four observatories, two botanical gardens and the Ordnance Survey of India, gives the sense of deliberate central policy. However, this capacity was built initially through responses to individual efforts and unanticipated opportunities, which were approved by the Madras or Bengal government before receiving belated consideration from the Court of Directors in London. Some of these initiatives resulted from the increasing presence of people with relevant skills and interests being posted or otherwise travelling to India. In part this was due to changes to education – including an increasing number of medical practitioners with training or degrees that included botany and chemistry – and a developing view that fostering natural knowledge or artificial practice was a route to self-improvement and professional opportunity. Such attitudes might be expressed as leisured, gentlemanly interests or through the ambitions of medical, military and naval officers, to either of which the Company might respond. In this fashion, from around 1783, the EIC 'came to support' Johann König, a former pupil of Linnaeus, who had come to India as a medical officer on a mission of the Danish East India Company and had been a naturalist to the Nawab of Arcot.[89]

Having begun the precedent of paying for a naturalist, after König's death in 1785 the Company agreed to transfer the allowance to Patrick Russell in 1787. He was an Edinburgh-trained surgeon, already in India for personal reasons, who had become a friend and collaborator of König.[90] He was, the Company's secretary Thomas Morton reported, 'employed in the execution of a Plan for rendering the Botanical improvements of the late Dr Konig more immediately subservient to the uses of life as well with regard to Medicinal applications as to Arts and Manufactures.' Writing from East India House to Joseph Banks, president of the Royal Society and a key advisor to government and other interests on economic botany, Morton, also conveyed the

87 Drayton, *Nature's Government*, 116–17 (quotation p. 117).

88 Ehrlich, *The East India Company*, 18, 37. See also Howes, who suggests that, by the nineteenth century, the artworks collected and displayed at East India House 'reflected the outward aims of modern corporate collections, such as education and philanthropy', aiming to create a positive image of the Company's activities in response to criticism: *The Art of a Corporation*, 3.

89 Drayton, *Nature's Government*, 118; Arthur MacGregor, 'European Enlightenment in India: An Episode of Anglo-German Collaboration in the Natural Sciences on the Coromandel Coast, Late 1700s–Early 1800s', in *Naturalists in the Field*, ed. MacGregor (Leiden: Brill, 2018), 365–92 (pp. 367–8).

90 MacGregor, 'European Enlightenment in India', 372–3.

Court of Directors' wish to have Banks's opinion on these proposals.[91] Banks ultimately received König's collection and oversaw the publication, by order of the EIC, of three volumes based on his work, begun by Russell and completed by William Roxburgh, another Edinburgh-trained surgeon with botanical interests and networks, already based in Madras.[92] Russell's preface noted that the volumes had an 'immediate view to utility', giving preference to 'subjects connected either with medicine, the arts, or manufactures', but carefully added that 'the liberality of the Honourable Court of Directors encourages the admission of new plants … although their qualities and uses may as yet remain unexplored'. [93]

Despite these indications of central support, it was, similarly, the initiative of an individual that led to the foundation of a botanical garden in Calcutta. It was proposed by Robert Kyd, an officer in the Bengal Engineers with botanical interests and, possibly, an Edinburgh medical training. He made a moral argument that some return should be offered for 'the accumulated Riches which have accrued to Great Britain' from India by the cultivation of plants that would protect against famine. However, he also noted that his plans 'ultimately tend to the extension of the national commerce and riches'.[94] His suggestions were approved by Lord Cornwallis, who sailed in 1786 to assume his appointment as governor-general, and received support from the Company and its Board of Control (a committee of six members of the Privy Council who oversaw matters relating to defence and foreign policy after the East India Act of 1784). The EIC was being urged to act in a more state-like manner, and economic botany was part of this equation. Sir George Yonge, Secretary at War, urged Cornwallis to return a Botanical Despatch and hoped that, now that the government in India was 'awakened to this object, the most extensive and salutary consequences may be hoped for'. The garden opened in 1787 but remained very much Kyd's individual effort. Although he was singled out for praise in communications to the Company from the governor-general and Banks, when he appealed for the donation of plants he was disappointed by a poor response.[95] The sanguine hopes that famine would be ended by British rulers were wildly misplaced.

Despite the kinds of interactions with Banks indicted in the paragraphs above, and the extent to which the EIC was becoming part of the still 'loosely connected network of institutions and influence which made up the British State', the Company's 'long tradition of jealously guarding its privileges' continued to be felt. Banks found

[91] Thomas Morton to Joseph Banks, 27 January 1787, in *The Indian and Pacific Correspondence of Sir Joseph Banks, 1768–1820, Volume 2*, ed. Neil Chambers (Abingdon and New York: Routledge, 2021), 154.

[92] MacGregor, 'European Enlightenment in India', 383–4; Ray Desmond, 'Roxburgh, William (1751–1815)', *ODNB*.

[93] Patrick Russell, 'Preface', in William Roxburgh, *Plants of the Coast of Coromandel: Selected from Drawings and Descriptions Presented to the Hon. Court of Directors of the East India Company*, vol. 1 (London: 1795), i.

[94] John Gascoigne, *Science in the Service of Empire: Joseph Banks, the British State and the Uses of Science in the Age of Revolution* (Cambridge: Cambridge University Press, 1998), 138; Adrian P. Thomas, 'The Establishment of Calcutta Botanic Garden: Plant Transfer, Science and the East India Company, 1786–1806', *Journal of the Royal Asiatic Society* 16, no. 2 (2006): 165–77 (p. 167).

[95] Ray Desmond, 'Kyd, Robert (1745–1793)', *ODNB*; Drayton, *Nature's Government* (quotation p. 118); Thomas, 'The Establishment of Calcutta Botanic Garden', 171.

it significantly more difficult to gain direct influence here than in either British state or scholarly institutions.[96] His common, encouraging refrain of national benefit and utility was sometimes echoed (the EIC secretary Thomas Morton asked Banks to share hemp and flax seeds with those qualified to experiment and comment, with the aim of growing hemp in England as 'an object of general utility'), but his mercantilist aim of avoiding dependency on products under foreign control did not always align with the EIC's business model and trade monopoly. Individual directors, such as Nathaniel Smith, might support Banks's views, and he was close to Henry Dundas, who dominated the Board of Control but 'the Company as a whole proved to be rather lukewarm about some of Banks's pet projects'. Kyd told him that the Company was, for example, reluctant to promote indigo dye manufacture and the impression is that they were unwilling to shift from their 'role as an importer of handicrafts to that of a supplier of raw materials'.[97] When the Company did offer support to scientific endeavours, commercial interests were always present, even if a reputation for enlightened disinterest was also useful. The £1,200 that the EIC granted the scientific survey voyage of Matthew Flinders was, according Banks, paid 'to Encourage the men of Science to discover such things as will be useful to the Commerce of India & to find new passages'.[98]

As we have seen, the EIC was ahead of the British Admiralty in some aspects of navigational practice, but here again individual enterprise was key. The Company did not make a decision to create a hydrographic office but did, in 1779, respond to Alexander Dalrymple's initiatives, including his proposing himself for a new position

> for examining the Ships' Journals from the earliest times that Notice may be given of every danger which has hitherto been discovered, and for publishing from time to time such Charts and Nautical Directions as a comparison of the various Journals and other Materials may enable him to do.[99]

Employed as a writer by the EIC in 1752, Dalrymple clearly believed that his technical and mathematical skills could be a route to advancement, either within the Company or elsewhere. In a portrait, probably taken in 1765, he chose to represent himself in the uniform of an East India Company captain with the tools of the trade: celestial globe, map, hydrographic charts and a pair of dividers (Fig. 7.6/Plate 13).[100] However, while he later called himself Hydrographer of the East India Company, this position was not recorded as such and was not salaried: payments were made quarterly 'For examining the Ships' Journals &ca'. Although Dalrymple received an impressive £500 per annum, the Company clearly did not want to commit to the creation of an office.[101] As Andrew

[96] Gascoigne, *Science in the Service of Empire*, 135–45 (pp. 135–6).

[97] Morton to Banks, 13 April 1785, Chambers, *Indian and Pacific Correspondence*, 89; Gascoigne, *Science in the Service of Empire*, 141–3.

[98] Quoted in Drayton, *Natures Government*, 109.

[99] Quoted in Cook, 'Establishing the Sea Routes', 129.

[100] Andrew S. Cook, 'Dalrymple, Alexander (1737–1808)', *ODNB*; Andrew S. Cook, 'Alexander Dalrymple (1737–1808), Hydrographer to the East India Company and to the Admiralty as Publisher' (PhD thesis, University of St Andrews, 1993), 19–21. In 1765, Dalrymple had returned to Britain to promote a trading settlement at Balambangan, following a series of exploratory voyages.

[101] Cook, 'Alexander Dalrymple', 105.

Figure 7.6 Portrait of Alexander Dalrymple, attributed to John Thomas Seaton and thought to have been painted at Newhailes House in 1765 (NMS: T.2008.168). © National Museums Scotland.

See also colour version, reproduced in plate section 13.

Cook has documented, the position was not only created at Dalrymple's urging, but the manner in which he set about compiling and publishing charts based on the materials in the Company's London archive was largely left to his own discretion.[102]

The Madras Observatory was also founded in response to proposals from a company officer, this time with the added benefit of a private donation. However, the fact that Michael Topping was in Madras from 1785 as Chief Marine Surveyor of Fort St George, and had been sent out to India with appropriate instruments, was an indication of the Company's increasing reliance on individuals with specialist skills. It is nevertheless telling that he was aided and supported by the presence of an interested amateur of science, William Petrie. Appointed as writer in 1765, Petrie had risen to senior merchant by 1776 and was owner of a substantial estate. There he established a private observatory in 1786, furnished with instruments and an assistant. Temporarily put to Company use in support of a coastal survey, Petrie then offered both the building and instruments to the Bengal Government in 1789. In 1790 the Court of

[102] Cook, 'Alexander Dalrymple', 26, 27, Chapter 4.

Directors accepted the offer, together with Topping's proposal to found a Company observatory, for the purpose of 'promoting the knowledge of Astronomy, Geography and Navigation in India'.[103] The Bengal Government provided commitment, investing in another building and additional instruments as well as establishing the post of Government Astronomer. Topping's proposal to found a Survey School, to produce properly trained assistants for surveying and cartographical work, came to fruition in 1794. It was closed in 1810 on financial grounds, but the ongoing need for technical training was such that it reopened in 1818.[104]

By this date, it is evident that there was, from the centre, a changing policy towards training and to the results of and wider access to the Company's knowledge-producing activities. We might note the opening of an establishment for training the Company's writers, equipping them for trade, administration and government, at Haileybury in 1806 and a Military Seminary at Addiscombe in 1809. Jessica Ratcliff suggests that the foundation of the Company's museum, inaugurated in the new East India House in 1801, reflected a new tendency of 'intervening in and controlling … what had long existed as an individual, unofficial practice'. [105] While, once again, the result of a proposal by an employee outside the Company's central governance, the museum created a visible and accessible distinction between the old and new East India House. Hastings expressed his support for the scheme in a letter read out before the Court of Directors. He paid tribute to 'the liberality, and no less to the wisdom' of a Company that, looking beyond

> pecuniary gain, has joined a desire to add the acquisition of knowledge (and wonderful will be the stores which the projected institution under such auspices will lay open to them) to the power, the riches, and the glory which its acts have already so largely contributed to the British Empire and Name …[106]

The suggestion that the 'acquisition of knowledge' was separable from 'the power, the riches, and the glory' gave an ennobling gloss to proceedings that was much needed: Hastings had only recently faced impeachment, in a trial that implicated Company rule more widely. Looking for rehabilitation, in the same year that the museum opened, Hastings was elected a fellow of the Royal Society. His election certificate described him as 'a gentleman of great and extensive knowledge in various branches of Science', 'highly deserving' of the honour of election.[107]

[103] S. M. Razaullau Ansari, 'Early Modern Observatories in India, 1792–1900', in Uma Das Gupta, *Science and Modern India: An Institutional History, c. 1784–1947* (Delhi et al.: Pearson Longman, 2011), 349–80 (pp. 351–2); R. K. Kochhar, 'The Growth of Modern Astronomy in India, 1651–1960', *Vistas in Astronomy* 34 (1991): 69–105 (pp. 78–9).

[104] John Bosco Lourdusamy, 'College of Engineering, Guindy, 1794–1947', in Gupta, *Science and Modern India*, 429–52 (pp. 431–2).

[105] Ratcliff, 'The Company Museum', 502.

[106] Warren Hastings to Stephen Lushington, 15 November 1799, inserted in the Court Book for 20 November 1799, printed in A. J. Arberry, *The Library of the India Office: A Historical Sketch* (London: The India Office, 1938), 23.

[107] Warren Hastings, certificate of election to the Royal Society, EC/1801/14 (Royal Society).

Conclusion

The information kept in East India House and access to it were, in the earliest years of the EIC, shaped by the Company's strong preference for manuscript over print as a tool for equipping its mariners with the navigational knowledge they needed. This was in line with their concern for security to protect their monopoly from possible rivals, including the private enterprises of their own investors and servants. However, relevant information, including that gleaned directly from Company voyages, was soon available in print, first as sailing instructions and later as charts. This was partly a reflection of the difficulty of keeping the knowledge acquired by a wide number of individuals associated with the Company secure, but also an acceptance of the benefits a flourishing commercial world of navigational skills and equipment could offer. The Company relied on this to provide books, charts, instruments and training to its navy – increasingly so in the second half of the eighteenth century, when it demanded further specific skills of its ships' masters and reached the peak of its long-range maritime activity. We have argued that not only did relevant businesses gravitate towards East India House and the nearby social and supportive space of the Jerusalem Coffee House, but that the scale and consistency of the demand helped form the novelty of the Navigation Warehouse. It was partly the extent to which individuals could and did invest in their own education, together with the changing nature of education for navigators (including mathematics and astronomy) and medical practitioners (including chemistry and botany), that increasingly led to people with scientific interests and knowledge-making careers within the India Service.

In the second half of the eighteenth century, the EIC's administrative tasks grew immensely more complex as it strove to come to terms with the need to manage a large and heterogeneous territorial empire. At the same time, the legitimacy of its monopoly and the sense of its obligation to some larger notion of the public good than the profit of its investors came under increasingly close scrutiny and, in some quarters, vocal criticism. Governing and exploiting the new territories meant the need for a precise sense of their extent, natural resources and productive capacity, and thus for experts, entrepreneurs and instruments to assess and test them. Despite largely responding rather than initiating projects, the Court of Directors strongly insisted on the centralization of the information, reports, maps, surveys and charts produced by these efforts, and on its own authority to appoint those responsible, specify the scope of their activity and judge their qualifications. It was, however, beginning to present itself as a space for making knowledge that would be available more widely, sponsoring publications and the work of individuals in botany, surveying, hydrography and astronomy.

The various pressures on the EIC – to manage its territories effectively and to find new ways of extracting profit from them, to maintain public goodwill and a productive relationship with government, and to assert its authority over its servants in India without forfeiting their efficiency and cooperation – produced varied and sometimes contradictory approaches to the collection, storage, distribution and use of scientific and technical information. Much of what the Company produced was commercially or militarily sensitive and needed to be tightly controlled; some of this information, such as new maps and charts, was urgently needed on the ground but could only be made useful by being codified and reproduced in London. On some occasions, a national as well as a Company interest was asserted. The Company, at times, appeared to embrace a more disinterested notion of the pursuit of natural and useful knowledge, and it collaborated with or assisted organizations such as the Board of Trade, the Royal Society or the Society of Arts with their enquiries. This, leading up to the formal establishment of the Company's Museum in 1800 and the more systematized provision of training and scientific establishments in the nineteenth century, was part of a strategy for appealing to a broader notion of the Company's contribution to the national life at a time when its function and conduct faced increasingly strict scrutiny and debates over its legitimacy in the later eighteenth.

8

Conclusion

This book's opening case studies placed us in the multi-institutional and multifunctional site of the medieval Tower of London to the east of the City. It was, as we have shown, a site of production, testing and training as well as of governance, management and display that had developed, often haphazardly, over centuries. It seems apt to conclude this study by considering a new multi-institutional site, built to the west. This was Somerset House, a public building erected in English Palladian style between 1776 and 1801. Designed and begun by William Chambers, it was continued by James Wyatt, brother of the architect of the new Trinity House. It brought naval administration and a series of public offices into the same purpose-built space as several learned societies, creating a 'national building' to rival or eclipse London's civic institutions. It presented an image of the city as an imperial capital, where knowledge supported power, and power supported knowledge. It combined several of the institutions and functions that we have met in previous chapters, but largely removed them from what had also been an industrial location. At the west end of the Strand, Somerset House was upstream of the 'range of wharfs, warehouses, and manufactures' that lined the river from Blackfriars to London Bridge.[1] Though easily reachable, with direct access to the river built into the fabric, Somerset House was further still from the new dockyards of the first decade of the nineteenth century – the London, West and East India Docks in Wapping, the Isle of Dogs and Blackwall – and the growing industrial site of Woolwich Arsenal.

This building, then, combined learning and government bureaucracy, intending to link in stone the kinds of productive informal and personal networks that we have shown to have worked across and between the offices and corporations explored in this book. Among the first institutions to be housed there were the Royal Academy, the Royal Society and the Society of Antiquaries, given rent-free accommodation in the north wing; the Navy Board and the Stamp Office in the south; the Tax and Exchequer Offices in the east; and subsidiary naval boards, as well as residences, in the west. While hands-on practice and productive activities were not absent, they appeared subordinate to a world of the mind and the pen. The building's decoration, by sculptors including Joseph Wilton and Agostino Carlini, repeated familiar themes, with representations

1 Rudolph Ackermann, *The Microcosm of London; or, London in Miniature*, 3 vols (London: 1808–10), vol. 3, 280.

of the royal family and Britain linked to personifications of virtues and symbols of maritime empire. While the Thames is represented on the south, riverside façade, the Strand front emphasized national concerns: the keystones of the nine arches represent the Ocean and the eight principal rivers of Great Britain (Thames, Humber, Mersey, Dee, Medway, Tweed, Tyne and Severn). A statue of George III with a reclining Father Thames in the courtyard (just visible in Fig. 8.1), by John Bacon sculptor of the pediment to the new East India House, combines imperial, martial, maritime and riverine symbols.[2] In debates about the cost of the project, arguments were advanced to support the appropriateness of the splendour of such a great public work, representing as it did the wealth and importance of London to the rest of the world.[3] The inclusion of learned societies was, like the architecture and artworks, a statement of liberal patronage by the Crown and state. Joseph Banks hoped it would 'be understood by all

Figure 8.1 The courtyard of Somerset House, including (a somewhat inaccurate rendering of) the statue of George III and Father Thames. From Thomas H. Shepherd, *London and Its Environs in the Nineteenth Century* (London, 1828). York University Library/Internet Archive.

[2] Raymond Needham and Alexander Webster, *Somerset House: Past and Present* (New York: E.P. Dutton & Co., 1906), 199, 203–4; Adriano Aymonino, 'Somerset House (1776–1801)', in *Companion to the History of Architecture*, ed. H. F. Muallgrave (John Wiley & Sons, 2017); https://doi.org/10.1002/9781118887226.wbcha074.

[3] Jocelyn Anderson, 'The Commission for Somerset House in the "Eye of the Public"', *The Georgian Group Journal* 24 (2016): 81–94.

Europe, as meant to confer on [the Royal Society] an external splendor' to reflect 'the consideration in which they have been held for more than a century'.[4]

Somerset House has largely been considered from an architectural point of view and would benefit from the kind of analysis that we have brought to bear on other sites within the city. Broadly speaking, however, it could be seen to be part of the shift away from the kind of hybrid spaces and 'trading zones' represented by the Tower in the early modern period. That the Royal Society, later followed by the Geological and Astronomical Societies, was placed within this icon of the British state 'recognised the close ties of organised science to the political, social and religious establishment' by the end of the eighteenth century. But the decision was also part of the process of separating 'science' from other kinds of activity.[5] Our case studies show that natural knowledge produced within metropolitan institutions had – for reasons of personal and institutional status and advancement as well as a response to newly codified approaches to a range of tasks – increasingly been reinterpreted or represented as contributing to and benefitting from the mathematical, experimental and natural sciences. In their chapter on 'Scientific London' in the first decades of the nineteenth century, Morus, Schaffer and Secord present readers with a 'new imperial scientific capital' undergoing a 'second scientific revolution' that drew on the products of commerce and empire, supplied and created by the merchant and royal navies. The newly founded London University, they point out, was advertised as an 'Emporium for the supply of intellectual goods' to the 'Queen of Cities – Empress of the commercial world'.[6] The scientific knowledge created was, however, increasingly distanced from its origins and repackaged and represented within a different set of spaces, aimed at new audiences.

The institutions explored in this book are not those traditionally included with histories of science but their inclusion provides a richer picture of London's cultures of knowledge and practice that is both broader and more diverse. In exploring their sites, we have emphasized how they brought together varied activities, materials and sources of information, as well as individuals with a range of skills and varying status. This came about as a result of pursuing the core functions of the institutions, in a collective, or at least corporate, manner, and responding to opportunities or requirements according to changing circumstances. These sites have been shown to be multifunctional, whether at the level of the busy Tower or Woolwich complexes, with their overlapping spaces, or of the courtyard houses within which guilds and companies supported a range of operations and the needs of different groups of people. Questions of access to spaces and secrecy surrounding particular processes or information have recurred throughout. Rules established levels of trust, marking insiders from outsiders, or reinforced internal hierarchies of power, yet in practice often operated with a considerable degree of flexibility. Thus, we have seen orders to keep Ordnance workers out of spaces dedicated to producing the nation's currency at

[4] D. C. Martin, 'Former Homes of the Royal Society', *Notes and Records of the Royal Society of London* 22, no. 1 (1967): 12–19 (p. 16).

[5] Morus, Schaffer and Secord, 'Scientific London', 130.

[6] Ibid., 136.

the Mint, to stop commercial rivals securing information about sea routes kept within East India House, and to halt the sharing of information about assaying with those who were not freemen of the Goldsmiths Company. Equally, for pragmatic or opportunistic reasons, institutional elites might turn a blind eye or choose to break their own rules.

The geographies of our sites have, then, been part of the story, acting both to unite and to divide those who belonged to or had business with the institutions and corporations we have explored. Part of a crowded city, each had ways of excluding or selectively receiving outsiders. Locations might be chosen for their proximity to other sites of interest – as with Trinity House's move to Tower Street Ward – or be changed as a result of the institution's presence. Thus, the metrology, surveying and gunnery undertaken within the Tower by officers and workers of the Mint and Ordnance were increasingly served by practical mathematicians (instrument makers, mathematical teachers and publishers) who established themselves within the Bulwark and on Tower Wharf. Similarly, we have argued that by choosing not to centralize training or provision of instruments, books and charts, and yet requiring individuals to be qualified in particular ways, the East India Company attracted commercial providers to the streets around its building and the coffee house frequented by its captains. The efficacy of this provision was certified by examinations. At East India House, Trinity House, the Barber Surgeons and elsewhere, these were a key means by which corporate control was asserted. Even when performed by a hired hand, as in the case of the mathematical examiner at Trinity House, the fact that the examinations always took place within the corporation's building signified collective agreement about their validity and re-affirmed institutional headquarters as sites of skilled, knowledgeable practice.

Sites and corporations could, however, attract controversy and confidence in their authority was not guaranteed. Corporations worked hard to host visitors or display goods in ways that supported their claims to ancient pedigree, well-grounded experience, particular skills and social status. There was always a risk that, if reputations – such as that of the Common Assayer at the Assay Office – were not carefully maintained, or procedures – such as those regarding dissection of bodies within Barber Surgeons' Hall – not followed, the corporation's privileges would be challenged or public confidence lost. Being too financially successful, whether because of the impact on others' livelihoods or on political power, could also attract criticism. Trinity's House's lucrative monopoly over the supply of ballast and the EIC's extractive and individually enriching practices in the eastern trade, both led to calls to provide more internal charity and to act in the wider public interest. Acting as patron to 'disinterested' knowledge was one important response to such criticism, and helps account for support given to, for example, formal experiments on lighthouse illumination or relatively open-ended enquiry in botany or astronomy. While the utility of scientific knowledge was strongly asserted, apparently 'useless' knowledge also had its uses.

This book has considered several different types of knowledge and practice in relation to the natural world. We have followed our sources to tell new stories about our chosen institutions, exploring what areas of their knowledge cultures were recognized to be important and require particular negotiation within these settings. Rather than privileging the knowledge and practice that became increasingly associated with universities, scientific societies, theory and book learning we have treated it equally, or seen it as intertwined, with that which was and sometimes remained artisanal, gained through hands-on and tacit experience. Indeed, this book has repeatedly demonstrated the fundamental importance to the development of metropolitan science of workers, artisans and technicians working within and between London's institutions. We have encountered assayers, goldsmiths, coiners, instrument makers, mariners, producers of weapons, medical practitioners, among others, engaged in trials, experiments and investigations into material and natural properties at institutional sites. Some of these craftspeople were well known to their contemporaries, like the loquacious John Reynolds at the Tower, but most were relatively anonymous, leaving mere traces in the archives, 'a whisper, in history'.[7] We have endeavoured to recover some of these individual and collective voices and contributions to skilled technical practices. Taking institutions as our primary focus has allowed us to show that they interacted across differing social and professional contexts and between purely pragmatic and more generalized kinds of knowledge. While mathematicized ballistics or navigation had a clear crossover with scholarship and book learning, they were nevertheless pursued within the overlapping 'working worlds' of our institutions to meet the same kinds of individual and collective objectives as trial-and-error melting, refining and coining. Within London's institutions, all of these activities were of potential interest to outsiders, from curious gentlemen to industrial spies.

As mathematical, experimental and natural philosophy gained cultural authority, in part through its claims to successful application in industrial processes and practical activities, it became an obvious resource for institutions to make use of, both in practice and for purposes of self-presentation. External innovation could be a challenge to ancient authority and experience but we have shown that it was often welcomed, incorporated, managed and co-opted by our institutions, whether profit-driven companies, government departments or craft guilds. Each survived or thrived into the nineteenth century and beyond, necessarily adapting to their changing circumstances. In the centuries before this, 'metropolitan science', understood broadly as the knowledge and practices developed within the networks of skills and expertise that were instantiated in and around London institutions, made its mark. Because these institutions were dedicated to other main purposes, metropolitan science was characterized by its thorough integration into commercial, civic, governmental and imperial projects.

7 Sugata Ray, 'A "Small" Story of the Jasmine Flower in the Age of Global Botany', *Crafting Enlightenment*, ed. Cannady and Ferng, 247–72 (p. 271).

Selected bibliography

Online and digital sources

Calendar of Treasury Books, British History Online (https://www.british-history.ac.uk/cal-treasury-books/)

Cambridge Digital Library: Royal Greenwich Observatory Archives (https://cudl.lib.cam.ac.uk/collections/longitude)

Historical Texts, including Early English Books Online and Eighteenth Century Collections Online (https://historicaltexts.jisc.ac.uk/)

History of Parliament Online (https://www.historyofparliamentonline.org/)

John Strype's Survey of London (https://www.dhi.ac.uk/strype/)

Old Bailey Proceedings Online (www.oldbaileyonline.org, version 8.0, 27 May 2022).

Oxford Dictionary of National Biography (https://www.oxforddnb.com/)

Oxford English Dictionary Online (www.newtonproject.ox.ac.uk)

Papers of the Board of Longitude (https://cudl.lib.cam.ac.uk/collections/rgo14/1)

Project Gutenberg (https://www.gutenberg.org/)

SIMON (Scientific Instrument Makers, Observations and Notes) database, Royal Museums Greenwich

The Newton Project (https://www.newtonproject.ox.ac.uk/)

Trinity House (https://www.trinityhouse.co.uk/lighthouses-and-lightvessels)

Webster Signature Database (http://historydb.adlerplanetarium.org/signatures/)

Object and image collections

British Museum (https://www.britishmuseum.org/collection)

London Picture Library (https://www.londonpicturearchive.org.uk/)

Museum of Archaeology and Anthropology (https://maa.cam.ac.uk/collections)

National Maritime Museum (https://www.rmg.co.uk/collections/objects)

National Museums Scotland (https://www.nms.ac.uk/explore-our-collections/search-our-collections/)

Royal Armories Collection (https://collections.royalarmouries.org/)

Royal Artillery Museum (https://royalartillerymuseum.com/collection/overview-2)

Royal Mint Museum (https://www.royalmintmuseum.org.uk/collection/)

Science Museum (https://collection.sciencemuseumgroup.org.uk/)

Trinity House (https://www.trinityhouse.co.uk/)

Whipple Museum (https://collections.whipplemuseum.cam.ac.uk/)

Archival sources

Barbers' Company Archive

British Library

Additional Manuscripts (Add. MS)
Harley Manuscripts (Harley MS)
India Office Records (IOR)
Lansdowne Manuscripts (Lansdowne MS)
Sloane Manuscripts (Sloane MS)
Cambridge University Library, Manuscripts
Newton Papers (MS.Add)
Royal Greenwich Observatory Archives (RGO)
Goldsmith's Hall Archive
Guildhall Library
Clockmakers Company (CLC/L/CD)
Spectacle Makers Company (CLC/L/SH)
London Metropolitan Archives
Corporation of Trinity House (CLC/526)
New York Public Library, Manuscripts and Archive Division
Records of the British Surveyor of Ordnance (MssCol)
Royal Society
Classified Papers (CLP)
Council Minutes (CMO)
Election Certificates (EC)
The National Archives
Admiralty Papers (ADM)
Probate Records (PROB)
Records created or inherited by the Ministry of Supply and successors, the Ordnance Board, and related bodies (SUPP)
Records of the Ordnance Office and its Successors (WO)
Royal Mint (MINT)
Treasury: William Lowndes (T 48)

Primary sources

Agricola, Georgius. *De re metallica*, translated by Herbert, and Lou Hoover. New York: Dover Publications, 1950.

Anon. *A Most Humble Mem[O]Randum from Peter Blondeau, Concerning the Offers Made to Him by This Commonwealth, for the Coyning of the Monie, by a New Invention, Not yet Practised in Any State of the World, the Which Will Prevent Counterfeiting, Casting, Washing, and Clipping of the Same*. London, 1653.

Anon. 'Directions for Observations and Experiments to Be Made by Masters of Ships, Pilots, and other Fit Persons in their Sea-Voyages', *Philosophical Transactions* 2, no. 24 (1667): 433–48.

Anon. 'Paul Sandby and His Times', *Library of the Fine Arts* 2 (1831): 338–45.

Baily, Francis. *An Account of the Revd. John Flamsteed, the First Astronomer-Royal*. London: Printed by Order of the Lords Commissioners of the Admiralty, 1835.

Banister, John. *The Workes of That Famous Chyrurgian, Mr Iohn Banester by Him Digested into Five Bookes*. London, 1633.

Beighton, H. 'A Description of the Water-works at London Bridge', *The Universal Magazine of Knowledge and Pleasure* 5 (1749): 241–4.

Birch, Thomas. *The History of the Royal Society of London for Improving of Natural Knowledge*. London: A. Milar in the Strand, 1756.

Birdwood, George, and William Foster (eds). *Register of Letters etc., of the Governor and Company of Merchants of London Trading into the East Indies, 1600–1619*. London: Bernard Quaritch, 1893.

Biringucci, Vannoccio. *The Pirotechnia of Vannoccio Biringuccio*, translated by Cyril Stanley Smith and Martha Teach Gnudi. New York: The American Institute of Mining and Metallurgical Engineers, 1942.

Clowes, William. *A Short and Profitable Treatise Touching the Cure of the Disease Called (Morbus Gallicus) by Unctions, Set Forth by William Clowes, of London, Chirurgion*. London, 1579.

Corporation of Moniers in the Mint. *The Answer of the Corporation of Moniers in the Mint, at the Tower of London, to Two False and Scandalous Libells Printed at London, and Lately Come Forth without Date*. London, 1653.

Cotton, Joseph. *Memoir on the Origin and Incorporation of the Trinity House of Deptford Strond*. London: J. Darling, 1818.

East India Company. *The Lawes or Standing Orders of the East India Company*. London, 1621.

Eldred, William. *The Gunners Glasse*. London, 1646.

Evelyn, John. *Memoirs, Illustrative of the Life and Writings of John Evelyn, Esq. F.R.S ....*, 2 vols, edited by William Bray. 2nd ed. London: Henry Colburn, 1819.

Evelyn, John. *The Diary of John Evelyn*, vol. 3, *Kalendarium, 1650–1672*, edited by E.S. de Beer. Oxford: Clarendon Press, 1955.

Forbes, Eric G., Lesley Murdin and Frances Willmoth (eds). *Correspondence of John Flamsteed, Volume 1: 1666–1682* and *Volume 2: 1682–1703*. Bristol and Philadelphia: Institute of Physics Publishing, 1995–97.

Goodcole, Henry. *Heavens Speedie Hue and Cry Sent after Lust and Murder*. London, 1635.

Harris, G.G. (ed). *Trinity House of Deptford Transactions, 1609–1635*, 2 vols. London: London Records Society, 1983. Available at British History Online http://www.british-history.ac.uk/london-record-soc/vol19.

Howes, Edmond. *The Annales, or Generall Chronicle of England, Begun First by Maister John Stow* ... London, 1615.

Lucar, Cyrian. *A Treatise Named Lucarsolace*. London: Richard Field for John Harrison, 1590.

Lyle, J.V. (ed). *Acts of the Privy Council of England Volume 35, 1616–1617*. London, 1927. Available at British History Online https://www.british-history.ac.uk/acts-privy-council/vol35/.

Maitland, William. *The History of London, from Its Foundation by the Romans to the Present Time*. London, 1739.

Malynes, Gerard de. *Consuetudo, vel, Lex mercatoria; or, The ancient-law merchant*, vol. 1. London, 1622. 2nd ed. London, 1636.

Norton, Robert. *The Gunner*. London: 1628.

Ogilby, John, *The Relation of His Majestie's Entertainment Passing through the City of London, to His Coronation*. London and Edinburgh, 1661.

Oresme, Nicholas. *The de moneta of Nicholas Oresme and English Mint Documents*, translated by Charles Johnson. London: Thomas Nelson and Sons, 1965.

Peele, George. *The Device of the Pageant Borne before Wolstan Dixie Lord Maior of the Citie of London*. London: Printed by Edward Allde, 1585.

Pepys, Samuel. *Samuel Pepys's Naval Minutes*, edited by J.R. Tanner. London: The Naval Records Society, 1926.

Pepys, Samuel. *The Diary of Samuel Pepys*, 11 vols, edited by Robert Latham and William Matthews. London: Bell, 1970–1985.

Platter, Thomas. *Thomas Platter's Travels in England 1599*, translated by Clare Williams. London: Jonathan Cape, 1937.

Prideaux, Walter Sherburne. *Memorials of the Goldsmiths' Company*, 2 vols. London: Eyre and Spottiswoode, 1896–97.

Purchas, Samuel. *Purchas His Pilgrimage*. London: 1626.

Read, Alexander. *The Chirurgicall Lectures of Tumors and Ulcers*. London, 1635.

Robins, Benjamin. *New Principles of Gunnery*, edited by Charles Hutton. London: 1805.

Rooke, Lawrence. 'Directions for Seamen, Bound for Far Voyages', *Philosophical Transactions* 1, no. 8 (1665): 140–3.

Rowe, Jacob. *Navigation Improved: In Two Books*. London: 1725.

Royal Military Academy, Woolwich. *Rules and Orders for the Royal Academy at Woolwich*. London, 1741.

Saumarez, Henry De. *An Account of the Proceedings of Henry De Saumarez of the Island of Guernsey*, Gent. London, 1717.

Sprat, Thomas. *The History of the Royal Society of London for the Improving of Natural Knowledge*. London, 1667.

Stow, John. *A Survey of London. Reprinted from the text of 1603*, vol. 1, edited by C.L. Kingsford. Oxford: Clarendon, 1908.

Strype, John. *A Survey of the Cities of London and Westminster*. London, 1720. Online edition, The Stuart London Project, 2007 https://www.dhi.ac.uk/strype/.

Tatham, John. *Aqua Triumphalis, being a True Relation of the Honourable the City of Londons Entertaining Their Sacred Majesties upon the River of Thames and Wellcoming Them from Hampton-Court to White-Hall Expressed and Set Forth in Severall Shews and Pageants the 23. Day of August, 1662*. London, 1662.

Whormby, John. *An Account of the Corporation of Trinity House, Deptford Strond, and of Sea Marks in General*. London, 1746.

Worsop, Edward. *A Discoverie of Sundrie Errors and Faults* … London, 1582.

Wybard, John. *Tactometria. Seu, Tetagmenometria. Or, The Geometry of Regulars Practically Proposed*. London, 1650.

Young, Sidney. *The Annals of the Barber-Surgeons of London; Compiled from Their Records and Other Sources, by Sidney Young*. London: Blades, East & Blades, 1890.

Secondary sources

Adams, Andrew, and Richard Woodman. *Light upon the Waters: The History of Trinity House 1514–2014*. London: The Corporation of Trinity House, 2013.

Agar, Jon. *Science in the Twentieth Century and Beyond*. Cambridge and Malden, MA: Polity Press, 2012.

Agar, Jon, and Crosbie Smith (eds). *Making Space for Science: Territorial Themes in the Shaping of Knowledge*. Basingstoke: Macmillan, 1998.

Alford, Stephen. *London's Triumph: Merchant Adventurers and the Tudor City*. London: Allen Lane, 2017.

Appleby, John H. 'A New Perspective on John Rowley, Virtuoso Master of Mechanics and Hydraulic Engineer', *Annals of Science* 53, no. 1 (1996): 1–27.

Archer, Ian W. *The Pursuit of Stability: Social Relations in Elizabethan London*. Cambridge: Cambridge University Press, 1991.

Archer, Ian W. 'The City of London and River Pageantry: 1400–1856'. In *Royal River: Power, Pageantry and the Thames*, edited by Susan Doran, 80–5. London: Scala, 2012.

Arnade, Peter, Martha Howell and Walter Simons. 'Fertile Spaces: The Productivity of Urban Space in Northern Europe', *The Journal of Interdisciplinary History* 32, no. 4 (2002): 515–48.

Ash, Eric H. *Power, Knowledge, and Expertise in Elizabethan England*. Baltimore and London: Johns Hopkins University Press, 2004.

Ash, Eric H. 'Introduction: Expertise and the Early Modern State', *Osiris* 25 (2010): 1–24.

Ash, Mitchell G. (ed). *Science in the Metropolis: Vienna in Transnational Context 1848–1914*. London: Routledge, 2020.

Ashworth, William J. *Customs and Excise: Trade, Production, and Consumption in England, 1640–1845*. Oxford: Oxford University Press, 2003.

Atkins, Samuel Elliott, and William Henry Overall. *Some Account of the Worshipful Company of Clockmakers of the City of London*. London: [Worshipful Company of Clockmakers], 1881.

Aubin, David. 'The Fading Star of the Paris Observatory in the Nineteenth Century: Astronomers' Urban Culture of Circulation and Observation', *Osiris* 18 (2003): 79–100.

Ayres, James. *Art, Artisans and Apprentices: Apprentice Painters and Sculptors in the Early Modern British Tradition*. Oxford: Oxbow Books, 2014.

Baker, Alexi. 'Symbiosis and Style: The Production, Sale and Purchase of Instruments in the Luxury Markets of Eighteenth-Century London'. In *How Scientific Instruments Have Changed Hands*, edited by Alison D. Morrison-Low, Sara J. Sechner and Paulo Brenni, 1–20. Leiden and Boston: Brill, 2016.

Baker, Alexi Shannon. '"This Ingenious Business": The Socio-Economics of the Scientific Instrument Trade in London, 1700–1750'. PhD thesis, University of Oxford, 2010.

Bantz, Dorothee. 'Metropolitan Natural Histories: Inventing Science, Building Cities, and Displaying the World'. In *Science in the Metropolis: Vienna in Transnational Context 1848–1914*, edited by Mitchell G. Ash, 25–42. London: Routledge, 2020.

Barbour, Richard. *The Third Voyage Journals: Writing and Performance in the London East India Company, 1607–10*. New York: Palgrave Macmillan, 2009.

Barker, N.P. 'The Architecture of the English Board of Ordnance: 1660–1750'. PhD thesis, University of Reading, 1985.

Barrett, C.R.B. *The Trinity House of Deptford Strond*. London: Lawrence & Bullen, 1893.

Barrett, Katy. 'The Wanton Line: Hogarth and the Public Life of Longitude'. PhD thesis, University of Cambridge, 2014.

Bayley, Justine, and Harriet White. 'Evidence for Workshop Practices at the Tudor Mint in the Tower of London'. In *The Renaissance Workshop*, edited by David Saunders, Marika Spring and Andrew Meek, 138–43. London: Archetype Publications, 2013.

Beeley, Philip. 'Practical Mathematicians and Mathematical Practice in Later Seventeenth-Century London', *British Journal for the History of Science* 52, no. 2 (2019): 225–48.

Bennett, J.A. 'The Mechanics' Philosophy and the Mechanical Philosophy', *History of Science* 24, no. 1 (1986): 1–28.

Bennett, J.A. 'Geometry and Surveying in Early-Seventeenth-Century England', *Annals of Science* 48, no. 4 (1991): 345–54.

Bennett, Jim. 'Shopping for Instruments in Paris and London'. In *Merchants and Marvels: Commerce, Science, and Art in Early Modern Europe*, edited by Pamela Smith and Paula Findlen, 370–95. London and New York: Routledge, 2002.

Bennett, Jim. '"The Rev. Mr. Nevil Maskelyne, F.R.S. and Myself": The Story of Robert Waddington'. In *Maskelyne: Astronomer Royal*, edited by Rebekah Higgitt, 59–88. London: Robert Hale, 2014.

Bennett, Jim. 'Mathematicians on Board: Introducing Lunar Distances to Life at Sea', *The British Journal for the History of Science* 52, no. 1 (2019): 65–83.

Bennett, Jim, and Rebekah Higgitt. 'London 1600–1800: Communities of Natural Knowledge and Artificial Practice', *The British Journal for the History of Science* 52, no. 2 (2019): 183–96.

Berlin, Michael. 'Broken All in Pieces': Artisans and the Regulation of Workmanship in Early Modern London'. In *The Artisan and the European Town, 1500–1900*, edited by Geoffrey Crossick, 75–91. Aldershot: Scolar, 1997.

Berlin, Michael. 'Guilds in Decline? London Livery Companies and the Rise of a Liberal Economy, 1600–1800'. In *Guilds, Innovation and the European Economy*, edited by S.R. Epstein and Maarten Prak, 316–41. Cambridge: Cambridge University Press, 2008.

Bertucci, Paola. 'Enlightened Secrets: Silk, Intelligent Travel, and Industrial Espionage in Eighteenth-Century France', *Technology and Culture* 54, no. 4 (2013): 820–52.

Bertucci, Paola. *Artisanal Enlightenment: Science and the Mechanical Arts in Old Regime France*. New Haven and London: Yale University Press, 2017.

Biggs, Norman. 'John Reynolds of the Mint: A Mathematician in the Service of King and Commonwealth', *Historia Mathematica* 48 (2019): 1–28.

Binnema, Ted. *Enlightened Zeal: the Hudson's Bay Company and Scientific Networks, 1670–1870*. Toronto, Buffalo and London: University of Toronto Press, 2014.

Bishop, Jennifer. 'The Clerk's Tale: Civic Writing in Sixteenth-Century London', *Past & Present* 230, no. 11 (2016): 112–30.

Bowen, H.V. *The Business of Empire: The East India Company and Imperial Britain, 1756–1833*. Cambridge: Cambridge University Press, 2006.

Brant, Clare, and Susan E. Whyman (eds). *Walking the Streets of Eighteenth-Century London: John Gay's Trivia (1716)*. Oxford: Oxford University Press, 2009.

Brantz, Dorothee. 'Metropolitan Natural Histories: Inventing Science, Building Cities, and Displaying the World'. In *Science in the Metropolis: Vienna in Transnational Context 1848–1914*, edited by Mitchell G. Ash, 25–42. London: Routledge, 2020.

Brantz, Dorothee, Sasha Disko and Georg Wagner-Kyora. *Thick Space: Approaches to Metropolitanism*. Transcript Verlag: Bielefeld, 2012.

Brewer, John. 'The Eighteenth-Century British State'. In *An Imperial State at War: Britain from 1689 to 1815*, edited by Lawrence Stone, 52–71. London: Routledge, 1994.

Brock, Aske Laursen, Guido Meersbergen and Edmond Smith (eds). *Trading Companies and Travel Knowledge in the Early Modern World*. Abingdon and New York: Routledge, 2022.

Brockbank, William. 'Old Anatomical Theatres and What Took Place Therein', *Medical History* 12, no. 4 (1968): 317–84.

Bryden, D.J. 'Evidence from Advertising for Mathematical Instrument Making in London, 1556–1714', *Annals of Science* 49, no. 4 (1992): 301–36.

Buchanan, Brenda J. '"The Art and Mystery of Making Gunpowder": The English Experience in the Seventeenth and Eighteenth Centuries'. In *The Heirs of Archimedes: Science and the Art of War through the Age of Enlightenment*, edited by Brett D. Steele and Tamera Dorland, 233–74. Cambridge, MA and London: MIT Press, 2005.

Bull, Stephen. '*The Furie of the Ordnance*': *Artillery in the English Civil Wars*. Woodbridge: The Boydell Press, 2008.

Challis, C.E. 'Mint Officials and Moneyers of the Tudor Period', *British Numismatic Journal* 45 (1975): 157–97.

Challis, Christopher E. *The Tudor Coinage*. Manchester: Manchester University Press, 1978.

Challis, Christopher E. *A New History of the Royal Mint*. Cambridge and New York: Cambridge University Press, 1992.

Chamberland, Celeste. 'Honor, Brotherhood and the Corporate Ethos of London's Barber-Surgeons' Company', *Journal of the History of Medicine and Allied Sciences* 64, no. 3 (2009): 300–32.

Chamberland, Celeste. 'Between the Hall and the Market: William Clowes and Surgical Self-Fashioning in Elizabethan London', *The Sixteenth Century Journal* 41, no. 1 (2010): 69–89.

Chamberland, Celeste. 'From Apprentice to Master: Social Disciplining and Surgical Education in Early Modern London, 1570–1640', *History of Education Quarterly* 53, no. 1 (2013): 21–44.

Chaudhuri, K.N. *The English East India Company: The Study of an Early Joint-Stock Company, 1600–1640*. London: Frank Cass and Co., 1965.

Clark, Peter (ed). *The Cambridge Urban History of Britain, Volume 2: 1540–1840*. Cambridge: Cambridge University Press, 2000.

Clifton, Gloria. *Dictionary of British Scientific Instrument Makers, 1550–1851*. London: Zwemmer in association with the National Maritime Museum, 1995.

Cock, Emily. '"Lead[ing] 'em by the Nose into Publick Shame and Derision": Gaspare Tagiacozzi, Alexander Read and the Lost History of Plastic Surgery, 1600–1800', *Social History of Medicine* 28, no. 1 (2014): 1–21.

Connor, R.D. *The Weights and Measures of England*. London: H.M. Stationary Office, 1987.

Convertito-Farrar, Cori, and Kenneth Cozens. 'Trinity House Ballast Office', Paper Given at the Thames Symposium, 2009.

Cook, Andrew S. 'Alexander Dalrymple, (1737–1808), Hydrographer to the East India Company and to the Admiralty as Publisher'. PhD thesis, University of St Andrews, 1993.

Cook, Andrew S. 'Establishing the Sea Routes to India and China: Stages in the Development of Hydrographical Knowledge'. In *The Worlds of the East India Company*, edited by H.V. Bowen, Margarette Lincoln and Nigel Rigby, 119–36. Woodbridge: Boydell Press, 2003.

Cook, B.J. 'Humphrey Cole at the Mint'. In *Humphrey Cole: Mint, Measurement and Maps in Elizabethan England*, edited by Silke Ackermann, 21–6. London: British Museum, 1998.

Cook, B.J. '"Stampt with Your Own Image": The Numismatic Dimension of Two Stuart Successions'. In *Stuart Succession Literature: Moments and Transformations*, edited by Paulina Kewes and Andrew McRae, 303–18. Oxford: Oxford University Press, 2018.

Cook, Harold J. *Matters of Exchange: Commerce, Medicine, and Science in the Dutch Golden Age*. New Haven and London: Yale University Press, 2007.

Cooper, Tarnya. *Citizen Portrait: Portrait Painting and the Urban Elite of Tudor and Jacobean England and Wales*. New Haven and London: Yale University Press, 2012.

Cormack, Leslie B., Steven A. Walton and John A. Shuster (eds). *Mathematical Practitioners and the Transformation of Natural Knowledge in Early Modern Europe*. Cham: Springer, 2017.

Craig, John. *Newton at the Mint*. Cambridge: Cambridge University Press, 1946.

Crawforth, M.A. 'Instrument Makers in the London Guilds', *Annals of Science* 44, no. 4 (1987): 319–77.

Cregan, Kate. *The Theatre of the Body: Staging Death and Embodying Life in Early Modern London*. Turnhout: Brepols, 2009.

Cregan, Kate. 'Teaching the Anatomical Body in Seventeenth-Century London', *Medicine Studies* 2 (2010): 21–36.

Cunningham, Andrew, and Perry Williams. 'De-Centring the "Big Picture": "The Origins of Modern Science" and the Modern Origins of Science', *The British Journal for the History of Science* 26, no. 4 (1993): 407–32.

Daunton, Martin. 'London and the World'. In *London – World City 1800–1840*, edited by Celina Fox, 21–38. New Haven and London: Yale University Press, 1992.

David, Andrew. 'Lieutenant Murdoch Mackenzie and His Survey of the Bristol Channel and the South Coast of England', *The Cartographic Journal* 40, no. 1 (2003): 69–78.

Davids, Karel. 'Cities, Long-Distance Corporations and Open Air Sciences: Antwerp, Amsterdam and Leiden in the Early Modern Period'. In *Knowledge and the Early Modern City: A History of Entanglements*, edited by Bert De Munck and Antonella Romano, 126–48. London: Routledge, 2020.

Davidson, Simon C. 'Marine Chronometers: The Rapid Adoption of New Technology by East India Captains in the Period 1770–1792 on Over 580 Voyages', *Antiquarian Horology* 40 (2019): 76–91.

Davies, Matthew, '"Monuments of Honour": Clerks, Histories and Heroes in the London Livery Companies'. In *Parliament, Personalities and Power: Papers Presented to Linda S. Clark*, edited by H. Kleineke, 143–66. Woodbridge: Boydell, 2011.

Dickinson, H.W. *Educating the Royal Navy: Eighteenth- and Nineteenth-Century Education for Officers*. London and New York: Routledge 2007

Dierge, Sven, Jens Lachmund and Andrew Mendelsohn (eds). 'Introduction: Toward an Urban History of Science', *Science in the City, Osiris* 18 (2003): 1–19.

Dixon, Jennim. 'The Toyshop, the Cabinet, and Eighteenth-Century Curiosity', *History of Retailing and Consumption* 5 (2019): 205–27.

Dobson, Jessie, and R. Milnes Walker. *Barbers and Barber-Surgeons of London: A History of the Barbers' and Barber-Surgeons' Companies*. Oxford: Blackwell Scientific for The Worshipful Company of Barbers, 1979.

Drayton, Richard. *Nature's Government: Science, Imperial Britain and the 'Improvement' of the World*. New Haven and London: Yale University Press, 2000.

Dunn, Richard, and Rebekah Higgitt. *Finding Longitude: How Ships, Clocks and Stars Helped Solve the Longitude Problem*. Glasgow: Collins, 2014.

Dunn, Richard, and Rebekah Higgitt (eds). *Navigational Enterprises in Europe and Its Empires, 1730–1850*. Basingstoke: Palgrave Macmillan, 2016.

Eamon, William. 'From the Secrets of Nature to Public Knowledge: The Origins of the Concept of Openness in Science', *Minerva* 23, no. 3 (1986): 321–47.

Eamon, William. *Science and the Secrets of Nature: Books of Secrets in Medieval and Early Modern Culture*. Princeton, NJ: Princeton University Press, 1994.

Edney, Matthew H. *Mapping an Empire: The Geographical Construction of British India, 1765–1843*. Chicago: University of Chicago Press, 1997.

Ehrlich, Joshua. *The East India Company and the Politics of Knowledge*. Cambridge: Cambridge University Press, 2023.

Epstein, S.R., and Maarten Prak (eds). *Guilds, Innovation and the European Economy*. Cambridge: Cambridge University Press, 2008.

Fara, Patricia. *Life after Gravity: Isaac Newton's London Career*. Oxford: Oxford University Press, 2021.

Felten, Sebastian. *Money in the Dutch Republic: Everyday Practice and Circuits of Exchange*. Cambridge: Cambridge University Press, 2022.

Ferrari, Giovanna. 'Public Anatomy Lessons and the Carnival: The Anatomy Theatre of Bologna', *Past and Present* 117 (1987): 50–106.

Ffoulkes, Charles. *The Gun-Founders of England: With a List of English and Continental Gun-Founders from the XIV to the XIX Centuries*. Cambridge: Cambridge University Press, 2011

Field, J.V., and Frank James (eds). *Renaissance and Revolution: Humanists, Scholars and Craftsmen*. Cambridge: Cambridge University Press, 1993.

Fissel, Mary E. 'The Marketplace of Print'. In *Medicine and the Market in England and Its Colonies*, edited by Mark S.R. Jenner and Patrick Wallis, 108–32. New York: Palgrave Macmillan, 2007

Forbes, Archibald. *A History of the Army Ordnance Services*, 3 vols. London: Medici Society, 1929.

Forbes, John S. *Hallmark: A History of the London Assay Office*. London: Unicorn, 1999.

Forbes, John S. 'Search, Immigration and the Goldsmiths' Company: A Study in the Decline of Its Powers'. In *Guilds, Society and Economy in London 1450–1800*, edited by Ian Anders Gadd, and Patrick Wallis, 115–25. London: Centre for Metropolitan History, 2002.

Foster, William. *The East India House: Its History and Associations*. London: John Lane the Bodley Head Ltd, 1924.

Fox, Celina (ed). *London – World City 1800–1840*. New Haven and London: Yale University Press, 1992.

Fox, Celina. *The Arts of Industry in the Age of Enlightenment*. New Haven: Yale University Press, 2009.

France, Catherine Ann. 'Gunnery and the Struggle for the New Science (1537–1687)'. PhD thesis, University of Leeds, 2014.

Freeman, Jessica. 'The Mistery of Coiners and the King's Moneyers of the Tower of London, *c.*1340– *c.*1530', *British Numismatic Journal* 70 (2000): 67–82.

Gadd, Ian Anders, and Patrick Wallis (eds). *Guilds, Society and Economy in London 1450–1800*. London: Centre for Metropolitan History, 2002.

Garrioch, David. 'House Names, Shop Signs and Social Organization in Western European Cities, 1500–1900', *Urban History* 21, no. 1 (1994): 20–48.

Gascoigne, John. *Science in the Service of Empire: Joseph Banks, the British State and the Uses of Science in the Age of Revolution*. Cambridge: Cambridge University Press, 1998.

Goldgar, Anne, and Robert I. Frost. *Institutional Culture in Early Modern Society*. Leiden and Boston: Brill, 2004.

Gowing, Laura. '"The Freedom of the Streets": Women and Social Space, 1560–1640'. In *Londinopolis: Essays in the Cultural and Social History of Early Modern London*, edited by Mark S.R. Jenner and Paul Griffiths, 130–51. Manchester: Manchester University Press, 2000.

Gowing, Laura. *Common Bodies: Women, Touch and Power in Seventeenth-Century England*. New Haven and London: Yale University Press, 2003.

Grier, Jason. 'Navigation, Commercial Exchange and the Problem of Long-Distance Control in England and the English East India Company, 1673–1755'. PhD thesis, York University, 2018.

Griffiths, Paul. 'Secrecy and Authority in Late Sixteenth- and Seventeenth-Century London', *The Historical Journal* 40, no. 4 (1997): 925–51.

Guggisberg, F.G. *'The Shop;' the Story of the Royal Military Academy*. London, Paris, New York and Melbourne: Cassell and Company, 1900.

Gupta, Uma Das. *Science and Modern India: An Institutional History, c. 1784–1947*. Delhi: Pearson Longman, 2011.

Hall, A. Rupert. 'Gunnery, Science and the Royal Society'. In *The Uses of Science in the Age of Newton*, edited by John G. Burke, 111–41. Berkeley and Los Angeles: University of California Press, 1983.

Hallett, Mark, Nigel Llewellyn and Martin Myrone (eds). *Court, Country, City: British Art and Architecture, 1660–1735*. New Haven and London: Yale University Press, 2016.

Harding, Vanessa. 'City, Capital, and Metropolis: The Changing Shape of Seventeenth-Century London'. In *Imagining Early Modern London: Perceptions and Portrayals of the City from Stow to Strype, 1596–1720*, edited by J.F. Merritt, 117–43. Cambridge: Cambridge University Press, 2001.

Harkness, Deborah. '"Strange" Ideas and "English" Knowledge: Natural Science Exchange in Elizabethan London'. In *Merchants and Marvels: Commerce, Science, and Art in Early Modern Europe*, edited by Pamela Smith and Paula Findlen, 137–60. London and New York: Routledge, 2002.

Harkness, Deborah. *The Jewel House: Elizabethan London and the Scientific Revolution*. New Haven and London: Yale University Press, 2007.

Harkness, Deborah, and Jean E. Howard. 'Introduction: The Great World of Early Modern London', *Huntington Library Quarterly* 71, no. 1 (2008): 1–9.

Hayes, John. 'The Trinity House Group Portrait', *Burlington Magazine* 106, no. 735 (1964): 309–16.

Hellawell, Philippa. '"The Best and Most Practical Philosophers": Seamen and the Authority of Experience in Early Modern Science', *History of Science* 58, no. 1 (2019): 28–50.

Higgitt, Rebekah. 'The Mathematical Examiners of Trinity House: Teaching and Examining Mathematics for Navigation in London during the Long Eighteenth Century'. In *Beyond the Academy: The Practice of Mathematics, 1600–1850*, edited by Philip Beeley and Christopher Hollings, 80–111. Oxford: Oxford University Press, 2024.

Higton, Hester. 'Portrait of an Instrument-Maker: Wenceslaus Hollar's Engraving of Elias Allen', *British Journal for the History of Science* 37, no. 2 (2004): 147–66.

Hilaire-Pérez, Liliane. 'Technology as a Public Culture in the Eighteenth Century: The Artisans' Legacy', *History of Science* 45, no. 2 (2007): 135–53

Hill, Tracey. *Pageantry and Power: A Cultural History of the Lord Mayor's Show, 1585–1639*. Manchester: Manchester University Press, 2010.

Hitchcock, Tim, and Heather Shore. *The Streets of London: From the Great Fire to the Great Stink*. London: Rivers Oram Press, 2003.

Hochadel, Oliver. 'Periphery and Metropolis: Some Historiographical Reflections on the Urban History of Science'. In *Science in the Metropolis: Vienna in Transnational Context 1848–1914*, edited by Mitchell G. Ash, 43–63. London: Routledge, 2020.

Howard, Jean E. *Theatre of a City: The Places of London Comedy, 1598–1642*. Philadelphia: University of Pennsylvania Press, 2007.

Howes, Jennifer. *The Art of a Corporation: The East India Company as Patron and Collector, 1600–1800*. London: Routledge India, 2023.

Howse, Derek. *Greenwich Observatory. Volume 3: The Buildings and Instruments*. London: Taylor and Francis, 1975.

Hunt, Arnold. 'The Early Modern Secretary and the Early Modern Archive'. In *Archives and Information in the Early Modern World*, edited by Alexandra Walsham and Liesbeth Corens, 105–31. Oxford: Published for The British Academy by Oxford University Press, 2018.

Iliffe, Rob. 'Material Doubts: Hooke, Artisan Culture and the Exchange of Information in 1670s London', *The British Journal for the History of Science* 28, no. 3 (1995): 285–318.

Iliffe, Rob. 'Mathematical Characters: Flamsteed and Christ's Hospital Mathematical School'. In *Flamsteed's Stars: New Perspectives on the Life and Work of the First Astronomer Royal, 1646–1719*, edited by Frances Willmoth, 115–44. Woodbridge: Boydell Press, 1997.

Jacobi, Lauren. *The Architecture of Banking in Renaissance Italy: Constructing the Spaces of Money*. Cambridge: Cambridge University Press, 2019.

Jardine, Boris. 'Henry Sutton's Collaboration with John Reynolds (Gauger, Assayer and Clerk at the Royal Mint)', *Bulletin of the Scientific Instrument Society* 130 (2016): 4–7.

Jardine, Boris. 'Instruments of Statecraft: Humphrey Cole, Elizabethan Economic Policy and the Rise of Practical Mathematics', *Annals of Science* 75, no. 4 (2018): 304–29.

Jasanoff, Sheila. 'The Idiom of Co-production'. In *States of Knowledge: the Co-production of Science and Social Order*, edited by Sheila Jasanoff, 1–12. London and New York: Routledge, 2004.

Jenner, Mark S.R. 'From Conduit Community to Commercial Network? Water in London, 1500–1725'. In *Londinopolis: Essays in the Cultural and Social History of Early Modern London*, edited by Jenner and Paul Griffiths, 250–72. Manchester: Manchester University Press, 2000.

Jenner, Mark S.R., and Paul Griffiths (eds). *Londinopolis: Essays in the Cultural and Social History of Early Modern London*. Manchester: Manchester University Press, 2000.

Jenner, Mark S.R., and Patrick Wallis (eds). *Medicine and the Market in England and Its Colonies*. New York: Palgrave Macmillan, 2007.

Jones, Neil, and Paul Ridgway. *Light through a Lens: An Illustrated Celebration of 500 Years of Trinity House*. London: Bloomsbury, 2014.

Jütte, Daniel. *The Age of Secrecy: Jews, Christians, and the Economy of Secrets, 1400–1800*. New Haven: Yale University Press, 2015.

Kanefsky, John, and John Robey. 'Steam Engines in 18th-Century Britain: A Quantative Assessment', *Technology and Culture* 21 (1980): 161–86.

Kilburn-Toppin, Jasmine. '"A Place of Great Trust to Be Supplied by Men of Skill and Integrity": Assayers and Knowledge Cultures in Late Sixteenth- and Seventeenth-Century London', *The British Journal for the History of Science* 52, no. 2 (2019): 197–223.

Kilburn-Toppin, Jasmine. *Crafting Identities: Artisan Culture in London, c.1550–1640*. Manchester: Manchester University Press, 2021.

Kilburn-Toppin, Jasmine. 'Writing Knowledge, Forging Histories: Metallurgical Recipes, Artisan-Authors and Institutional Cultures in Early Modern London', *Cultural and Social History* 18, no. 3 (2021): 297–314.

Kilburn-Toppin, Jasmine. 'Instrument Makers, Shops, and Expertise in Eighteenth-Century London'. In *Spaces of Enlightenment Science*, edited by Gordon McOuat and Larry Stewart, 74–90. Leiden: Brill, 2022.

Klein, Ursula. 'Technoscience Avant la Lettre', *Perspectives on Science* 13, no. 2 (2005): 226–66.

Klein, Ursula, and E.C. Spary (eds). *Materials and Expertise in Early Modern Europe: Between Market and Laboratory*. Chicago and London: University of Chicago Press, 2010.

Knights, Mark. 'A City Revolution: The Remodelling of the London Livery Companies in the 1680s', *English Historical Review* 112, no. 449 (1997): 1141–78.

Kochhar, R.K. 'The Growth of Modern Astronomy in India, 1651–1960', *Vistas in Astronomy* 34 (1991): 69–105.

Kumar, Deepak. 'The Evolution of Colonial Science in India: Natural History and the East India Company'. In *Imperialism and the Natural World*, edited by John M. MacKenzie, 51–66. Manchester: Manchester University Press, 1990.

La Roncière, Monique de. 'Manuscript Charts by John Thornton, Hydrographer of the East India Company (1669–1701)', *Imago Mundi* 19, no. 1 (1965): 46–50.

Lefebvre, Henri. *The Production of Space*, translated by Donald Nicholson-Smith. Oxford: Basil Blackwell, 1991.

Leong, Elaine. 'Learning Medicine by the Book: Reading and Writing Surgical Manuals in Early Modern London', *BJHS Themes* 5 (2020): 93–110.

Leslie, J.H. 'The Arms of the Board of Ordnance', *Journal for the Society for Army Historical Research* 1, no. 5 (1922): 210–13.

Lincoln, Margarette. *London and the Seventeenth Century: The Making of the World's Greatest City*. New Haven and London: Yale University Press, 2021.

Linebaugh, Peter. 'The Tyburn Riot against the Surgeons'. In *Albion's Fatal Tree: Crime and Society in Eighteenth Century England*, edited by Dougals Hay, Peter Linebaugh, John G. Rule, E.P. Thompson and Cal Winslow, 65–117. New York: Pantheon, 1975.

Livingstone, David. *Putting Science in Its Place: Geographies of Scientific Knowledge*. Chicago and London: University of Chicago Press, 2003.

Long, Pamela O. 'The Openness of Knowledge: An Ideal and Its Context in Sixteenth-Century Writings on Metallurgy', *Technology and Culture* 32, no. 2 (1991): 318–55.

Long, Pamela O. *Openness, Secrecy, Authorship: Technical Arts and the Culture of Knowledge from Antiquity to the Renaissance*. Baltimore and London: Johns Hopkins University Press, 2001.

Long, Pamela O. *Artisan/Practitioners and the Rise of the New Sciences, 1400–1600*. Corvallis: Oregon State University Press, 2011.

Long, Pamela O. 'Trading Zones in Early Modern Europe', *Isis* 106, no. 4 (2015): 840–7.

Lucas, Anya, and Henry Russell. *The Livery Halls of the City of London*. London: Merrell in Association with the Worshipful Company of Chartered Architects, 2018.

Luu, Lien Bich. *Immigrants and the Industries of London*. Aldershot and Burlington, VT: Ashgate, 2005.

Lux, David S., and Harold J. Cook. 'Closed Circles or Open Networks? Communicating at a Distance during the Scientific Revolution', *History of Science* 36, no. 2 (1998): 179–211.

MacGregor, Arthur. '"A Magazin of all Manner of Inventions": Museums in the Quest for "Salomon's House" in Seventeenth-Century England', *Journal of the History of Collections* 1, no. 2 (1989): 207–12.

MacGregor, Arthur (ed). *Naturalists in the Field*. Leiden: Brill, 2018.

Manderson, Lenore, Mark Davis, Chip Colwell and Tanja Ahlin. 'On Secrecy, Disclosure, the Public, and the Private in Anthropology', *Current Anthropology* 56, no. 12 (2015): 183–90.

Mandressi, Rafael. 'Affected Doctors', *Osiris* 31 (2016): 119–36.

Marples, Alice. 'The Science of Money: Isaac Newton's Mastering of the Mint', *Notes and Records* 76, no. 3 (2022): 507–25.

Marshall, Douglas W. 'Military Maps of the Eighteenth-Century and the Tower of London Drawing Room', *Imago Mundi* 32 (1980): 21–44.

Matthews, Anya. 'Honour, Ornament, and Frugality: The Reconstruction of London's Livery Halls after the Great Fire'. In *Court, Country, City: British Art and Architecture, 1660–1735*, edited by Mark Hallett, Nigel Llewellyn and Martin Myrone, 73–104. New Haven and London: Yale University Press, 2016.

Mauskopf, Seymour H. 'Gunpowder and the Chemical Revolution', *Osiris* 4 (1988): 93–118.

Mauskopf, Seymour H. 'Chemistry in the Arsenal: State Regulation and Scientific Methodology of Gunpowder in Eighteenth-Century England and France'. In *The Heirs of Archimedes: Science and the Art of War through the Age of Enlightenment*, edited by Brett D. Steele and Tamera Dorland, 293–330. Cambridge, MA and London: MIT Press, 2005.

Mauskopf, Seymour H. 'The Crisis of English Gunpowder in the Eighteenth Century'. In *Materials and Expertise in Early Modern Europe: Between Market and Laboratory*, edited by Ursula Klein and E.C. Spary, 288–320. Chicago and London: University of Chicago Press, 2010.

McAleer, John. 'Exhibiting the "Strangest of All Empires": The East India Company, East India House, and Britain's Asian Empire'. In *The MacKenzie Moment and Imperial History*, edited by Stephanie Barczewski and Martin Farr, 25–45. Cham: Springer International Publishing, 2019.

Mendelsohn, J. Andrew. 'Introduction: Civic Medicine'. In *Civic Medicine: Physician, Polity, and Pen in Early Modern Europe*, edited by Annemarie Kinzelbach and Ruth Schilling, 1–6. Abingdon and New York: Routledge, 2020.

Mercer, Malcolm. 'Shaping the Ordnance Office Collections at the Tower of London: The Impact of Colonial Expansion, Diplomacy, and Donation in the Early Nineteenth Century', *Museum History Journal* 9, no. 2 (2016): 153–67.

Millburn, John R. 'The Office of Ordnance and the Instrument-Making Trade in the Mid-Eighteenth Century', *Annals of Science* 45, no. 3 (1988): 221–93.

Millburn, John R. 'William Deane and His Ordnance Bills', *Bulletin of the Scientific Instrument Society* 45 (1995): 12–18.

Millburn, John R. 'Some English Military Instrument Makers of the Late 17th Century', *Bulletin of the Scientific Instrument Society* 68 (2001): 2–5.

Miller, David P. 'Longitude Networks on Land and Sea'. In *Navigational Enterprises in Europe and Its Empires, 1730–1850*, edited by Richard Dunn and Rebekah Higgitt, 223–47. Basingstoke: Palgrave Macmillan, 2016.

Miller, David Philip. 'The Revival of the Physical Sciences in Britain, 1815–1840', *Osiris* 2 (1986): 107–34.

Mishra, Rupali. *A Business of State: Commerce, Politics, and the Birth of the East India Company*. Cambridge, MA: Harvard University Press, 2018.

Mitchell, David. *Silversmiths in Elizabethan and Stuart London: Their Lives and Marks*. Woodbridge: The Boydell Press/The Goldsmiths' Company, 2017.

Morrissey, L.J. 'English Pageant-Waggons', *Eighteenth-Century Studies* 9, no. 3 (1976): 353–74.

Morus, Iwan, Simon Schaffer and Jim Secord. 'Scientific London'. In *London – World City 1800–1840*, edited by Celina Fox, 129–42. New Haven and London: Yale University Press, 1992.

Mörzer Bruyns, W.F.J. 'Navigation on Dutch East India Company Ships around the 1740s', *The Mariner's Mirror* 78 (1992): 143–54.

Mosley, Adam. 'Objects, Texts and Images in the History of Science', *Studies in the History and Philosophy of Science A* 38, no. 2 (2007): 289–302.

Moxham, Noah. 'Natural Knowledge, Inc.: The Royal Society as a Metropolitan Corporation', *The British Journal for the History of Science* 52, no. 2 (2019): 249–71.

Munck, Bert De, and Antonella Romano (eds). *Knowledge and the Early Modern City: A History of Entanglements*. London: Routledge, 2020.

Newman, William R. 'Alchemy, Assaying, and Experiment'. In *Instruments and Experimentation in the History of Chemistry*, edited by Frederic L. Holmes and Trevor H. Levere, 35–54. Cambridge, MA and London: MIT Press, 2000.

Newman, William R., and Lawrence M. Principe. *Alchemy Tried in the Fire: Starkey, Boyle, and the Fate of Helmontian Chymistry*. Chicago and London: University of Chicago Press, 2002.

Nummedal, Tara. *Alchemy and Authority in the Holy Roman Empire*. Chicago and London: University of Chicago Press, 2007.

Nummedal, Tara E. 'Words and Works in the History of Alchemy', *Isis* 102, no. 2 (2011): 330–7.

Ogborn, Miles. *Spaces of Modernity: London's Geographies 1680–1780*. New York and London: Guilford Press, 1998.

Ogilvie, Sheilagh. 'Rehabilitating the Guilds: A Reply', *Economic History Review* 61, no. 1 (2008): 175–82.

Ogilvie, Sheilagh. *The European Guilds: An Economic Analysis*. London and Princeton: Princeton University Press, 2019.

Ophir, Adi, and Steven Shapin. 'The Place of Knowledge: A Methodological Survey', *Science in Context* 4, no. 1 (1991): 3–22.

O'Sullivan, Tanya. *Geographies of City Science: Urban Lives and Origin Debates in Late Victorian Dublin*. Pittsburgh: Pittsburgh University Press, 2019.

Outram, Dorinda. 'New Spaces in Natural History'. In *Cultures of Natural History*, edited by Nicholas Jardine, James A. Secord and Emma C. Spary, 249–65. Cambridge: Cambridge University Press, 1995.

Park, Katherine. 'The Criminal and the Saintly Body: Autopsy and Dissection in Renaissance Italy', *Renaissance Quarterly* 47, no. 1 (1994): 1–33.

Parnell, Geoffrey. 'The Buildings and Works of the Office of Ordnance at the Tower of London, 1660–1722'. PhD thesis, University of London, 1996.

Parnell, Geoffrey. 'The Ordnance Drawing Room 1716–52', *English Heritage Historical Review* 9, no. 1 (2014): 122–79.

Pastorino, Cesare. 'The Mine and the Furnace: Francis Bacon, Thomas Russell, and Early Stuart Mining Culture', *Early Science and Medicine* 14, no. 5 (2009): 630–60.

Pearl, Valerie. 'Change and Stability in Seventeenth-Century London', *The London Journal* 5, no. 1 (1979): 3–34.

Peck, Linda Levy. *Consuming Splendor: Society and Culture in Seventeenth-Century England*. Cambridge and New York: Cambridge University Press, 2005.

Pelling, Margaret, with Frances White. *Medical Conflicts in Early Modern London: Patronage, Physicians, and Irregular Practitioners, 1550–1640*. Oxford: Clarendon Press; New York: Oxford University Press, 2003.

Peters, Kate, Alexandra Walsham and Liesbeth Corens (eds). *Archives and Information in the Early Modern World*. Oxford: Published for The British Academy by Oxford University Press, 2018.

Pham, Charlotte Minh Hà. 'European Navigation, Nautical Instructions and Charts of the Cochinchinese Coast (16th–19th Centuries)', *Moussons. Recherche En Sciences Humaines Sur l'Asie Du Sud-Est* 27 (2016): 101–29.

Plumb, Christopher. 'Exotic Animals in Eighteenth-Century Britain'. PhD thesis, University of Manchester, 2010.

Plumb, Christopher. *The Georgian Menagerie: Exotic Animals in Eighteenth-Century London*. London: I.B. Tauris, 2015.

Porter, Whitworth. *History of the Corps of Engineers*. London and New York: Longmans, Green, and Co., 1889.

Principe, Lawrence M., and William R. Newman. 'Some Problems with the Historiography of Alchemy'. In *Secrets of Nature: Astrology and Alchemy in Early Modern Europe*, edited by William R. Newman and Anthony Grafton, 385–431. Cambridge, MA. and London: MIT Press, 2001.

Raj, Kapil. *Relocating Modern Science: Circulation and the Construction of Knowledge in South Asia and Europe, 1650–1900*. London: Palgrave Macmillan, 2007.

Ratcliff, Jessica. 'The East India Company, the Company's Museum, and the Political Economy of Natural History in the Early Nineteenth Century', *Isis* 107, no. 3 (2016): 495–517.

Reddaway, Thomas Fiddean. *The Rebuilding of London after the Great Fire*. London: Jonathan Cape, 1940.

Reddaway, T.F., and Lorna Walker. *The Early History of the Goldsmiths Company, 1327–1509*. London: Arnold, 1975.

Rhynas Brown, Ruth. 'Notes on the Wealdon Furnaces in the Records of the Board of Ordnance, 1660–1700', *Wealdon Iron: Bulletin of the Wealdon Iron Research Group* 2, no. 13 (1993): 20–30.

Roberts, Lissa. 'Mapping Steam Engines and Skill in Eighteenth-Century Holland'. In *The Mindful Hand: Inquiry and Invention from the Late Renaissance to Early Industrialisation*, edited by Lissa Roberts, Simon Schaffer and Peter Dear, 197–218. Amsterdam: Royal Netherlands Academy of Arts and Sciences, 2007.

Roberts, Lissa, Simon Schaffer, and Peter Dear (eds). *The Mindful Hand: Inquiry and Invention from the Late Renaissance to Early Industrialisation*. Amsterdam: Royal Netherlands Academy of Arts and Sciences, 2007.

Robischon, Mary Margaret. 'Scientific Instrument Makers in London during the Seventeenth and Eighteenth Centuries'. PhD thesis, University of Michigan, 1983.

Sachse, William L. 'The Journal of Nathan Prince, 1747', *The American Neptune* 16, no. 2 (1956): 81–97.

Sandman, Alison. 'Mirroring the World: Sea Charts, Navigation, and Territorial Claims in Sixteenth-Century Spain'. In *Merchants and Marvels: Commerce, Science, and Art in Early Modern Europe*, edited by Pamela Smith and Paula Findlen, 83–108. New York and London: Routledge, 2002.

Sandman, Alison. 'Controlling Knowledge: Navigation, Cartography, and Secrecy in the Early Modern Spanish Atlantic'. In *Science and Empire in the Atlantic World*, edited by James Delbourgo and Nicholas Dew, 31–52. London: Routledge, 2008.

Sawday, Jonathan. *The Body Emblazoned: Dissection and the Human Body in Renaissance Culture*. London: Routledge, 1995.

Schaffer, Simon. '"The Charter'd Thames": Naval Architecture and Experimental Spaces in Georgian Britain'. In *The Mindful Hand: Inquiry and Invention from the Late Renaissance to Early Industrialisation*, edited by Lissa Roberts, Simon Schaffer and Peter Dear, 278–305. Amsterdam: Royal Netherlands Academy of Arts and Sciences, 2007.

Schneer, Jonathan. *The Thames: England's River*. New Haven and London: Yale University Press, 2005.

Schofield, John. *Medieval London Houses*. New Haven and London: Yale University Press, 1994.
Schotte, Margaret E. *Sailing School: Navigating Science and Skill, 1550–1800*. Baltimore: Johns Hopkins University Press, 2019.
Schwoerer, Lois G. *Gun Culture in Early Modern England*. Charlottesville: University of Virginia Press, 2016.
Secord, James A. 'Knowledge in Transit', *Isis* 95, no. 4 (2004): 654–72.
Selwood, Jacob. *Diversity and Difference in Early Modern London*. Farnham: Ashgate, 2010.
Shapin, Steven. 'The House of Experiment in Seventeenth-Century England', *Isis* 79, no. 3 (1988): 373–404.
Shapin, Steven. *A Social History of Truth: Civility and Science in Seventeenth-Century England*. Chicago and London: Chicago University Press, 1994.
Shapin, Steven, and Simon Schaffer. *Leviathan and the Air-Pump: Hobbes, Boyle, and the Experimental Life*, 2nd ed. Princeton, NJ: Princeton University Press, 2011.
Simmons, Anna. 'Medicines, Monopolies and Mortars: The Chemical Laboratory and Pharmaceutical Trade at the Society of Apothecaries in the Eighteenth Century', *Ambix* 53, no. 3 (2006): 221–36.
Simmons, Anna. 'Stills, Status, Stocks and Science: The Laboratories at Apothecaries' Hall in the Nineteenth Century', *Ambix* 61, no. 2 (2014): 141–61.
Simmons, Anna. 'Trade, Knowledge and Networks: The Activities of the Society of Apothecaries and Its Members in London, *c.*1670–*c.*1800', *The British Journal for the History of Science* 52, no. 2 (2019): 273–96.
Slack, Paul. *The Invention of Improvement: Information and Material Progress in Seventeenth-Century England*. Oxford: Oxford University Press, 2014.
Sloan, Kim. 'Thomas Weston and the Academy at Greenwich', *Transactions of the Greenwich and Lewisham Antiquarian Society* 9, no. 6 (1984): 313–33.
Smith, Edmond. *Merchants: The Community that Shaped England's Trade and Empire*. New Haven and London: Yale University Press, 2021.
Smith, Pamela H. *The Body of the Artisan: Art and Experience in the Scientific Revolution*. Chicago: University of Chicago Press, 2004.
Smith, Pamela H. 'Laboratories'. In *The Cambridge History of Science. Volume 3, Early Modern Science*, edited by Katherine Park, 290–305. Cambridge: Cambridge University Press, 2006.
Smith, Pamela H. 'Vermillion, Mercury, Blood, and Lizards: Matter and Meaning in Metalworking'. In *Materials and Expertise in Early Modern Europe: Between Market and Laboratory*, edited by Ursula Klein and E.C. Spary, 29–49. Chicago and London: University of Chicago Press, 2010.
Smith, Pamela H. 'Why Write a Book? From Lived Experience to the Written World in Early Modern Europe', *Bulletin of the German Historical Institute* 47 (2010): 25–50.
Smith, Pamela H. 'What Is a Secret? Secrets and Craft Knowledge in Early Modern Europe'. In *Secrets and Knowledge in Medicine and Science, 1500–1800*, edited by Elaine Leong and Alisha Rankin, 47–66. Farnham and Burlington, VT: Ashgate, 2011.
Smith, Pamela H., and Paula Findlen (eds). *Merchants and Marvels: Commerce, Science, and Art in Early Modern Europe*. London and New York: Routledge, 2002.
Smith, Pamela H., Amy R.W. Meyers and Harold J. Cook (eds). *Ways of Making and Knowing: The Material Culture of Empirical Knowledge*. Ann Arbor: University of Michigan Press, 2013.

Steele, Brett D. 'Muskets and Pendulums: Benjamin Robins, Leonhard Euler, and the Ballistics Revolution', *Technology and Culture* 35, no. 2 (1994): 348–82.

Steele, Brett D., and Tamera Dorland (eds). *The Heirs of Archimedes: Science and the Art of War through the Age of Enlightenment*. Cambridge, MA and London: MIT Press, 2005.

Stern, P.J. *The Company-State: Corporate Sovereignty and the Early Modern Foundation of the British Empire in India*. Oxford: Oxford University Press, 2011.

Stevenson, Christine. 'Occasional Architecture in Seventeenth-Century London', *Architectural History* 49 (2006): 35–74.

Stevenson, Christine. *The City and the King: Architecture and Politics in Restoration London*. New Haven and London: Yale University Press, 2013.

Stevenson, Christine. 'Making Empire Visible at the Second Royal Exchange, London'. In *Court, Country, City: British Art and Architecture, 1660–1735*, edited by Mark Hallett, Nigel Llewellyn and Martin Myrone, 51–72. New Haven and London: Yale University Press, 2016.

Stewart, Larry. *The Rise of Public Science: Rhetoric, Technology, and Natural Philosophy in Newtonian Britain, 1660–1750*. Cambridge: Cambridge University Press, 1992.

Stewart, Larry. 'Other Centres of Calculation, or, Where the Royal Society Didn't Count: Commerce, Coffee-Houses and Natural Philosophy in Early Modern London', *The British Journal for the History of Science* 32, no. 2 (1999): 133–53.

Stewart, Larry, and Paul Weindling. 'Philosophical Threads: Natural Philosophy and Public Experiment among the Weavers of Spitalfields', *The British Journal for the History of Science* 28, no. 1 (1995): 37–62.

Stielau, Allison. '"Adding to the Good Silver with Other Trickery": Purity and Contamination in Clement VII's Emergency Currency'. In *Contamination and Purity in Early Modern Art and Architecture*, edited by Lauren Jacobi and Danile M. Zolli, 157–82. Amsterdam: Amsterdam University Press, 2021.

Stobart, Jon, Andrew Hann and Victoria Morgan. *Spaces of Consumption: Leisure and Shopping in the English Town, c.1680–1830*. London: Routledge, 2007.

Tanner, J.R. 'Samuel Pepys and the Trinity House', *The English Historical Review* 44, no. 176 (1929): 573–87.

Taub, Liba. 'Introduction: Reengaging with Instruments', *Isis* 102, no. 4 (2011): 689–96.

Tittler, Robert. 'Portrait Collection and Display in the English Civic Body, *c.*1540–1640', *Journal of the History of Collections* 20, no. 2 (2008): 161–72.

Tittler, Robert. 'Portraiture, Precedence and Politics amongst the London Liveries *c.* 1540–1640', *Urban History* 35, no. 3 (2008): 349–62.

Tittler, Robert. *The Face of the City: Civic Portraiture and Civic Identity in Early Modern England*. Manchester: Manchester University Press, 2013.

Tomlinson, H.C. *Guns and Government: The Ordnance Office under the Later Stuarts*. London: Royal Historical Society, Swift Printers, 1979.

Tomory, Leslie. *The History of the London Water Industry, 1580–1820*. Baltimore: Johns Hopkins University Press, 2017

Tracey, Kevin. 'Calculating Value: Using and Collecting the Tools of Early Modern Mathematics'. PhD thesis, Swansea University, 2019.

Turner, G.L'E. 'Queen Elizabeth's Instrument Makers'. In *Humphrey Cole: Mint, Measurement and Maps in Elizabethan England*, edited by Silke Ackermann, 3–9. London: British Museum, 1998.

Walker, Matthew. 'Architecture, Anatomy and the New Science in Early Modern London: Robert Hooke's College of Physicians', *Journal of the Society of Architectural Historians* 72, no. 4 (2013): 475–502.

Wallace, Anthony F.C. *The Social Context of Innovation: Bureaucrats, Families, and Heroes in the Early Industrial Revolution, as Foreseen in Bacon's New Atlantis*. Lincoln and London: University of Nebraska Press, 2003.
Wallis, Patrick. 'Controlling Commodities: Search and Reconciliation in the Early Modern Livery Companies'. In *Guilds, Society and Economy in London 1450–1800*, edited by Ian Anders Gadd and Patrick Wallis, 85–100. London: Centre for Metropolitan History, 2002.
Wallis, Patrick. 'Consumption, Retailing, and Medicine in Early-Modern London', *The Economic History Review* 61, no. 1 (2008): 26–5.
Wallis, Patrick. 'Between Apprenticeship and Skill: Acquiring Knowledge outside the Academy in Early Modern England', *Science in Context* 32 (2019): 155–70.
Wallis, Patrick, and Catherine Wright. 'Evidence, Artisan Experience, and Authority in Early Modern England'. In *Ways of Making and Knowing: The Material Culture of Empirical Knowledge*, edited by Pamela H. Smith, Amy R.W. Meyers and Harold J. Cook, 138–63. Ann Arbor: University of Michigan Press, 2013.
Walsh, Claire. 'Shop Design and the Display of Goods in Eighteenth-Century London', *Journal of Design History* 8, no. 3 (1995): 157–6.
Walsh, Claire. 'The Design of London Goldsmiths' Shops in the Early Eighteenth Century'. In *Goldsmiths, Silversmiths and Bankers: Innovation and the Transfer of Skill, 1550 to 1750*, edited by David Mitchell, 96–111. Stroud: Alan Sutton, 1995.
Walton, Steven A. 'The Art of Gunnery in Renaissance England'. PhD thesis, University of Toronto, 1999.
Walton, Steven A. 'The Tower Gunners and the Artillery Company in the Artillery Garden before 1630', *Journal of the Ordnance Society* 18 (2006): 53–66.
Ward, Joseph P. *Metropolitan Communities: Trade Guilds, Identity, and Change in Early Modern London*. Stanford, CA: Stanford University Press, 1997.
Ward, Joseph P. 'The Taming of the Thames', *Huntington Library Quarterly* 71, no. 1 (2008): 55–75.
Wardhaugh, Benjamin. *Gunpowder & Geometry: The Life of Charles Hutton, Pitt Boy, Mathematician and Scientific Rebel*. London: William Collins, 2019.
Warner, Deborah Jean. 'What Is a Scientific Instrument, When Did It Become One, and Why?', *The British Journal for the History of Science* 23, no. 1 (1990): 83–93.
Wear, Andrew. *Knowledge and Practice in English Medicine, 1550–1680*. Cambridge: Cambridge University Press, 2016.
Werrett, Simon. 'Congreve's Rational Rockets', *Notes & Records of the Royal Society* 63, no. 1 (2009): 35–56.
Werrett, Simon. *Fireworks: Pyrotechnic Arts and Sciences in European History*. Chicago: University of Chicago Press, 2010.
Werrett, Simon. 'Matter and Facts: Material Culture and the History of Science'. In *Material Evidence: Learning from Archaeological Practice*, edited by Alison Wylie and Robert Chapman, 339–52. New York: Routledge, 2014.
Werrett, Simon. *Thrifty Science: Making the Most of Materials in the History of Experiment*. Chicago and London: University of Chicago Press, 2019.
Westfall, Richard. *Never at Rest: A Biography of Isaac Newton*. Cambridge: Cambridge University Press, 1981.
Williams, J. 'Mathematics and the Alloying of Coinage 1202–1700: Part II', *Annals of Science* 52, no. 3 (1995): 235–63.
Willmoth, Frances. 'Mathematical Sciences and Military Technology: The Ordnance Office in the Reign of Charles II'. In *Renaissance and Revolution: Humanists, Scholars*

and Craftsmen, edited by J.V. Field and Frank James, 117–32. Cambridge: Cambridge University Press, 1993.
Willmoth, Frances. *Sir Jonas Moore: Practical Mathematics and Restoration Science*. Woodbridge: Boydell & Brewer, 1993.
Winterbottom, Anna. *Hybrid Knowledge in the Early East India Company World*. Basingstoke: Palgrave Macmillan, 2016.
Winterbottom, Anna. 'An Experimental Community: The East India Company in London, 1600–1800', *British Journal for the History of Science* 52, no. 2 (2019): 323–43.
Withington, Phil. 'Public Discourse, Corporate Citizenship, and State Formation in Early Modern England', *The American Historical Review* 112, no. 4 (2007): 1016–38.
Wolfe, Heather, and Peter Stallybrass. 'The Material Culture of Record-Keeping in Early Modern England'. In *Archives and Information in the Early Modern World*, edited by Kate Peters, Alexandra Walsham and Liesbeth Corens, 179–209. Oxford: Published for The British Academy by Oxford University Press, 2018.
Wortham, Simon. 'Sovereign Counterfeits: The Trial of the Pyx', *Renaissance Quarterly* 49, no. 2 (1996): 334–59.
Yale, Elizabeth (ed). 'Focus: The History of Archives and the History of Science', *Isis* 107, no. 1 (2016): 74–120.

Index